Lecture Notes in Computer Science   **10467**

Commenced Publication in 1973
Founding and Former Series Editors:
Gerhard Goos, Juris Hartmanis, and Jan van Leeuwen

More information about this series at http://www.springer.com/series/7407

Robert Brijder · Lulu Qian (Eds.)

DNA Computing and Molecular Programming

23rd International Conference, DNA 23
Austin, TX, USA, September 24–28, 2017
Proceedings

 Springer

Editors
Robert Brijder (iD)
Hasselt University
Diepenbeek
Belgium

Lulu Qian (iD)
California Institute of Technology
Pasadena, CA
USA

ISSN 0302-9743 ISSN 1611-3349 (electronic)
Lecture Notes in Computer Science
ISBN 978-3-319-66798-0 ISBN 978-3-319-66799-7 (eBook)
DOI 10.1007/978-3-319-66799-7

Library of Congress Control Number: 2017951307

LNCS Sublibrary: SL1 – Theoretical Computer Science and General Issues

Printed on acid-free paper

This Springer imprint is published by Springer Nature
The registered company is Springer International Publishing AG
The registered company address is: Gewerbestrasse 11, 6330 Cham, Switzerland

Preface

This volume contains papers presented at DNA 23: The 23rd International Conference on DNA Computing and Molecular Programming. The conference was held at The University of Texas at Austin during September 24–28, 2017, and organized under the auspices of the International Society for Nanoscale Science, Computation and Engineering (ISNSCE). A workshop day, consisting of the Education Workshop and the Synthetic Biology Workshop, was held on September 24, 2017. The DNA conference series aims to draw together mathematics, computer science, physics, chemistry, biology, and nanotechnology to address the analysis, design, and synthesis of information-based molecular systems.

Papers and presentations were sought in all areas that relate to biomolecular computing, including, but not restricted to: algorithms and models for computation with biomolecular systems; computational processes in vitro and in vivo; molecular switches, gates, devices, and circuits; molecular folding and self-assembly of nanostructures; analysis and theoretical models of laboratory techniques; molecular motors and molecular robotics; studies of fault-tolerance and error correction; software tools for analysis, simulation, and design; synthetic biology and in vitro evolution; applications in engineering, physics, chemistry, biology, and medicine.

For DNA 23 we received 109 submissions: 23 Track A submissions, 10 Track B submissions, and 76 Track C submissions. Each Track A and B submission was reviewed by at least three reviewers, with an average of more than four reviews per submission. The Program Committee accepted 16 submissions in Track A and 8 submissions in Track B. Also, 8 submissions in Track C were selected for both poster and breaking-news short oral presentations. The papers accepted for Track A are published in these conference proceedings.

In addition to the accepted submissions, the scientific program included invited talks from David Doty, Eric Klavins, Cristopher Moore, Vincent Noireaux, Yannick Rondelez, and Karin Strauss. The Education Workshop included talks from Eric Klavins, Carlos Castro, and Brian Korgel, and The Synthetic Biology Workshop included talks from Reinhard Heckel, Sri Kosuri, and James Diggans.

We thank all invited speakers and all authors who submitted their work to this conference. Finally, we thank the members of the Program Committee and the additional reviewers for their hard work in reviewing the submissions and taking part in post-review discussions.

July 2017

Robert Brijder
Lulu Qian

Organization

DNA 23 was organized by the University of Texas at Austin in cooperation with the International Society for Nanoscale Science, Computation and Engineering (ISNSCE).

Organizing Committee

Andrew Ellington (Co-chair)	University of Texas at Austin, USA
David Soloveichik (Co-chair)	University of Texas at Austin, USA
Tiffo Carmichael	University of Texas at Austin, USA
Curtis White	University of Texas at Austin, USA

Steering Committee

Anne Condon (Co-chair)	University of British Columbia, Canada
Natasha Jonoska (Co-chair)	University of South Florida, USA
Luca Cardelli	Microsoft Research, Cambridge, UK
Masami Hagiya	University of Tokyo, Japan
Lila Kari	University of Western Ontario, Canada
Satoshi Kobayashi	University of Electro-Communication, Japan
Chengde Mao	Purdue University, USA
Satoshi Murata	Tohoku University, Japan
John Reif	Duke University, USA
Grzegorz Rozenberg	Leiden University, The Netherlands
Nadrian Seeman	New York University, USA
Friedrich Simmel	Technical University of Munich, Germany
Andrew Turberfield	Oxford University, UK
Hao Yan	Arizona State University, USA
Erik Winfree	California Institute of Technology, USA

Program Committee

Robert Brijder (Co-chair)	Hasselt University, Belgium
Lulu Qian (Co-chair)	California Institute of Technology, USA
Ebbe Sloth Andersen	Aarhus University, Denmark
Luca Cardelli	Microsoft Research, Cambridge, UK
Anne Condon	University of British Columbia, Canada
David Doty	University of California, Davis, USA
Andrew Ellington	University of Texas at Austin, USA
Andre Estevez-Torres	Pierre and Marie Curie University, France

Elisa Franco	University of California at Riverside, USA
Cody Geary	California Institute of Technology, USA
Kurt Gothelf	Aarhus University, Denmark
Hongzhou Gu	Yale University, USA
Natasha Jonoska	University of South Florida, USA
Ming-Yang Kao	Northwestern University, USA
Lila Kari	University of Western Ontario, Canada
Ibuki Kawamata	Tohoku University, Japan
Yonggang Ke	Georgia Institute of Technology, USA
Chenxiang Lin	Yale University, USA
Yan Liu	Arizona State University, USA
Jan Manuch	University of British Columbia, Canada
Chengde Mao	Purdue University, USA
Pekka Orponen	Aalto University, Finland
Andrew Phillips	Microsoft Research, Cambridge, UK
John Reif	Duke University, USA
Yannick Rondelez	CNRS/ESPCI Paris, France
Robert Schweller	University of Texas Rio Grande Valley, USA
Shinnosuke Seki	University of Electro-Communications, Japan
David Soloveichik	University of Texas at Austin, USA
Darko Stefanovic	University of New Mexico, USA
Chris Thachuk	California Institute of Technology, USA
Andrew Turberfield	University of Oxford, UK
Damien Woods	Inria, France
Bernard Yurke	Boise State University, USA

Additional Reviewers

Nathanael Aubert-Kato
Keenan Breik
Cameron Chalk
Steven Graves
Hiroaki Hata
Matthew R. Lakin
Luca Laurenti
Trent Rogers

Sponsors

Savran Technologies
BIOO Scientific
Promega Corporation
Integrated DNA Technologies

International Society of Nanoscale Science, Computing, and Engineering
The Office of Naval Research

Army Research Office
The National Science Foundation

The following departments of the University of Texas at Austin:
Center for Systems and Synthetic Biology
Department of Molecular Biosciences
Assistant VP for Research
Cockrell School of Engineering
Electrical & Computer Engineering
Center for Infectious Disease

Contents

Automated, Constraint-Based Analysis
of Tethered DNA Nanostructures

Matthew R. Lakin[1,2,3]([✉]) and Andrew Phillips[4]

[1] Department of Computer Science, University of New Mexico,
Albuquerque, NM, USA
mlakin@cs.unm.edu
[2] Center for Biomedical Engineering, University of New Mexico,
Albuquerque, NM, USA
[3] Department of Chemical and Biological Engineering, University of New Mexico,
Albuquerque, NM, USA
[4] Microsoft Research, 21 Station Rd, Cambridge CB1 2FB, UK
aphillip@microsoft.com

Abstract. Implementing DNA computing circuits using components tethered to a surface offers several advantages over using components that freely diffuse in bulk solution. However, automated computational modeling of tethered circuits is far more challenging than for solution-phase circuits, because molecular geometry must be taken into account when deciding whether two tethered species may interact. Here, we tackle this issue by translating a tethered molecular circuit into a constraint problem that encodes the possible physical configurations of the components, using a simple biophysical model. We use a satisfaction modulo theories (SMT) solver to determine whether the constraint problem associated with a given structure is satisfiable, which corresponds to whether that structure is physically realizable given the constraints imposed by the tether geometry. We apply this technique to example structures from the literature, and discuss how this approach could be integrated with a reaction enumerator to enable fully automated analysis of tethered molecular computing systems.

1 Introduction

Molecular computing using solution-phase DNA circuits is a powerful method for implementing nanoscale information processing systems. In particular, DNA strand displacement has emerged as a proven method of engineering networks of programmable computational elements [1,2] and a powerful theoretical framework has built up around it [3,4]. In the computational modeling of DNA strand displacement networks, the assumption that all components are freely diffusing in bulk solution means that the physical conformations of interacting molecular species may be neglected when enumerating the possible reactions. This assumption is heavily exploited in existing modeling systems for solution-phase DNA strand displacement networks, such as the DSD programming language [5] and the associated Visual DSD software [6].

© Springer International Publishing AG 2017
R. Brijder and L. Qian (Eds.): DNA 23 2017, LNCS 10467, pp. 1–16, 2017.
DOI: 10.1007/978-3-319-66799-7_1

However, the fact that any pair of components in a freely diffusing molecular circuit may potentially diffuse into proximity and interact means that the entire circuit must be designed so as to eliminate crosstalk from all possible undesired interactions. Thus, even if the circuit is designed in a modular fashion, each instance of a given module must be implemented using different nucleotide sequences, which limits the scalability of circuits. Indeed, the largest DNA computing circuit built to date is the "square root" circuit reported by Qian and Winfree [7], which required 74 initial DNA species (excluding input signals) for a total of 130 different DNA strands. Reducing the number of distinct molecular species required to implement complex computational systems is therefore a crucial direction for future research in molecular computing.

Advances in the field of structural DNA nanotechnology, in particular, the development of DNA origami [8] as a reliable method for building DNA tile nanostructures [9], has raised the possibility of implementing molecular computers using components that are tethered to a DNA origami tile surface. This would enable nucleotide sequences to be shared between components that are tethered far enough apart to not interact, which means that large-scale circuits could be constructed using many copies of standardized components. Furthermore, tethering circuit components in close proximity increases the speed of circuit computation because components do not need to diffuse in three dimensions prior to interacting [10]. Indeed, several proposals for implementing molecular computers using surface-bound DNA strand displacement networks based on hairpin-opening reactions have been previously published [11,12]. A further motivation for studying molecular circuits tethered to DNA tiles is that DNA nanostructures show great promise as a vehicle to deliver theranostic molecular devices to cells and tissues [13].

From a modeling perspective, however, the move to implementing molecular computers using components that are tethered to a surface presents some challenges. In particular, tethering components to the surface at particular locations makes it necessary to consider the geometry of the system when deciding whether two components can actually interact with each other. For example, it might be the case that two components contain complementary single-stranded domains, but are tethered to the surface too far apart to actually be able to hybridize. In certain cases it may be possible to derive simple expressions to compute whether domains can interact [12], but this is not possible in the general case. Our previous work on modeling tethered strand displacement networks [14] avoided this issue by relying on the programmer to specify which components could interact, however, in many cases this requires additional biophysical calculations to be made separately. Thus, in order to build an automated compiler for tethered molecular circuits, we require an automated, general purpose method for computing whether the geometry of two tethered molecular species permits them to interact.

In this paper, we present an automated solution to the problem of reaction enumeration in tethered strand displacement systems. We translate tethered structures into sets of arithmetic constraints on variables that represent the physical locations of their components. If the constraint problem associated with a given structure is *satisfiable*, it means that there is a plausible physical configuration of the components that can be adopted by the system. Conversely, if the constraint problem is *unsatisfiable*, there is no physical configuration of the components that can produce that structure. By solving the constraints using an off-the-shelf satisfaction modulo theories (SMT) solver, we obtain a procedure that allows us to automatically detect whether a structure is physically plausible, given the constraints imposed by the tethers. To our knowledge, this is the first fully automated procedure for analyzing molecular geometry in the context of a tethered molecular computing system.

2   Localized Processes

In this section, we present our language of localized processes for representing tethered DNA structures. Definition 1 extends the process calculus-based presentation of DNA nanostructures from our previous work [15] with syntax for tethers on strands that attach them to a surface (representing, e.g., a DNA origami tile) at particular coordinates, and annotations to represent the physical coordinates of different parts of the structure.

Definition 1 (Localized processes). *The syntax of localized processes, P, is expressed in terms of* domain names a, b, ..., bond names i, j, k, ..., variables x, y, z, ..., *and* real-valued constants $r \in \mathbb{R}$. *Then, the grammar of localized processes is as follows.*

Coordinate $c ::= x$		*Variable*
$\mid r$		*Real-valued constant*
Tether $t ::= tether$		*Tethered*
$\mid \varepsilon$		*Untethered*
Position $\pi ::= t(c_1, c_2, c_3)$		*Position with coordinates*
Domain $d ::= a$		*Domain name*
$\mid a^*$		*Complemented domain name*
(Un)bound domain $o ::= d$		*Free domain*
$\mid d!i$		*Bound domain with bond i*
Strand $S ::= <\pi_0 \ o_1 \ \pi_1 \ \cdots \ \pi_{M-1} \ o_M \ \pi_M>$		*Strand with M domains, $M \geq 1$*
Process $P ::= (S_1 \mid \cdots \mid S_N)$		*Multiset of N strands, $N \geq 0$*

We write $strands(P)$, $bonds(P)$, $posns(P)$, and $vars(P)$ for the sets of *strands, bonds, positions,* and *variables* that occur in P, respectively. We consider processes *equal* up to re-ordering of strands, renaming of bonds, and renaming of variables, and we only consider processes that are *well-formed*, by which we mean that each bond in $bonds(P)$ appears *exactly twice* and is shared between complementary domains, and that each variable in $vars(P)$ appears *exactly once*.

We assume the existence of a predicate *toehold* such that $toehold(a)$ returns true if and only if a is a toehold domain. For convenience, we also write $a\hat{\ }$ iff $toehold(a)$ holds. We also assume the existence of a function *len* such that $len(a)$ returns the length in nucleotides of the DNA sequence for the domain name a (and thus returns a positive integer). We assume that these functions are also lifted to possibly-complemented domains d, so that $f(a^*) = f(a)$, for $f \in \{toehold, len\}$. Similarly, we lift the complementation syntax to possibly-complemented domains, by defining $(a^*)^* = a$.

Between each pair of domains on every strand, and on each strand terminus, is a *position* that represents the location of that part of the structure in 3D space. The position $\varepsilon(c_x, c_y, c_z)$, which we may abbreviate as simply (c_x, c_y, c_z), means that part of the structure is untethered and located at x-coordinate c_x, y-coordinate c_y, and z-coordinate c_z. (Each of the coordinates may be a variable that can be assigned a value during the constraint solving procedure, or a real-valued constant that represents a specific location.) Similarly, the position $tether(c_x, c_y, c_z)$ represents a tether that attaches that part of the strand to the underlying surface at the location (c_x, c_y, c_z) where, typically, $c_z = 0$ and c_x and c_y are real-valued constants. In practice, we may elide certain positions from the process syntax altogether, with the understanding that the physical position of junction between the domains on either side (or the strand terminus, if at the end of the strand) will be represented by a freshly generated position with syntax $(c_x^\dagger, c_y^\dagger, c_z^\dagger)$, where $c_x^\dagger$, $c_y^\dagger$, and $c_z^\dagger$ are unique, freshly generated variables.

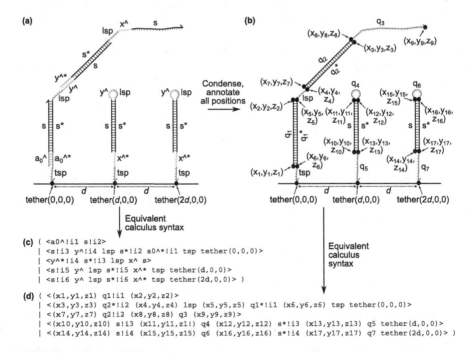

Fig. 1. Example of representing a tethered structure. **(a)** Secondary structure of a three-stator transmission line after [11], with the three stators arranged in a straight line. **(b)** Condensed version of the structure from (a), with all positions fully annotated. Here, the q domains are the freshly generated domain names. **(c)** Calculus syntax representing the initial structure from (a). **(d)** Calculus syntax representing the condensed structure from (b).

Note that we use the term "process" here for historical reasons. Although the long-term goal of our work is to enable automated enumeration of interactions in localized processes, in this paper we focus solely on the problem of analyzing the geometry of a single state of the localized process, and we use the process syntax solely as a means of representing the structure in question. Figure 1 presents graphical and syntactic representations of the tethered system that will serve as our running example throughout the paper—a three-stator transmission line after [11].

3 Biophysical Model

Before we present our translation of localized processes into constraint sets, we first present the assumptions about the biophysics of DNA that will underlie the translation. Crucially, we will model double-stranded DNA duplexes as rigid rods and single-stranded DNA as infinitely flexible, freely jointed chains. We also assume that joints (nicks) between double-stranded duplexes are infinitely flexible. We will neglect the thickness of DNA strands, and will also neglect the length of the bonds between complementary bases by requiring the ends of the two strands that make up a duplex to be positioned at exactly the same point in space. In computing whether structural components may interact, we will also not account for steric effects that could, for example, eliminate solutions to the constraint problem that would require part of the structure to pass through another to form a catenane. This model is clearly simplified, and we refer the user to Sect. 7 for further discussion of these assumptions and how this model might be made more realistic.

4 Condensing Localized Processes

When deriving geometric constraints from localized processes, we will model every double-stranded duplex as a single rigid rod, as described in Sect. 3 above, even when that duplex consists of multiple sequence domains. Thus, before converting a localized process into a set of constraints we must first *condense* the process by combining all domains on a given strand that represent a continuous duplex into a single extended domain. To reduce the size of the resulting constraint problem, we will also condense continuous single-stranded regions into extended domains. The condensing process is a straightforward binary relation $\longrightarrow_{condense}$ on localized processes that can be defined by the following rewrite rules:

$$(<\cdots \ \pi_j \ d_1!i_1 \ \pi_{j+1} \ d_2!i_2 \ \pi_{j+2} \ \cdots> \ | \ <\cdots \ \pi_k \ d_2{}^*!i_2 \ \pi_{k+1} \ d_1{}^*!i_1 \ \pi_{k+2} \ \cdots> \ | \ P)$$
$$\longrightarrow_{condense} (<\cdots \ \pi_j \ q!i_1 \ \pi_{j+2} \ \cdots> \ | \ <\cdots \ \pi_k \ q^*!i_1 \ \pi_{k+2} \ \cdots> \ | \ P)$$

$$(<\cdots \ \pi_j \ d_1!i_1 \ \pi_{j+1} \ d_2!i_2 \ \pi_{j+2} \ \cdots \ \pi_k \ d_2{}^*!i_2 \ \pi_{k+1} \ d_1{}^*!i_1 \ \pi_{k+2} \ \cdots> \ | \ P)$$
$$\longrightarrow_{condense} (<\cdots \ \pi_j \ q!i_1 \ \pi_{j+2} \ \cdots \ \pi_k \ q^*!i_1 \ \pi_{k+2} \ \cdots> \ | \ P)$$

$$(<\cdots \ \pi_j \ d_1 \ \pi_{j+1} \ d_2 \ \pi_{j+2} \ \cdots> \ | \ P) \longrightarrow_{condense} (<\cdots \ \pi_j \ q \ \pi_{j+2} \ \cdots> \ | \ P)$$

where, in each case, q is a freshly-chosen domain name, for which we assume that $len(q) = len(d_1) + len(d_2)$. We also require that any position removed by the condensing process has the form $\varepsilon(x, y, z)$, that is, the position is untethered and is specified solely by variables and not by constants. In the rule definitions, these positions are referred to as π_{j+1} and π_{k+1}. This condition essentially says that no information is lost by discarding that position, because it was not previously constrained to a specific location in space. If this condition is violated, the condensing process fails and we cannot proceed. However, for practical purposes this is a reasonable condition to impose, since to our knowledge there are no proposed or implemented designs for tethered molecular circuits that include a tether part way along a DNA duplex.

Thus, a single condensing step takes two neighbouring bound domains, either on two strands (the first rule) or on a single strand (the second rule), or two neighbouring single-stranded domains on the same strand (the third rule) and converts them into a single domain with a freshly-generated name (which avoids conflicts with other domain names in the system) whose length is the sum of the lengths of the two domains it replaces. If the domains were initially bound, they remain so in the condensed version of the process. Furthermore, the coordinates that represent the positions of the endpoints of the new bound domains are the same as those that represented the endpoints of the neighbouring bound domains that were replaced. Figure 1 presents an example of condensing, as applied to our running example system.

It is not hard to see that the condensing process preserves well-formedness of processes (since, in the first two rules, both occurrences of i_2 are removed and i_1 connects two complementary domains in the resulting process), that it is terminating (since every rule application removes a pair of neighbouring bound domains) and that it produces a unique normal form when no more rule applications are possible (modulo the choice of freshly-named new domains that are introduced).

Henceforth we will assume that all processes have been condensed via this procedure, so that all duplexes are represented by a single domain on each bound strand and all neighbouring single-stranded domains have been similarly collapsed into a given domain. This ensures that all domain junctions in the process correspond to points where the structure is flexible.

5   Geometric Constraints for Localized Processes

This section details the main technical contribution of the paper, where we define a translation of a (condensed) localized process into a *set* of arithmetic constraints that represent the possible geometry of the structure according to the biophysical model outlined in the previous section. This set of constraints will collectively encode the geometry of the structure, and our intention is that the set structure represents an implicit conjunction, so that the set of constraints is satisfied iff *all* constraints within the set are simultaneously satisfied (under some mapping of variables to values).

We will use a 3D Cartesian coordinate system that allows us to locate each domain junction with respect to the surface to which the tethered components are attached. For simplicity, we assume that all tethered components are attached to the *same* surface, though that restriction could be straightforwardly relaxed if we assume that there are no interactions between components tethered to different surfaces. By convention, we will assume that the surface (which could model, e.g., a DNA origami tile) lies in the $z = 0$ plane, and that all components are tethered to the same side of the surface and protrude on the positive z side. We now begin to define the different classes of constraints that will make up the constraint-based representation of structure geometry. First, we define constraints that represent the physical length of each domain in a process.

Definition 2 (Domain length constraints). *For a localized process P we define the corresponding domain length constraints, $\mathsf{dlc}(P)$, as follows:*

$$\mathsf{dlc}(P) = \bigcup_{S \in strands(P)} \mathsf{dlc}(S)$$

where $\mathsf{dlc}(<\pi_0 \; o_1 \; \pi_1 \; \cdots \; \pi_{N-1} \; o_N \; \pi_N>) =$
$\mathsf{dlc}(\pi_0, o_1, \pi_1) \cup \mathsf{dlc}(\pi_1, o_2, \pi_2) \cup \cdots \cup \mathsf{dlc}(\pi_{N-1}, o_N, \pi_N)$

and $\mathsf{dlc}(t(c_x, c_y, c_z), o, t'(c'_x, c'_y, c'_z)) =$
$$\begin{cases} \{(c'_x - c_x)^2 + (c'_y - c_y)^2 + (c'_z - c_z)^2 = (len(a) \times \mathsf{L_{ds}})^2\} & if \; o = a!i \\ \{(c'_x - c_x)^2 + (c'_y - c_y)^2 + (c'_z - c_z)^2 \leq (len(a) \times \mathsf{L_{ss}})^2\} & if \; o = a. \end{cases}$$

In this definition, we must distinguish between single-stranded and double-stranded domains. Thus, $len(a) \times \mathsf{L_{ds}}$ is the (fixed) length of a double-stranded domain a and $len(a) \times \mathsf{L_{ss}}$ is the *maximum* extended length of a single-stranded domain a. The constraints then specify the fixed distance between the two ends of a double-stranded domain (which we model as a rigid rod) or the maximum distance between the two ends of a single-stranded domain (which we model as an infinitely flexible freely-jointed chain), in terms of the variables used to represent the coordinates of each end of the domain. We now define the constraints that encode hybridization between domains.

Definition 3 (Hybridization constraints). *For a localized process P we define the corresponding hybridization constraints, $\mathsf{hc}(P)$, as follows:*

$$\mathsf{hc}(P) = \bigcup_{i \in bonds(P)} \mathsf{hc}_P(i)$$

where $\mathsf{hc}_P(i) = \{c_x = c'''_x, c_y = c'''_y, c_z = c'''_z, c'_x = c''_x, c'_y = c''_y, c'_z = c''_z\}$
and $<\cdots t(c_x, c_y, c_z) \; a!i \; t'(c'_x, c'_y, c'_z) \cdots> \in strands(P)$
and $<\cdots t''(c''_x, c''_y, c''_z) \; a^*!i \; t'''(c'''_x, c'''_y, c'''_z) \cdots> \in strands(P).$

As one of the key simplifications in our biophysical model, we assume that the diameter of a DNA duplex is negligible. Thus, if the two (complementary) domains are hybridized, the 5' end of one domain must be colocated with the 3' end of the other, and vice versa. Note that, in the above definition, since $i \in bonds(P)$ and since we assume that P is well-formed, it follows that there exists precisely one pair of complementary domains connected via the bond i, which will satisfy the side conditions.

Finally, we define constraints that situate the structure with respect to the tile surface. This includes the constraints imposed by tethers, which anchor one end of a domain at a fixed location on the tile surface, and constraints that stipulate that no part of the structure may pass through the tile surface.

Definition 4 (Tile constraints). *For a localized process P we define the corresponding tile constraints, $\mathsf{tc}(P)$, as follows:*

$$\mathsf{tc}(P) = \bigcup_{\pi \in posns(P)} \mathsf{tc}(\pi)$$

where $\quad \mathsf{tc}(tether(c_x, c_y, c_z)) = \{c_z = 0\}$
and $\quad \mathsf{tc}(\varepsilon(c_x, c_y, c_z)) = \{c_z \geq 0\}.$

Thus, positions on a strand where a tether appears (in practice this is typically at the end of the strand, though our model does not require this) are constrained to specific coordinates, with a z coordinate of 0. Non-tethered positions are simply constrained to have a non-negative z coordinate, i.e., they cannot be "below" the tile. We now combine the different kinds of constraint defined above to create a constraint-based model that encodes the possible geometric configurations of a localized process.

Definition 5 (Constraint-based biophysical models). *For a condensed, localized process P, we write $\mathcal{M}_P$ for the full, constraint-based model, which is defined as follows:*
$$\mathcal{M}_P = \mathsf{dlc}(P) \cup \mathsf{hc}(P) \cup \mathsf{tc}(P).$$

Thus, all constraints from domain lengths, hybridization, and tethers are unioned together to produce a single set of constraints that represents the possible geometric configurations of the structure of the localized process.

Definition 6 (Plausibility of localized processes). *We say that a localized process P is plausible iff its associated constraint set is satisfiable, that is, if there exists a mapping from the variables to real numbers such that all constraints in the constraint set are simultaneously satisfied when the mapping is applied. A localized process that is not plausible is implausible.*

If a localized process is implausible, this means there is no way to arrange the domains in space to form the specified structure without breaking one or more of the geometric constraints imposed by that structure. For example, in the case of a tethered DNA circuit, this may mean that the tethered locations of two components may leave them too far apart to interact with each other.

Thus, we can enumerate reactions between tethered components while respecting the tether geometry by checking plausibility of the localized process P that represents the candidate product structure, i.e., by checking satisfiability of the corresponding constraint problem $\mathcal{M}_P$.

6   Results

6.1   Prototype Implementation

We implemented a prototype system for checking satisfiability of constraint-based geometric models using the Z3 SMT solver [16]. The prototype is written in Python, and takes as input a localized process, expressed in the syntax from Definition 1. This is then condensed and converted into the corresponding constraint problem (as defined in Definition 5) using the Z3 Python API which can then be checked for satisfiability.

Our preliminary experiments represented domain positions using real-valued variables, because Z3 includes a complete algorithm for solving non-linear constraints over real variables exactly [17]. However, while this algorithm is complete it can be very computationally expensive, and even just changing the values of certain constants within a constraint problem (without changing the structure of the problem) can cause the time taken to solve the constraints to increase from microseconds to many hours. Therefore, we adopted an alternative approach to encode the constraints using floating-point variables, which can be represented within Z3 as bit-vectors of a fixed width. This makes the time taken to solve the constraints far more predictable, on the order of thirty seconds to one minute depending on the particular problem. However, floating-point representations are inexact, which introduces the possibility of constraints being incorrectly deemed unsatisfiable if any of the required variable assignments cannot be represented exactly using the floating-point representation in question. (The width of the bit vector, and the numbers of bits used to represent the significand and exponent, can be tuned to reduce numerical inaccuracy at the cost of increased compute time to solve the constraints.) We addressed this issue by modifying our constraints slightly, by introducing a *tolerance* parameter ε, so that variables will be considered equal if their values fall within that range. This addresses the issue of numerical inaccuracy because, for a judicious choice of ε, it is highly likely that there will be representable values within the tolerance range that satisfy the constraints. To implement this (approximate) solving algorithm, we simply modify the generated constraints in the model as follows, where e is an arithmetic expression and c is a constant. Thus:

$\cdot\ e \oplus c$ becomes $e \oplus (c - \varepsilon)$, for $\oplus \in \{>, \geq\}$;
$\cdot\ e \oplus c$ becomes $e \oplus (c + \varepsilon)$, for $\oplus \in \{<, \leq\}$; and
$\cdot\ e = c$ becomes $\{e \geq (c - \varepsilon), e \leq (c + \varepsilon)\}$.

This transformation from exact to inexact constraints causes the number of constraints to double at most, and does not increase the number of variables at all. Furthermore, the terms of the form $c \pm \varepsilon$ are constants that can be precomputed.

6.2   Example

We tested our approach using an example taken from the repeater system on hairpin-based circuits on DNA origami tiles [11], introduced in Fig. 1. We are particularly interested in the reaction between two tethered species, i.e., when the first stator hairpin has been opened, the freely diffusing fuel hairpin has bound and opened, and the opened fuel hairpin is trying to bind to the second stator hairpin, which is still closed. Because the stator hairpins share common nucleotide sequences, there are two possible interactions, which are shown in Fig. 2(a) and (b). Figure 2(a) shows the desired interaction, where the opened fuel hairpin binds to the second stator hairpin (S_1), and Fig. 2(b) shows an undesirable interaction, where the opened fuel hairpin binds to the third stator hairpin (S_2).

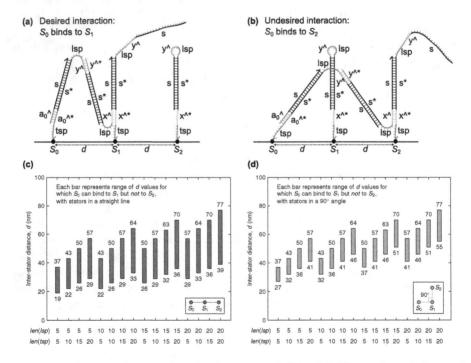

Fig. 2. Results on constraint-based structure modeling. **(a)** Illustration of the product structure for a three-stator transmission line after [11], when S_0 binds to S_1, as desired. **(b)** Illustration of the product structure for the undesirable interaction of the three-stator transmission line, when S_0 binds to S_2. **(c)** Results from constraint-based analysis of the transmission line system when all three stators are arranged in a straight line (bird's eye view shown in inset). Each bar shows the range of integer values of d for which S_0 can bind to S_1 but *not* to S_2, as required for correct signal transmission, for various combinations of loop spacer (lsp) and tether spacer (tsp) domain lengths. **(d)** Results from constraint-based analysis of the transmission line system when all three stators are arranged in a 90° angle (bird's eye view shown in inset), with data analysis carried out as specified in part (c).

In our presentation, every hairpin has a "loop spacer" domain (lsp), and every *tethered* hairpin has a "tether spacer" domain (tsp). Adjusting the lengths of these domains and the inter-stator distance (d) changes the possible behaviors of the system, and the goal of our analysis is to find the parameter sets that enable correct signal transmission, i.e., so that the opened stator S_0 can bind to S_1, resulting in the desired product structure shown in Fig. 2(a), but *not* to S_2, which would result in the undesired product structure shown in Fig. 2(b).

Results from constraint-based analysis of this example system with the stators arranged in a straight line are presented in Fig. 2(c). As described above, we used a floating-point representation for our constraint variables, with an 8 bit exponent and an 8 bit significand. The ε value was 10^{-5}. Values for L_{ds} and L_{ss} were as used in previous work [12], that is, $L_{ds} = 0.34$ nm and $L_{ss} = 0.68$ nm. The lengths of all toehold domains were 5 nucleotides and the lengths of all long domains were 30 nucleotides. For each combination of loop spacer (lsp) and tether spacer (tsp) lengths (5, 10, 15, and 20 nucleotides each) we used our prototype system to construct the condensed, localized process representation of the candidate structures formed by interactions between S_0 and S_1 and between S_0 and S_2, convert them to our constraint representation, and test plausibility of each using Z3, for values of the inter-stator distance (d) chosen at 1 nm intervals ranging from 1 nm to 100 nm.[1] Each bar in Fig. 2(c) represents the range of d values for which S_0 can bind to S_1 but *not* to S_2, as required for correct signal transmission. Below the bar, the stators are close enough together that S_0 can bind directly to S_2, and above the bar, the stators are so far apart that S_0 cannot even reach S_1. These results show that the range of acceptable values for d increases as the lengths of the lsp and tsp domains increase, as we would expect.

We also analyzed a similar example system that used the same structures, except that the three stator hairpins are arranged to form a 90° angle on the tile surface. In the straight line system, we added extra constraints that all y-coordinates equal zero, as this decreases solving time, but this was *not* done for the 90° angle system. Hence, this example illustrates the applicability of our approach to processes whose components occur in non-trivial geometric arrangements. Results from constraint-based analysis of the 90° angle system are presented in Fig. 2(d). These results show that, for identical loop spacer (lsp) and tether spacer (tsp) domain lengths, the maximum safe inter-stator distances are the same in both the straight-line and 90° angle cases, but the minimum safe distance is larger in the 90° angle case. This is because S_2 can be reached diagonally from S_0 when the stators are not arranged in a straight line. Thus, our constraint-based analysis can determine the feasible range of inter-stator distances that enable correct signal transmission without skipping any stators in the sequence, and thereby serves as a proof of concept for automated analysis of tethered molecular circuits.

---

[1] See the Supporting Information (available from the first author's web page) for full details of the examples and corresponding constraints.

7   Discussion

The main contribution of this paper is the development of a constraint-based methodology for *automatically* analyzing molecular geometry to determine if certain interactions between species are possible, given the physical constraints imposed by tethers that attach components to a surface (e.g., a DNA origami tile) at specific locations. Our approach therefore offers a principled, general-purpose alternative to the somewhat *ad hoc* rules adopted to handle molecular geometry in other reaction enumerators, e.g., our strand graph system [15] or the DyNAMiC Workbench [18].

7.1   Reaction Enumeration

The key advantage of our fully automated analysis is that it could be used in the main loop of a compiler for tethered molecular reaction networks. In this vein, Algorithm 1 presents a pseudocode algorithm for enumerating the state space of a tethered molecular reaction network, in which our constraint-based analysis of the plausibility of tethered molecular structures is used as a filter to prevent the addition of a transition to the state space, even if the domains match, if the reaction would yield a product process whose structure is implausible. We note that, strictly speaking, the constraint-solving process is only required for interactions between different parts of a single tethered structure since, for bimolecular reactions where one or both of the reactants is not tethered, one may

---

Algorithm 1: Pseudocode algorithm for constructing the reachable state space of a localized process. This pseudocode assumes the existence of several functions: *unprocessed_states*(S), which returns the set of states in S that have not yet been processed by the algorithm; *candidate_reactions*(P), which returns the set of possible reactions in the process P when considering only sequence complementarity and *not* molecular geometry; and *is_plausible*(P), which decides whether the localized process P is geometrically plausible, using our constraint-based technique.

---

begin
    initialize state space S with localized process P_{init} as the initial state;
    mark state P_{init} as unprocessed;
    while *unprocessed_states*$(S) \neq \varnothing$ **do**
        let $P_{reactant}$ be some element of *unprocessed_states*(S);
        for $(P_{reactant} \rightarrow P_{product}) \in$ *candidate_reactions*$(P_{reactant})$ **do**
            if *is_plausible*$(P_{product})$ **then**
                if $P_{product} \notin S$ **then**
                    ⌊ add state $P_{product}$ to S;
                add transition $(P_{reactant} \rightarrow P_{product})$ to S;
                mark state $P_{reactant}$ as processed;
end

---

assume that the non-tethered species can diffuse and adopt any conformation required to react.

7.2 Biophysical Model

The procedure outlined here is for computing *whether* two domains may bind, and an extension of this problem would be to determine *at what rate* that binding reaction may occur. In solution-phase systems, one approach to approximate binding rates for particular domains is to use free energy models to compute the free energy of a bound toehold domain, and then make the assumption that binding rates of that domain are dominated by diffusion, which allows an approximation of the off-rate to be computed. However, in tethered systems we cannot necessarily assume that on-rates are primarily influenced by diffusion. An alternative is to use the local concentration approach [19], which corresponds to calculating the probability that the domains will be in a conformation such that they may bind. It is possible that the output from the constraint solver could be used to estimate the number of different coordinate variable instantiations and hence the size of the parameter space in which the constraints are satisfied, which one might be able to use to approximate the rate. This would require us to consider not just the maximum possible extension of single-stranded domains, but also the probability that they are extended to given lengths e.g., using a worm-like chain biophysical model [19].

Indeed, the biophysical model that we have used in this paper is deliberately simple, so that the structures can be compiled directly to tractable constraint problems. Some of our assumptions seem necessary for the technique to work, such as the assumption that hybridized domains are co-localized in space, because to do otherwise would require additional constraints on the angles of certain domains to ensure that they are parallel, which would greatly complicate the constraint problems to be solved. Other assumptions, such as the condensing of single-stranded domains, make sense as a simple optimization which reduces the number of variables present in the constraint problem. An interesting direction for future research will be to consider further optimizations that could speed up the analysis of structures and enable larger structures to be analyzed: one possibility might be to prune constraints that arise from parts of the structure that are not actually essential to determining whether the structure is plausible or not, another might be to remove domain length constraints that are technically redundant because the length of those domains has already been constrained by hybridization to some other domain. The Z3 solver itself includes facilities for the simplification of constraint problems, which could be directed in specific ways to optimize solving the type of problems produced by our translation process. Another possibility might be to explore alternative methods for deciding plausibility that do not involve SMT solving, e.g., employing sampling, search, or genetic algorithms over the space of variable instantiations to try to find an instantiation that satisfies all of the constraints. These alternative approaches could enable larger constraint problems, corresponding to more sophisticated molecular structures, to be tackled.

While our biophysical model is clearly not suited to in-depth biophysical investigations of nucleic acid dynamics, it *is* useful as a simplified model for (automated) preliminary investigations, that can be used to rule out certain designs and to guide more in-depth analyses, e.g., using coarse-grained simulation models [20]. The key advantage of an automated structural analysis tool such as ours is that it may significantly reduce the number of iterative experimental design cycles that must be carried out in order to successfully implemented a tethered molecular circuit, thereby reducing wasted labour and reagents expended on unsuccessful designs.

A fruitful avenue for further research will be to determine how much more realistic our biophysical model can be made without sacrificing ease-of-use or constraint solving performance. One possibility would be to include additional constraints, e.g., to enforce a minimum extension on single-stranded domains, or to require that the lengths of double-stranded domains must be within some factor of the true value, to model slight fluctuations in the duplex structure. (The latter constraint could be imposed at no extra cost because our floating-point constraint expansion already includes an error term.) Another possibility would be to constrain the angles that can be formed at the junctions between domains, but this could adversely impact performance.

It would be an interesting future research direction to investigate whether our biophysical model is conservative, in the sense that if two strands cannot interact in our model, then they will not be able to interact in a more realistic model. It seems likely that certain assumptions would need to be made for this statement to be true, such as rigidity of the underlying origami tile. Furthermore, the biophysical constants, such as the internucleotide distances, would need to be chosen judiciously. It is worth pointing out that the converse may not hold, i.e., our system may permit certain reactions that a more detailed model might rule out, e.g., due to steric effects that might prevent the DNA from actually adopting the required conformation.

7.3   Extensions

The discussion above is phrased entirely in terms of tethered structures, however, similar techniques could be used in non-tethered but non-trivial DNA structures, e.g., those studied in our previous work on strand graphs [15]. In this context, we would simply pick an arbitrary position from the structure to serve as the origin and would omit the $tc(P)$ term from Definition 5 (which would also remove the constraints that prohibit negative z-coordinates). This would enable us to use constraint solving to find reachable domain bindings in a complex DNA nanostructure, with the caveat that issues such as potential steric hindrance between strands would not be modelled. Existing free energy approaches allow calculation of base-pairing probabilities [21] for nucleic acid nanostructures that include certain restricted classes of pseudoknots in polynomial time. As before, this thermodynamic model will be more physically accurate than our model, as it is based on thermodynamic experiments at the individual base-pair level.

However, our work could similarly be used as a precursor to a more detailed analysis using free energy methods.

It is worth noting that, in all of the examples presented above, it is the case that for any position $tether(c_x, c_y, c_z)$, then $c_x = r_x$, $c_y = r_y$, and $c_z = 0$, for some constants r_x and r_y. It is also the case that for any position $\varepsilon(c_x, c_y, c_z)$, then $c_x = x$, $c_y = y$, and $c_z = z$, for some variables x, y, and z. That is, the locations of all tethers are fixed and are always in the $z = 0$ plane, and all untethered domain junctions are completely free to vary (subject to the other constraints). However, our approach is more general than this—in principle, we could leave certain tether coordinates unspecified and let the constraint solver search for satisfying instantiations. Thus, this work lays the foundation for the development of further automated tools for tethered molecular computing systems, such as circuit synthesis tools that take a logic function as input and return a layout for a molecular circuit that will implement that logic function without crosstalk. In this context, our constraint-based system would be used to determine the positioning and spacing of components that would be required to make the products of the desired reactions plausible but the products of the undesired reactions implausible. Ideally, any circuit synthesis algorithm of this type would keep the constraint problems that must be solved as small as possible for maximum efficiency, and might use some kind of global optimization procedure or genetic algorithm to converge towards a suitable design. We could also allow non-zero z-coordinates for tether locations, which would correspond to the underlying surface being a 3D nanostructure as opposed to a flat tile.

Acknowledgments. This material is based upon work supported by the National Science Foundation under grants 1525553, 1518861, and 1318833. The authors thank Neil Dalchau and Rasmus Petersen for productive discussions.

References

1. Zhang, D.Y., Seelig, G.: Dynamic DNA nanotechnology using strand-displacement reactions. Nat. Chem. **3**(2), 103–113 (2011)
2. Chen, Y.-J., Dalchau, N., Srinivas, N., Phillips, A., Cardelli, L., Soloveichik, D., Seelig, G.: Programmable chemical controllers made from DNA. Nat. Nanotechnol. **8**, 755–762 (2013)
3. Soloveichik, D., Seelig, G., Winfree, E.: DNA as a universal substrate for chemical kinetics. PNAS **107**(12), 5393–5398 (2010)
4. Cook, M., Soloveichik, D., Winfree, E., Bruck, J.: Programmability of chemical reaction networks. In: Algorithmic Bioprocesses, pp. 543–584. Springer, Heidelberg (2009)
5. Lakin, M.R., Youssef, S., Cardelli, L., Phillips, A.: Abstractions for DNA circuit design. JRS Interface **9**(68), 470–486 (2012)
6. Lakin, M.R., Youssef, S., Polo, F., Emmott, S., Phillips, A.: Visual DSD: a design and analysis tool for DNA strand displacement systems. Bioinformatics **27**(22), 3211–3213 (2011)
7. Qian, L., Winfree, E.: Scaling up digital circuit computation with DNA strand displacement cascades. Science **332**, 1196–1201 (2011)

8. Rothemund, P.W.K.: Folding DNA to create nanoscale shapes and patterns. Nature **440**, 297–302 (2006)

9. Tikhomirov, G., Petersen, P., Qian, L.: Programmable disorder in random DNA tilings. Nat. Nanotechnol. **12**, 251–259 (2017)

10. Bui, H., Miao, V., Garg, S., Mokhtar, R., Song, T., Reif, J.: Design and analysis of localized DNA hybridization chain reactions. Small **13**(12), 1602983 (2017)

11. Muscat, R.A., Strauss, K., Ceze, L., Seelig, G.: DNA-based molecular architecture with spatially localized components. In: Proceedings of ISCA 13 (2013)

12. Dalchau, N., Chandran, H., Gopalkrishnan, N., Phillips, A., Reif, J.: Probabilistic analysis of localized DNA hybridization circuits. ACS Synth. Biol. **4**(8), 898–913 (2015)

13. Walsh, A.S., Yin, H., Erben, C.M., Wood, M.J.A., Turberfield, A.J.: DNA cage delivery to mammalian cells. ACS Nano **5**(7), 5427–5432 (2011)

14. Lakin, M.R., Petersen, R., Gray, K.E., Phillips, A.: Abstract modelling of tethered DNA Circuits. In: Murata, S., Kobayashi, S. (eds.) DNA 2014. LNCS, vol. 8727, pp. 132–147. Springer, Cham (2014). doi:10.1007/978-3-319-11295-4_9

15. Petersen, R.L., Lakin, M.R., Phillips, A.: A strand graph semantics for DNA-based computation. Theor. Comput. Sci. **632**, 43–73 (2016)

16. Moura, L., Bjørner, N.: Z3: an efficient SMT solver. In: Ramakrishnan, C.R., Rehof, J. (eds.) TACAS 2008. LNCS, vol. 4963, pp. 337–340. Springer, Heidelberg (2008). doi:10.1007/978-3-540-78800-3_24

17. Jovanović, D., Moura, L.: Solving non-linear arithmetic. In: Gramlich, B., Miller, D., Sattler, U. (eds.) IJCAR 2012. LNCS (LNAI), vol. 7364, pp. 339–354. Springer, Heidelberg (2012). doi:10.1007/978-3-642-31365-3_27

18. Grun, C., Werfel, J., Zhang, D.Y., Yin, P.: DyNAMiC Workbench: an integrated development environment for dynamic DNA nanotechnology. JRS Interface **12**, 20150580 (2015)

19. Genot, A.J., Zhang, D.Y., Bath, J., Turberfield, A.J.: Remote toehold: a mechanism for flexible control of DNA hybridization kinetics. J. Am. Chem. Soc. **133**, 2177–2182 (2011)

20. Doye, J.P.K., Ouldridge, T.E., Louis, A.A., Romano, F., Šulc, P., Matek, C., Snodin, B.E.K., Rovigatti, L., Schreck, J.S., Harrison, R.M., Smith, W.P.J.: Coarse-graining DNA for simulations of DNA nanotechnology. Phys. Chem. Chem. Phys. **15**, 20395–20414 (2013)

21. Dirks, R.M., Pierce, N.A.: An algorithm for computing nucleic acid base-pairing probabilities including pseudoknots. J. Comput. Chem. **25**, 1295–1304 (2004)

DNA-Templated Synthesis Optimization

Bjarke N. Hansen, Kim S. Larsen$^{(\boxtimes)}$, Daniel Merkle$^{(\boxtimes)}$, and Alexei Mihalchuk

Department of Mathematics and Computer Science,
University of Southern Denmark, Odense, Denmark
{kslarsen,daniel}@imada.sdu.dk

Abstract. In chemistry, synthesis is the process in which a target compound is produced in a step-wise manner from given base compounds. A recent, promising approach for carrying out these reactions is DNA-templated synthesis, since, as opposed to more traditional methods, this approach leads to a much higher effective molarity and makes much desired one-pot synthesis possible. With this method, compounds are tagged with DNA sequences and reactions can be controlled by bringing two compounds together via their tags. This leads to new cost optimization problems of minimizing the number of different tags or strands to be used under various conditions. We identify relevant optimization criteria, provide the first computational approach to automatically inferring DNA-templated programs, and obtain optimal and near-optimal results.

1 Introduction

The first instance where DNA has been used to execute an algorithm in order to solve a combinatorial optimization problem dates back to 1994. In [1], Adleman demonstrated how a small instance of the Hamiltonian Path Problem could be solved using DNA sequences. Since then, DNA nanotechnology has been used as a powerful tool for a wide variety of research and engineering questions. Examples include polyhedral mesh rendering, where DNA sequences are designed such that they fold into predefined complex three-dimensional structures [3], and design of DNA-based molecular motors that can be used to transport cargo molecules [16]. Appealing features of DNA-based designs is their programmability, the inherent concurrency, the predictability, and the fact that DNA sequences are relatively cheap and easy to synthesize. The number of approaches utilizing DNA-based chemistry as a source for the discovery and the design of novel drug-like molecules has increased rapidly in recent years [7]. Basically all large pharmaceutical companies have already started utilizing this technology. DNA-based chemistry approaches include a method called DNA-templated organic synthesis [12], where the goal is to synthesize an organic compound in a step-wise manner. In an individual step of a synthesis plan [11], either two compounds are combined (affixation reaction) or a single compound is modified (cyclization reaction). This information can be captured in a rooted unary-binary tree, though often cyclization reactions can be ignored from a combinatorial point of view, making the tree binary. Chemists are aiming at *efficient* synthesis (the yield of all reactions and

© Springer International Publishing AG 2017
R. Brijder and L. Qian (Eds.): DNA 23 2017, LNCS 10467, pp. 17–32, 2017.
DOI: 10.1007/978-3-319-66799-7_2

therefore the yield of the overall process should be high) and *one-pot* synthesis (for instance, avoiding complicated separation and purification processes based on contaminating compounds that require subsequent extraction of a specific product from a mixture of compounds).

In DNA-templated synthesis, the base compounds are "tagged" with DNA sequences. These tags are used to bring the compounds in close vicinity (and thereby react). This is done by adding a complementary DNA strand, called an instruction strand, which is a concatenation of the complementary strands of the two tags that are attached to the base compounds. In contrast to classical synthesis approaches, DNA-templated synthesis allows for much lower concentrations of reactants due to the tagging, which leads to a dramatically increased effective molarity. We refer to [8,12] for in-depth reviews and specifically [10,13] for examples of successful, non-trivial, multi-step DNA-templated molecule syntheses.

The synthesis tree together with a specification of how to tag the base compounds and according to which topological ordering of the tree the reactions should be carried out defines a so-called DNA-templated program. While high-level formalisms for DNA computational structures have been studied [4,14] before, there are no prior attempts to automatically inferring DNA-templated programs based on a given synthesis tree. In [2], graph rewriting approaches have been used for verifying correctness of given DNA-templated programs, but neither were programs automatically inferred nor optimization questions answered. With careful choice of tagging and topological ordering, it is possible to use the same tags and strands repeatedly, which leads to the optimization problems we consider. To avoid unintended interference, tags and strands that should be different must be some mimimum edit distance away from each other. If one uses too many different tags or strands, these must be made longer in order to obtain this, leading to higher production costs.

Another cost stems from the tagging of chemical compounds, which is a somewhat sophisticated chemical procedure. Thus, while it is interesting to minimize the use of different tags and strands in general, it is also interesting just to minimize the number of different tags used on the base compounds.

We present (i) optimal or near-optimal methods for minimizing the number of strands, (ii) a somewhat more involved method for minimizing the number of strands and subsidiarily the number of tags, (iii) a method for minimizing the number of strands when only two different tags are allowed on base compounds, but longer programs using blocking are allowed, and, finally, (iv) a generic ILP formulation of the optimization problems which is then without time complexity bounds.

2   Modeling DNA-Templated Synthesis

The goal of this section is to present a model for DNA-templated synthesis such that we can work with these issues in a combinatorial manner. We identify some basic operations and restrictions on how these can be applied, with the goals in mind. We would like to emphasize that we do not make any simplifying assumptions, preventing our solutions from leading to programs that can be realized chemically. However, there could be other choices of computational units and goals, and our focus is on presenting an initial model that is as simple as possible while still capturing the fundamental chemical intricacies. Our description will lead to a definition of the input, available operations, constraints, and a number of optimization objectives.

From a chemist, we get *a synthesis tree*, which we assume is binary, where the *leaves represent compounds*. We refer to these as the *base* compounds. The tree can be interpreted as a recipe in the following manner. Each leaf of the tree represents an existing base compound. Now, we bring compounds to react in an order respecting the tree structure. Thus, first the compounds corresponding to two leaves are made to react, resulting in a new compound, which we refer to as an *intermediate* compound. We keep going until we reach the root, and have at that point produced our final *target* compound. The order of combining the compounds should simply be a topological ordering of the tree. We draw the trees with the root at the bottom, as it is usually done for synthesis trees, and hope our fellow computer scientists can accept this normality.

We detail the operations below. Our textual description is complete and self-contained, but it might be helpful to refer to the appendix for an example program. In order for two compounds to react, they must be in close proximity, and two compounds do not react if they are distant enough. To obtain proximity, the compounds are equipped (tagged) with DNA sequences, and the compound is at one of the two ends of the sequence. We refer to such a DNA sequence on a compound as a *tag* and choose an orientation so that we can refer to the left and right ends of a tag. Assume x and y are the tags of compounds X and Y, respectively, and X is at the right end of x and Y at the left end of y. If we add the *complementary* strand of the concatenation of x and y, denoted $\overline{xy}$, x and y will attach to the $\overline{x}$-part and $\overline{y}$-part of $\overline{xy}$, respectively, bringing X and Y close together and the reaction of X and Y takes place. We refer to such a strand as an *instruction* strand and the process as a *react* operation. In the above, and in the rest of the paper, when we refer to a strand, it can always be thought of as the concatenation of two tags. The resulting intermediate compound will lose one of its tags in the process of the reaction and will thus afterwards be tagged with either x or y in a deterministic fashion decided by the compounds, i.e., along with the synthesis tree, a chemist will tell us, for every internal node, which of the two tags from the child nodes will be the tag of the produced intermediate compound. We say that the node *inherits* the tag from the child in question and we may use a bold edge to indicate this. This annotated tree forms our input from the chemist. Note that the compounds and what they become when they react is not important to us; only the tags (and how they are attached) and strands are relevant to our computation.

After a reaction has been carried out using the $\overline{xy}$ instruction strand, a complementary *release* strand, xy, is added to release the compound, and, thus, prepare for further reactions. We will not need to consider this in the algorithms, but technically, this is obtained by really using a strand $\overline{x'y'}$ for the process, where $\overline{x'y'}$ is different from $\overline{xy}$, but similar enough for the process to work, and then the release is carried out using $x'y'$, such that $\overline{x'y'}$ and $x'y'$ combine and never react with anything else again. Thus, the reason the strand $\overline{x'y'}$ must be a little different from $\overline{xy}$ is to ensure that the later release it nearly 100% effective. Similarly, if x is the tag on the resulting compound, after the release, we add the complement of y in exact matching quantity so that they will combine with all the y tags, now flowing freely in the pot, making them inert such that they can be ignored in the remaining process. This necessary use of different but similar strands further increases the need for a large edit distance between tags (when viewed as sequences) as discussed earlier. Any mismatch in quantity will result in a proportional drop in the yield.

We disallow simultaneous releases, since they lead to a low overall yield as we explain now. Releasing two compounds using $\overline{xy}$ implies that one released compound must be tagged with an x and the other with a y. Otherwise (that is, if both compounds have the same tag), we cannot control subsequent operations. But this implies the presence of free-flowing y strands and x strands from the first and second reactions, respectively. These may attach to any later $\overline{xy}$ strand, resulting in a reduced overall yield.

A final chemical possibility we shall use as an operation in one section is the ability to temporary *block* a compound. A compound tagged with a strand x can be blocked by adding a strand $\overline{xy}$ or $\overline{yx}$, and can be released again in the same manner as described above.

We use blocking in Sect. 3, but otherwise simply delay the release of compounds while working on others, with the aim of producing a *one-pot* program. Compounds, corresponding to the leaves, may be added gradually, but we do not allow ourselves to produce compounds corresponding to subtrees separately and add them later.

Our computational choices are the following. Given the annotated synthesis tree, we must decide on tags for the leaves and a topological ordering, including when to add, when to release, and in one algorithm also when to block and which strand to block with. Recall that given tags on the leaves, the annotation determines the tags on internal nodes. Since we most often use delayed release to avoid interference, we will frequently label internal nodes with the instruction strand, i.e., the sequence of two tags. The tag attached to the intermediate compound produced at that node is always one of the tags the strand consists of, and which one it is, is determined by the inheritance information provided by the chemist.

In summary (Table 1), a program is a sequence of operations (tag, react, release, block), where *tag* attaches a specified tag to a base compound, *react* combines two intermediate compounds, *release* releases the resulting intermediate compound, and *block* blocks a compound. To be chemically feasible, left and right input compounds to any react operation must have the compound placed to the right and left, respectively, the react operations must form a topological ordering of the tree, compounds (unreleased as well as possibly blocked) must

Table 1. Operations of a DNA-templated program: note, that (i) the tag operation allows for attaching the compound to the left or right end of the tag, (ii) the inheritance for the react operation is given as input from the chemist, (iii) the release operation assumes an addition of complementary tags in order to handle waste, (iv) the blocking operation can bind the tagged compound to the left or right part of the added strand.

tag	(diagram)
react	(diagram)
release	(diagram)
block	(diagram)

be released (unblocked) before they are used again, any block operation must use a strand matching the compound tag to the left or right, all unreleased (and blocked) compounds in the pot at any given time must be unreleased (and blocked) with unique (at the time) strands, and if there are compounds in the pot with the same tag, all but one must be unreleased or blocked.

This is implied by the above, but just for emphasis, we cannot use strands of the form xx in a controlled process, so if we use τ different tags, we have at most $\tau(\tau - 1)$ different strands at our disposal.

We illustrate some of these restrictions now, using the smallest possible interesting synthesis trees. First note that because compounds are at one end of a tag, we cannot have an unreleased compound with an $\overline{ab}$ strand while using ba at the root of the other subtree. This is because when we release using ab, then (without loss of generality) the released compound is tagged with a and the compound is at the right end. Thus, later, it must react with a compound tagged with a b where the compound is at the left end. Thus, the strand from that subtree would have to have the form xb for some x; see Fig. 1.

The reader may have wondered if the reverse sequence of x is any different from x in a pot, or if xy could interfere with yx. Starting with the latter, breaking the sequences into their nucleotides, $\alpha_1\alpha_2 \cdots \alpha_{n_\alpha}\beta_1\beta_2 \cdots \beta_{n_\beta}$ is different from $\beta_1\beta_2 \cdots \beta_{n_\beta}\alpha_1\alpha_2 \cdots \alpha_{n_\alpha}$, and they are not the reverse of each other. Obviously, x cannot be distinguished from its reverse sequence in a pot. However, compounds are attached to one of the ends, so everything has an orientation.

Finally, to give a clean initial presentation, we do not consider the option of adding multiple strands simultaneously. Computationally it does not add anything and for most problems where the objective is to use for smallest number of different strands, it is counter-productive. However, in a lab, it could be desirable to know when this is an option. One could also lift the restriction of one-pot synthesis. However, since this would lead to a multi-criteria problem, we prefer to focus on the cleaner one-pot problem.

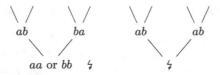

Fig. 1. Illustration of disallowed strand assignments. Left: Using strands ab and ba for two children requires the parent to be assigned either the strands aa or bb, which will result in a reduced overall yield, as with a probability of 50%, the corresponding compounds do not get in close proximity and therefore will not react. Right: Assume one subtree is already computed and the compound has to be unreleased with the complementary strand $\overline{ab}$. The corresponding unrelease needs to make the waste inert with $\overline{a}$ or $\overline{b}$, depending on which tag is now flowing freely in the pot. However, due to the disallowed simultaneous release of the other subtree, the release operation of the last of the two subtrees would accidentally make tagged compounds inert.

Some of the algorithms in this paper and graphical illustrations of the chemical processes can be inspected via a prototype implementation [9].

3 Minimizing the Number of Tags

In this section, our objective is to minimize the number of tags used on base compounds (the leaves), and as our second priority, we want to minimize the total number of tags used.

It turns out that, with appropriate blocking, it is always possible to arrive at a program using only two tags on base compounds, and clearly, for any two neighboring leaves with the same parent, the tags must be different. We refer to the two tags as a and b. Using the following recursively defined function, $\left\lceil \frac{\text{MNT}(\text{ROOT},0,0)}{2} \right\rceil$ will compute the minimum number of tags needed to block intermediate compounds when the basic compounds are tagged using only a and b.

Let t_a and t_b denote the subtrees of a tree t where the compound is tagged with an a and b, respectively. We keep track of tags used with a and with b separately, counting using c_a and c_b.

$$
\text{MNT}(t, c_a, c_b) =
\begin{cases}
\max(c_a, c_b) & \text{if } t \text{ is a leaf} \\
\min\left(\begin{array}{l} \max\left(\begin{array}{l} \text{MNT}(t_a, c_a, c_b), \\ \text{MNT}(t_b, c_a + 1, c_b) \end{array} \right), \\ \max\left(\begin{array}{l} \text{MNT}(t_a, c_a, c_b + 1), \\ \text{MNT}(t_b, c_a, c_b) \end{array} \right) \end{array} \right) & \text{otherwise}
\end{cases}
\tag{1}
$$

We discuss correctness and the derived program in the following (see the appendix for a simple example calculation). First, we decide arbitrarily between a and b for the final tag on the target compound that the root represents. If we use only the two tags a and b on base compounds, then we can determine all tagging recursively, since the chemist has informed us, for each node in the subtree, from which child we inherit the tag, i.e., if a node has a given tag, then a specific child of that node must have the same tag, and then the other child must be given the other tag (of the two tags a and b).

Since compounds have one of only two tags, any reaction involves both tags, so anything else in the pot must be blocked. In algorithms to be presented later, leaving them unreleased can also be an option, but in this particular case with only two tags on compounds, this would lead to the disallowed simultaneous release; see the earlier Fig. 1.

As a consequence, for any node with two non-leaf subtrees, we must decide which subtree to synthesize first, and then block while we work on the other subtree. In the subtree we synthesize first, we must block other compounds (corresponding to subtrees) recursively. We find the best subtree to block using the minimization in the formula above. The first entry in the minimization corresponds to first synthesizing and then blocking the subtree t_a. This requires no further resources while synthesizing that subtree, but while later synthesizing t_b, the compound from t_a must be blocked using a tag that has not been used for blocking subtrees on the way from the root to this node. Actually, when using some tag x to block a, for instance, this can be done (unconstrained) as ax or xa. Thus, each such tag x can be used twice, which accounts for the fraction $\frac{1}{2}$ in the final result, $\left\lceil \frac{\text{MNT}(\text{ROOT},0,0)}{2} \right\rceil$.

The best values can be computed using dynamic programming. If the tree is of height h, then each of the variables a and b in the expression can take on at most h different values, so if the tree has size n, then $O(nh^2)$ is an upper bound on the number of values to be computed and each value in a given node can be computed in constant time from values in the node's subtrees, so $O(nh^2)$ is also an upper bound on the computation time. A program can easily be extracted from the computed values by simply checking if the various minima are obtained from the left or right. An example program is shown in the appendix.

4   Minimizing the Number of Strands

In this section, we consider the problem when it is undesirable to use blocking, so that is disallowed, and our objective is to minimize the number of strands used. We allow for an arbitrary number of tags. As any instruction strand requires a unique complementary release strand, they will not be counted separately. It turns out that it is necessary and sufficient to use $\mathcal{S}(t)-1$ different strands, where $\mathcal{S}(t)$ is the (Horton-)Strahler number [15] of the synthesis tree t. Referring to the previous section, where we restricted ourselves to only using two different tags on base compounds, the Strahler number many strands would not in general be sufficient. The result in this section is accomplished without using blocking.

Definition 1. *The Strahler number $\mathcal{S}(t)$ of a tree t is defined as follows: If t is a leaf, then $\mathcal{S}(t) = 1$, and if t has two subtrees t_l and t_r, then*

$$\mathcal{S}(t) = \max(\min(\mathcal{S}(t_l), \mathcal{S}(t_r)) + 1, \max(\mathcal{S}(t_l), \mathcal{S}(t_r)))$$

$\mathcal{S}(t)$ is also referred to as the register number, i.e., the minimum number of registers required for evaluating a given arithmetic expression [6].

We are given a synthesis tree and information regarding from which child a node inherits its tag. To explain the tagging, it is easiest for us first to reorder the subtrees so that tags are inherited from subtrees according to a specific pattern. By a *layer* in a tree, we denote all the nodes of the same distance from the root. Given the synthesis tree, we order the subtrees such that when considering any layers from the left to the right, the tag is inherited alternately either from the left or from the right child, and we start by inheriting from the left; see bold edges in Fig. 2.

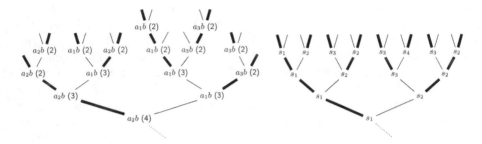

Fig. 2. Left: Illustration of the labeling algorithm that uses $\mathcal{S}(t) - 1$ many strands $a_1b, a_2b, \ldots$. Note that this is only one of the possible labelings, since strands are simply chosen from an available set, though we have consistently chosen the smallest indexed a_i available. The Strahler number is given in parenthesis. Right: Illustration of the labeling algorithm for complete binary trees: $s_1, s_2, \ldots$ is an antipath of strands, inheritance of tags is illustrated by bold lines.

Now, we explain how we label each node in our synthesis tree, excluding the leaves that contain the base compounds. For the labeling, we use the set $I = \{a_1b, a_2b, a_3b, \ldots\}$. This set contains strands that have pairwise different tags as their first parts (a_i) and identical tags as their second parts (b).

Recursively for a subtree t of the synthesis tree with ordered children as described above, we first compute the subtree with the larger Strahler number. In case of identical Strahler numbers, we choose the left subtree first. The strand assignment is done as follows: In case the subtrees have identical Strahler numbers, the subtree computed first will require a strand for the release operation. This strand cannot be used for any operation in the other subtree. If the Strahler numbers are different, this constraint will not apply. However, in all cases, neighboring operations need to use different strands. During the recursion, we keep track of the set of forbidden strands (this set grows by one element for the right subtree in the case of identical Strahler numbers) and the sibling reaction strand. Note that the constraint for the sibling reaction *only* applies to the sibling reaction. The pseudocode is given in Algorithm 1 and an illustration with an example of the labeling for a tree with Strahler number 4 is given in Fig. 2(Left). With regards to the number of strands, it is clear that the forbidden set F grows with the Strahler number,

Algorithm 1. Strahler Number Strands

Given: Synthesis tree t ▷ ordered children according to text description
       Set $A = \{a_1b, a_2b, \ldots, a_{\mathcal{S}(t)-1}b\}$ ▷ A: set of strands with $|A| = \mathcal{S}(t) - 1$

```
 1: function ASSIGNSTRAND(Tree t, Set F, Strand sibling)    ▷ F: forbidden strands
 2:     t_l, t_r ← LeftSubtree(t), RightSubtree(t)
 3:     if both t_l and t_r are base compounds (leaves) then
 4:         choose strand s from A \ (F ∪ {sibling})
 5:     else if one of t_l and t_r is a base compound (a leaf) then
 6:         t_x ← arg max_{t_i ∈ {t_l,t_r}} S(t_i)              ▷ t_x is the non-leaf tree
 7:         s ← ASSIGNSTRAND(t_x, F, _)
 8:     else
 9:         if S(t_l) > S(t_r) then
10:             s ← ASSIGNSTRAND(t_l, F, _)
11:             ASSIGNSTRAND(t_r, F, s)
12:         else if S(t_l) < S(t_r) then
13:             s_r ← ASSIGNSTRAND(t_r, F, _)
14:             s ← ASSIGNSTRAND(t_l, F, s_r)
15:         else                                              ▷ S(t_l) = S(t_r)
16:             s ← ASSIGNSTRAND(t_l, F, _)
17:             ASSIGNSTRAND(t_r, F ∪ {s}, s)
18:     assign s to t
19:     return s
20: ASSIGNSTRAND(t, ∅, _)
```

---

so if it was not for the temporary restriction given by the sibling, we use $\mathcal{S}(t) - 1$ strands. Recall that a leaf (with a compound) has Strahler number one, so the smallest subtree we assign a strand to has Strahler number two. With regards to the restriction, when the number of available strands is at least two, the temporary restriction does not matter, since we still have a strand we can choose. Thus, the only possible problem is when we recur from a tree with Strahler number three to smaller subtrees. If the subtrees have different Strahler numbers, there is no problem, since the restriction is imposed on the smaller one. If they have the same Strahler number, the sibling restriction coincides with the growing forbidden set, so only one strand option disappears, and the one required strand can be found.

With regards to chemical feasibility, siblings have different strands by construction, and b has its compound at the left and the compound coming from the right subtree will always be tagged with b. The opposite holds for the a_is, so the strands listed in the internal nodes indicate instruction strands fulfilling all requirements.

The upper bound just given is the interesting one. The lower bound that $\mathcal{S}(t) - 1$ different strands are necessary follows directly from the equivalent result for arithmetic expressions [6]; it is simply a matter of having to store at least that many intermediate results.

Strahler examples, as the ones produced in this section, can be found in [9].

5   Complete Binary Trees

The two problems of minimizing the number of strands used (Sect. 4) and minimizing the number of tags used under the constraint that all base compounds are tagged by one out of two tags (Sect. 3) can both be solved optimally in an efficient manner. In this section, we restrict the topology of the synthesis plan to complete binary trees and present an approach that minimizes the overall number of strands *as well as* bounds the overall number of tags to the optimal, possibly plus one. We accomplish this without using blocking.

The approach will employ so-called antipaths [5], which is a sequence of adjacent edges in a digraph, where every visited edge has opposite direction of the previously visited edge; and we will need some further restrictions defined below.

Definition 2. *An antipath in a digraph is a finite sequence of edges (u_i, u_j) having one of the following forms:*

$$(u_1, u_2), (u_3, u_2), (u_3, u_4), \ldots \quad or \quad (u_2, u_1), (u_2, u_3), (u_4, u_3), \ldots$$

An antipath is called return-free *if for any two successive edges (u, v) and (u', v'), $\{u, v,\} \neq \{u', v'\}$ and* non-overlapping *if no edge is used twice.*

In our construction, we will need return-free, non-overlapping antipaths as long as possible (each edge will correspond to a strand) from digraphs with as few vertices as possible (each vertex will correspond to a tag). The proof of the theorem below is available in the full version of our paper.

Theorem 1. *In a complete digraph $G_n = (V, E)$ over $n \geq 2$ vertices, the length of a longest return-free, non-overlapping antipath is $n(n-1)$ if n is odd and $n(n-2) + 1$ if n is even.*

As in all the other sections, we are given a synthesis tree and information regarding from which child a node inherits its tag. We reorder subtrees with regards to inheritance as in the previous section.

Separate from the tree structure, assume that we let each tag that we use represent a vertex in a digraph. Thus, a directed edge in the digraph is an ordered pair of tags, which we can interpret as a strand. We choose a longest antipath $s_1, s_2, s_3, \ldots$ in such a digraph, writing them as s_i for the ith strand. The number of tags (equal to the number of vertices in the digraph) we use depends on how long an antipath we need for the construction below.

First, we explain how we label each node in our synthesis tree, excluding the leaves that contain the base compounds. The root is labeled s_1 and, for ease of the definition below, artificially assume that the root has a parent, and that we moved left to get to the root. Moving from the root towards a leaf, we label each node with the same label as its parent (below it in our illustrations) if we move in the same direction as from the grandparent to our parent, and we label it with the next label (index one larger) if we change direction; see Fig. 2(Right).

Afterwards, the base compounds in the leaves can be tagged in the obvious manner, tagging the left (right) leaf with the left (right) part of its parent's strand.

From the labeled synthesis tree, we can define the program recursively. For a given node, we first compute the subtree whose root has the strand with the smallest index, leaving it unreleased while the other subtree is computed, after which the first subtree is released and the instruction of the node is carried out.

We now argue that the labeling algorithm produces a chemically feasible program. With regards to the reactions, due to the definition of the inheritance, a simple inductive argument establishes that at any node, the two input compounds stem from the left-most and right-most leaves of the subtree of the node. Thus, the compounds are tagged with the correct orientation for a reaction. With regards to the interference, the definition of the program explicitly states that the subtree whose root is labeled with the smallest indexed strand is computed first, and by the labeling algorithm, that strand is not used in the other subtree. Thus, no release operation can unintendedly release more than one compound.

Theorem 2. *The labeling algorithm uses the minimal number of strands and at most one more than the minimal number of tags.*

Proof. A complete binary tree of height h has Strahler number $h+1$, so we know from Sect. 4 that h is the optimal number of strands. The maximal number of direction changes from the root to the level next to the leaves is $h-1$, so, since the root is labeled s_1, the maximal label index is $1+(h-1)=h$.

Assume that it is somehow possible to make a program using the optimal number of tags τ. Observe that we can make at most $\tau(\tau-1)$ different strands from τ tags, so if τ is the optimal number tags, this must mean that this hypothetical program uses at most $\tau(\tau-1)$ strands.

If we allow for $\tau+1$ tags in our program, we know from Theorem 1 that an antipath of length at least $(\tau+1)(\tau-1)+1$ exists. Since we use the optimal number of strands and $(\tau+1)(\tau-1)+1\geq\tau(\tau-1)$ for any positive integer τ, the theorem follows. □

We remark that the construction is actually optimal also with regards to the number of tags in many cases. In fact, for all heights up to 25, we know that we are optimal, except for the heights 10–12. An example argument that the method is optimal for height 13 (in fact, the same argument works up to height 20) goes as follows. We know we need 13 different strands. With 4 tags, we can make only $4\cdot3=12$ different strands, so 5 tags are necessary for any program, and with 5 tags we can find antipaths of lengths up to $5\cdot4=20$. Similarly, for height 9 (in fact, down to height 7), we need four tags to have enough strands, and with four tags, we can make antipaths of lengths up to $4\cdot2+1=9$. It is the slightly limited lengths of antipaths for an even number of tags that prevents us from extending this optimality argument throughout the range 10–12.

Finally, the algorithm runs in linear time. The recursive definition of the longest antipath one can extract from the theorem is constructive and easily implemented in linear time in the number of strands needed for the synthesis tree algorithm, the labeling is a linear-time pre-order traversal, and the extraction of the program is a linear-time depth-first traversal.

6 Concluding Remarks

For small synthesis trees, one might consider *all* possible programs, i.e., all topological orderings of the synthesis tree with optional blocking at any node. For all such programs, one can find an optimal assignment of tags and strands. In the full version of the paper, we specify an integer linear program that does that.

Directly related to the questions we consider, it would be interesting to settle the near-optimality issue for complete binary trees, where we have provably optimal results for heights up to 25, except for heights 10–12. It may be necessary to loosen the constraint of using antipaths for the labeling slightly, but it requires great care to still ensure correctness. Also in relation to the complete binary tree algorithm, solutions could be used as the basis for solutions for trees that are not complete. For instance, adding long paths to a complete binary tree need not result in a higher cost in terms of number of tags and strands. It seems that for trees in general, the largest induced complete binary tree is the key to the cost and a formal extension from complete binary trees to trees in general exploiting this kernelization-like idea would be nice.

A quite different direction is to explore concurrency. If one uses more tags and strands than the bare minimum, some subtrees may become independent and even one-pot synthesis could allow for concurrency. Trade-off results between concurrency maximization and tag/strand minimization would be interesting.

Acknowledgment. The second and third authors were supported in part by the Danish Council for Independent Research, Natural Sciences, grants DFF-1323-00247 and DFF-7014-00041.

Appendix: DNA Program Example

We consider an example synthesis tree with four base compounds. The actual names of the compounds is not used in any of our algorithms, but for illustration, assume the base compounds are A, B, C, and D. Furthermore, we assume that the tagged compound A reacts with the tagged compound B $(A + B \rightarrow E)$, and that E will have the tag of B. The complete assumptions are

$$A + B \rightarrow E, E \text{ will inherit the tag of } B$$
$$C + D \rightarrow F, F \text{ will inherit the tag of } C$$
$$E + F \rightarrow X, X \text{ will inherit the tag of } E$$

and we demonstrate one possible program computing the target compound X as a one-pot synthesis.

We first tag the base compounds A at the left end of the tag a and B at the right end of the tag b. The tag a (respectively b) is depicted as a red (respectively blue) line in the following.

```
1   tag(A, a, left)
2   tag(B, b, right)
```

The state is as follows:

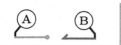

We add the complementary strand $\overline{ba}$ in order to bring A and B in close vicinity and they react to produce E. In this process, A loses its tag.

```
3   react(b̄ā)
```

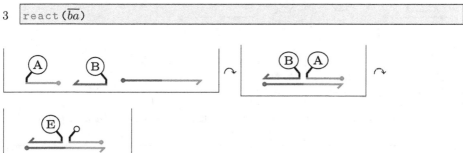

We release the produced tagged compound E with the strand ba and E is now tagged with b. The tag a is now unattached and we add the complementary tag $\overline{a}$ such that in the subsequent operations, it can be ignored.

```
4   release(ba)
```

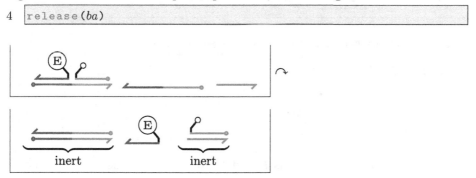

Since they are no longer relevant, we will not depict the inert strands in the following.

In order to avoid unintended interference, we block the tagged compound E with a strand $\overline{bc}$ (c shown in orange).

```
5   block(b̄c̄)
```

We proceed with the base compounds C and D in a similar manner. Note that C is tagged with a and D is tagged with b, i.e., adding them to the pot in the beginning would have led to unintended interference. By adding $\overline{ba}$, the tagged compounds C and D react to produce F, and D loses its tag.

```
6   tag(C, a, left)
7   tag(D, b, right)
8   react(ba)
```

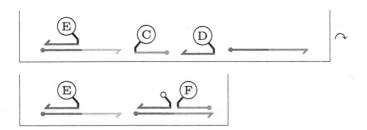

We then release the tagged compound F using the strand ba and pacify the tag b.

```
9   release(ba)
```

The blocked tagged compound E is released with the strand bc.

```
10   release(bc)
```

Finally, the tagged compounds E and F are brought in close vicinity using the strand $\overline{ba}$, producing X, and F loses its tag.

```
11   react(ba)
```

In the very last step, the target compound is released using strand ba, which finalizes the synthesis.

12 | `release(ba)`

The only non-inert tag is the tag attached to compound X, which makes it chemically easy to extract the compound from the pot. The synthesis required three different tags and two different strands (and their corresponding complementary tags and strands).

The given example also illustrates the minimization of the number of tags for blocking, when assuming that only two tags on the compounds are used (see Eq. 1), and the number of tags for blocking is to be minimized. Without loss of generality, we choose the goal compound X to be tagged with b. Given that decision, and given that we have restricted ourselves to using only two different tags on the compounds, there are no further choices for tagging: The tagging of all nodes in the tree is simply inferred as follows. The nodes A, C, and F need to be tagged with an a, and B, D, and E with a b. In this example, the subtree of the root X corresponding to $A + B \rightarrow E$ is synthesized before the subtree corresponding to $C + D \rightarrow F$. As we need to block the result of the former synthesis, we need an additional tag for blocking for the subtree E. With respect to Eq. 1, this corresponds to the recursive calculations for the inference $\max(\text{MNT}(E, 0, 0), \text{MNT}(F, 1, 0))$ (the choice to synthesize the subtree $C + D \rightarrow F$ first would, in this specific example, lead to the same overall result). This leads to the following base cases for the leaves: $\text{MNT}(A, 0, 0) = 0$ and $\text{MNT}(B, 0, 0) = 0$, and for the other subtree $\text{MNT}(C, 1, 0) = 1$ and $\text{MNT}(D, 1, 0) = 1$. Obviously, $\text{MNT}(E, 0, 0) = 0$ and $\text{MNT}(F, 1, 0) = 1$, leading to $\text{MNT}(X, 0, 0) = \min(\max(\text{MNT}(E, 0, 0), \text{MNT}(F, 1, 0)), \ldots) = 1$. Thus, only one additional tag is needed for blocking.

References

1. Adleman, L.: Molecular computation of solutions to combinatorial problems. Science **5187**, 1021–1024 (1994)
2. Andersen, J.L., Flamm, C., Hanczyc, M.M., Merkle, D.: Towards optimal DNA-templated computing. Int. J. Unconventional Comput. **11**(3–4), 185–203 (2015)
3. Benson, E., Mohammed, A., Gardell, J., Masich, S., Czeizler, E., Orponen, P., Högberg, B.: DNA rendering of polyhedral meshes at the nanoscale. Nature **523**, 441–444 (2015)
4. Cardelli, L.: Two-domain DNA strand displacement. In: DCM, pp. 47–61 (2010)
5. de Werra, D., Pasche, C.: Paths, chains, and antipaths. Networks **19**(1), 107–115 (1989)

6. Flajolet, P., Raoult, J., Vuillemin, J.: The number of registers required for evaluating arithmetic expressions. Theor. Comput. Sci. **9**(1), 99–125 (1979)
7. Goodnow, R.A., Dumelin, C.E., Keefe, A.D.: DNA-encoded chemistry: enabling the deeper sampling of chemical space. Nat. Rev. Drug Discov. **16**, 131–147 (2017)
8. Gorska, K., Winssinger, N.: Reactions templated by nucleic acids: more ways to translate oligonucleotide-based instructions into emerging function. Angew. Chem. Int. Ed. **52**(27), 6820–6843 (2013)
9. Hansen, B.N., Mihalchuk, A.: DNA-templated computing. Master's thesis, University of Southern Denmark, Denmark (2015). http://cheminf.imada.sdu.dk/dna/. Accessed 30 March 2017
10. He, Y., Liu, D.R.: A sequential strand-displacement strategy enables efficient six-step DNA-templated synthesis. J. Am. Chem. Soc. **133**(26), 9972–9975 (2011)
11. Hendrickson, J.B.: Systematic synthesis design. 6. Yield analysis and convergency. J. Am. Chem. Soc. **99**, 5439–5450 (1977)
12. Li, X., Liu, D.R.: DNA-templated organic synthesis: nature's strategy for controlling chemical reactivity applied to synthetic molecules. Angew. Chem. Int. Ed. **43**, 4848–4870 (2004)
13. Meng, W., Muscat, R.A., McKee, M.L., Milnes, P.J., El-Sagheer, A.H., Bath, J., Davis, B.G., Brown, T., O'Reilly, R.K., Turberfield, A.J.: An autonomous molecular assembler for programmable chemical synthesis. Nat. Chem. **8**(6), 542–548 (2016). doi:10.1038/nchem.2495
14. Phillips, A., Cardelli, L.: A programming language for composable DNA circuits. J. R. Soc. Interface **6**(Suppl. 4), S419–S436 (2009)
15. Strahler, A.: Hypsometric (area-altitude) analysis of erosional topography. Bull. Geol. Soc. Am. **63**, 1117–1142 (1952)
16. Wickham, S., Bath, J., Katsuda, Y., Endo, M., Hidaka, K., Sugiyama, H., Turberfield, A.: A DNA-based molecular motor that can navigate a network of tracks. Nat. Nanotechnol. **7**, 169–173 (2012)

Ruleset Optimization on Isomorphic Oritatami Systems

Yo-Sub Han and Hwee Kim[(✉)]

Department of Computer Science, Yonsei University,
50 Yonsei-Ro, Seodaemum-Gu, Seoul 03722, Republic of Korea
{emmous,kimhwee}@yonsei.ac.kr

Abstract. RNA cotranscriptional folding refers to the phenomenon in which an RNA transcript folds upon itself while being synthesized out of a gene. The oritatami system (OS) is a computation model of this phenomenon, which lets its sequence of beads (abstract molecules) fold cotranscriptionally by the interactions between beads according to its ruleset. We study the problem of reducing the ruleset size while maintaining the terminal conformations geometrically same. We first prove the hardness of finding the smallest ruleset, and suggest two approaches that reduce the ruleset size efficiently.

1 Introduction

In nature, a one-dimensional RNA sequence folds itself autonomously and gives rises to a highly-dimensional tertiary structure. It has been a challenging question to predict the tertiary structure from a primary structure. Based on experimental observations, researchers established various RNA structure prediction models including RNAfold [15], Pknots [9], mFold [14] and KineFold [13] (Fig. 1).

Recently, biochemists showed that the kinetics—the step-by-step dynamics of the reaction—plays an essential role in the geometric shape of the RNA foldings [1], since the folding caused by intermolecular reactions is faster than the RNA transcription rate [7]. By controlling cotranscriptional foldings, researchers succeeded in assembling a rectangular tile out of RNA, which is called RNA Origami [4]. This cotranscriptional folding was observed even at a single-nucleotide resolution [12]. From this kinetic point of view, Geary et al. [3] proposed a new folding model called the oritatami system (OS). An OS consists of a sequence of beads (which is the transcript) and a set of rules for possible intermolecular reactions between beads. An OS folds its bead sequence as follows: For each bead, the OS determines the best location that maximizes the number of possible interactions using a lookahead of a few upcoming beads and place the current bead at the location. Then it reads the next bead and repeat the same procedure until there is no more bead to place. The lookahead represents the reaction rate of the cotranscriptional folding and the number of interactions represents the energy level. In OS, we call the secondary structure *the conformation*, and the resulting secondary structure *the terminal conformation*. Figure 2 gives an analogy between RNA origami and oritatami systems.

© Springer International Publishing AG 2017
R. Brijder and L. Qian (Eds.): DNA 23 2017, LNCS 10467, pp. 33–45, 2017.
DOI: 10.1007/978-3-319-66799-7_3

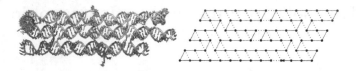

Fig. 1. (Left) An example of an RNA tile generated by RNA Origami [4]. (Right) A conformation representing the RNA tile in OS. The directed solid line represents a path, dots represent beads, and dotted lines represent interactions.

RNA Origami	Oritatami System
(A set of) Nucleotides	Beads
Transcript	Sequence of beads connected by a line
h-bonds between nucleotides	Interactions
Cotranscriptional folding rate	Delay
Resulting secondary structure	Conformation

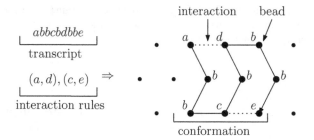

Fig. 2. An analogy between RNA origami and oritatami systems

Geary et al. implemented an OS counting in binary [2] and an OS simulating a cyclic tag system [3]. Han et al. [5] implemented an OS to solve the DNF tautology problem, and proved hardness of the OS equivalence problem. Han et al. [6] proposed the problem of removing self-attracting rules, and proved upper and lower bounds for number of required copies of bead types to remove self-attracting rules.

Since an OS folds its transcript according to its own ruleset, with more rules, it becomes more difficult to realize the system in experiments. In experiments, biochemists studied methods that synthesize the desired structure using a smaller number of basic components [10]. This motivates us to consider the problem of reducing the size of the alphabet and the ruleset from a theoretical point of view. Since an OS folds its transcript on the triangular lattice, it is important to preserve its geometric properties including the transcript path and interactions between beads while reducing the ruleset. Geary et al. [2] proved that given a set of paths and a transcript, it is NP-complete to find a ruleset that folds the transcript to the set of paths. Ota and Seki [8] proved that given a path, a transcript and a set of interactions, it is NP-complete to find a ruleset that folds the transcript to the path according to the given interactions. However, there is no research on reducing and optimizing the ruleset of a given OS while preserving all its geometric properties.

We say that two OSs are *isomorphic* if both have the same geometric properties. We first prove that, given an OS, it is NP-hard to find the smallest ruleset of an isomorphic OS in general. Then we propose two practical approaches to the problem: (1) We propose the bead type merging method—merge two bead types that have the same interaction with other bead types. (2) We propose representative fuzzy ruleset construction—a set of rulesets that results in the same set of terminal conformations. We design efficient algorithms to find a representative fuzzy ruleset from a given OS, reduce the size of the fuzzy ruleset by a modified bead type merging, and construct a reduced ruleset from the fuzzy ruleset.

2   Preliminaries

Let Σ be a finite set of types of abstract molecules, or *bead types*. By Σ^* (respectively Σ^ω), we denote the set of finite (one-way infinite) sequences of bead types in Σ. A sequence w in Σ^* can be represented as $w = b_1 b_2 \cdots b_n$ for some $n \geq 0$ and bead types $b_1, b_2, \ldots, b_n \in \Sigma$, where n is the *length* of w and denoted by $|w|$; in other words, a sequence w is a string over Σ. The sequence of length 0 is denoted by λ. For $1 \leq i \leq j \leq n$, the subsequence of w ranging from the i-th bead to j-th bead is denoted by $w[i : j]$, that is, $w[i : j] = b_i b_{i+1} \cdots b_j$. This notation is simplified as $w[i]$ when $j = i$, referring to the i-th bead of w. For $k \geq 1$, $w[1 : k]$ is a *prefix* of w. We use $w = w_1 \cdot w_2$, or simply $w_1 w_2$ to denote the catenation of two strings w_1 and w_2.

An undirected graph $G = (V, E)$ consists of a finite nonempty set V of nodes and a set E of unordered pairs of nodes of V. Each pair $e = \{u, v\}$ of nodes in E is an edge of G and e is said to join u and v. An weighted graph $G = (V, E)$ is a graph where each edge $e = (u, v)$ has an assigned weight $w(e)$. We denote an edge between u and v with a weight w in a weighted graph by $e = (\{u, v\}, w)$.

Oritatami systems fold their transcript, a sequence of beads, over the triangular lattice cotranscriptionally by letting nascent beads form as many hydrogen-bond-based interactions (*h-interactions*, or simply *interactions*) as possible according to a given set of rules. Let $\mathbb{T} = (V, E)$ be the triangular grid graph. A directed simple path $P = p_1 p_2 \cdots$ in $\mathbb{T}$ is a possibly-infinite sequence of pairwise-distinct points (vertices). Let $P[i]$ be the i-th point p_i and $|P|$ be the number of points in P. A *ruleset* $\mathcal{H} \subset \Sigma \times \Sigma$ is a symmetric relation over the set of pairs of bead types such that, for all bead types $a, b \in \Sigma$, $(a, b) \in \mathcal{H}$ implies $(b, a) \in \mathcal{H}$.

A conformation instance, or *configuration*, is a triple (P, w, H) of a directed path P in $\mathbb{T}$, $w \in \Sigma^* \cup \Sigma^\omega$, and a set $H \subseteq \{(i, j) \mid 1 \leq i, i + 2 \leq j, \{P[i], P[j]\} \in E\}$ of interactions. This is to be interpreted as the sequence w being folded while its i-th bead $w[i]$ is placed on the i-th point $P[i]$ along the path and there is an interaction between the i-th and j-th beads if and only if $(i, j) \in H$. Configurations (P_1, w_1, H_1) and (P_2, w_2, H_2) are *congruent* provided $w_1 = w_2$, $H_1 = H_2$, and P_1 can be transformed into P_2 by a combination of a translation, a reflection, and rotations by $60°$ degrees. The set of all configurations congruent to a configuration (P, w, H) is called the *conformation* of the configuration and denoted by $C = [(P, w, H)]$. We call w a *primary structure* of C. Let $\mathcal{H}$ be a ruleset. A rule $(a, b) \in \mathcal{H}$ is *useful* in the conformation $C = [(P, w, H)]$ if there exists $(i, j) \in H$ such that $w[i] = a$ and $w[j] = b$ or vice versa. Otherwise, the rule is *useless* in the conformation. An interaction $(i, j) \in H$ is *valid with respect to* $\mathcal{H}$, or simply $\mathcal{H}$-*valid*, if $(w[i], w[j]) \in \mathcal{H}$. We say that a conformation C is $\mathcal{H}$-*valid* if all of its interactions are $\mathcal{H}$-valid. For an integer $\alpha \geq 1$, C is *of arity* α if the maximum number of interactions per bead is α, that is, if for any $k \geq 1$, $\left|\{i \mid (i, k) \in H\}\right| + \left|\{j \mid (k, j) \in H\}\right| \leq \alpha$ and this inequality holds as an equation for some k. By $\mathcal{C}_{\leq\alpha}$, we denote the set of all conformations of arity at most α.

Oritatami systems grow conformations by elongating them under their own ruleset. For a finite conformation C_1, we say that a finite conformation C_2 is an *elongation* of C_1 by a bead $b \in \Sigma$ under a ruleset $\mathcal{H}$, written as $C_1 \xrightarrow{\mathcal{H}}_b C_2$, if there exists a configuration (P, w, H) of C_1 such that C_2 includes a configuration $(P \cdot p, w \cdot b, H \cup H')$, where $p \in V$ is a point not in P and $H' \subseteq \{(i, |P|{+}1) \mid 1 \leq i \leq |P| - 1, \{P[i], p\} \in E, (w[i], b) \in \mathcal{H}\}$. This operation is recursively extended to the elongation by a finite sequence of beads as follows: For any conformation C, $C \xrightarrow{\mathcal{H}^*}_\lambda C$; and for a finite sequence of beads w and a bead b, a conformation C_1 is elongated to a conformation C_2 by $w \cdot b$, written as $C_1 \xrightarrow{\mathcal{H}^*}_{w \cdot b} C_2$, if there is a conformation C' that satisfies $C_1 \xrightarrow{\mathcal{H}^*}_w C'$ and $C' \xrightarrow{\mathcal{H}}_b C_2$.

An *oritatami system (OS)* is a 6-tuple $\varXi = (\Sigma, w, \mathcal{H}, \delta, \alpha, C_\sigma = [(P_\sigma, w_\sigma, H_\sigma)])$, where $\mathcal{H}$ is a *ruleset*, $\delta \geq 1$ is a *delay*, and C_σ is an $\mathcal{H}$-valid initial *seed* conformation of arity at most α, upon which its *transcript* $w \in \Sigma^* \cup \Sigma^\omega$ is to be folded by stabilizing beads of w one at a time and minimize energy collaboratively with the succeeding $\delta - 1$ nascent beads. The energy of a conformation $C = [(P, w, H)]$ is $U(C) = -|H|$; namely, the more interactions a conformation has, the more stable it becomes. The set $\mathcal{F}(\varXi)$ of conformations *foldable* by this system is recursively defined as follows: The seed C_σ is in $\mathcal{F}(\varXi)$; and provided that an elongation C_i of C_σ by the prefix $w[1 : i]$ be foldable (i.e., $C_0 = C_\sigma$), its further elongation C_{i+1} by the next bead $w[i+1]$ is foldable if

$$C_{i+1} \in \underset{\substack{C \in \mathcal{C}_{\leq\alpha} \text{ s.t.} \\ C_i \xrightarrow{\mathcal{H}}_{w[i+1]} C}}{\text{argmin}} \ \min\left\{U(C') \,\Big|\, C \xrightarrow{\mathcal{H}^*}_{w[i+2:i+k]} C', k \leq \delta, C' \in \mathcal{C}_{\leq\alpha}\right\}. \quad (1)$$

Once we have C_{i+1}, we say that the bead $w[i+1]$ and its interactions are *stabilized* according to C_{i+1}. A conformation foldable by $\varXi$ is *terminal* if none of its elongations is foldable by $\varXi$. We use $C = [(P_\sigma P, w_\sigma w, H_\sigma \cup H)]$ to denote a terminal conformation, where w is folded along the path P with interactions in H. An OS is *deterministic* if, for all i, there exists at most one C_{i+1} that satisfies (1). Namely, a deterministic OS folds into a unique terminal conformation.

Figure 3 illustrates an example of an OS with delay 2, arity 2 and the ruleset $\{(a, b), (b, f), (d, f), (d, e)\}$; in (a), the system tries to stabilize the first bead a

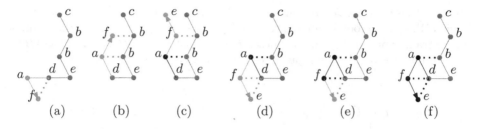

Fig. 3. An example OS with delay 2 and arity 2. The seed is colored in red, and the stabilized beads and interactions are colored in black. (Color figure online)

of the transcript, and the elongation in (a) gives 1 interaction. However, it is not the most stable one since the elongation in (b) gives 2 interactions in total. Thus, the first bead a is stabilized according to the location in (b). In (c), the system tries to stabilize the second bead f, and the elongation in (c) gives 1 interaction for the primary structure fe. However, the elogation in (d) gives 2 interactions in total. Thus, the second bead f is stabilized according to the location in (d). Note that f is not stabilized according to the location in (b), although the elongation in (b) is used to stabilize the first bead a.

Conformations C_1 and C_2 are *isomorphic* if there exist an instance (P_1, w_1, H_1) of C_1 and an instance (P_2, w_2, H_2) of C_2 such that $P_1 = P_2$ and $H_1 = H_2$. For two sets $\mathcal{C}_1$ and $\mathcal{C}_2$ of conformations, we say that two sets are isomorphic if there exists an one-to-one mapping $C_1 \in \mathcal{C}_1 \rightarrow C_2 \in \mathcal{C}_2$ such that C_1 and C_2 are isomorphic. We say that two oritatami systems are isomorphic if they fold the isomorphic set of foldable terminal conformations. A rule (a, b) is *useful* in an OS if the rule is useful in one of the terminal conformations of the system.

3 Hardness of Ruleset Optimization on Isomorphic Oritatami Systems

We first define the ruleset optimization problem on isomorphic OSs.

Problem 1 (RSOPT-Isomorphic). Given an OS $\varXi = (\varSigma, w, \mathcal{H}, \delta, \alpha, C_\sigma = [(P_\sigma, w_\sigma, H_\sigma)])$, find an isomorphic OS $\varXi' = (\varSigma', w', \mathcal{H}', \delta, \alpha, C'_\sigma = [(P_\sigma, w'_\sigma, H_\sigma)])$ where $|\mathcal{H}'|$ is minimum (See Fig. 4(a) and (c)).

We can think of the problem as follows: Suppose we are given a delay δ, an arity α, a path P_σ, a set H_σ of interactions and a set $\{(P_i, H_i)\}$, where P_i is a path and H_i is a set of interactions on P_i. Then, the problem is to find a seed primary structure w'_σ, a transcript w' and a smallest ruleset $\mathcal{H}'$, where the OS $\varXi' = (\varSigma', w', \mathcal{H}', \delta, \alpha, C'_\sigma = [(P_\sigma, w'_\sigma, H_\sigma)])$ successfully folds the set $\{[(P_\sigma P_i, w'_\sigma w', H_\sigma \cup H_i)]\}$ of terminal conformations (See Fig. 4(b) and (c)). Before we tackle the problem, we revisit the following problem.

Problem 2 (RSD-UniqConformation [8]). Given a finite conformation $C = [(P, w, H)]$, an alphabet Σ, an arity α, a delay δ, a seed $C_\sigma = [(P_\sigma, w_\sigma, H_\sigma)]^1$, and a finite transcript $w \in \Sigma^*$, find a ruleset $\mathcal{H}$ such that $C' = [(P_\sigma P, w_\sigma w, H_\sigma \cup H)]$ is the unique terminal conformation of the OS $\Xi = (\Sigma, w, \mathcal{H}, \delta, \alpha, C_\sigma)$.

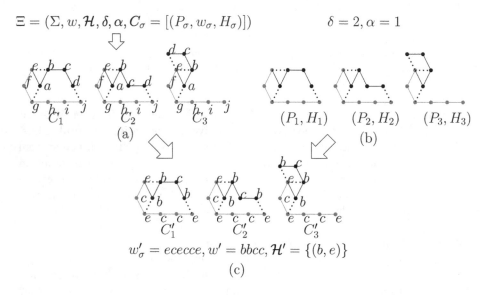

(a)

(b)

(c)

$w'_\sigma = ececce$, $w' = bbcc$, $\mathcal{H}' = \{(b, e)\}$

Fig. 4. An Illustration of two representations of Problem 1. The seed is colored in red. (Color figure online)

The problem is NP-hard when $\alpha, \delta \geq 2$ or $\delta \geq 3$. Note that the RSD-Isomorphic problem is different from the RSOPT-UniqConformation problem, since we should minimize the ruleset size by using an arbitrary transcript. Now, we propose another problem (Problem 3) and prove its hardness based on the proof for the RSD-UniqConformation problem [8]. Then, we prove the hardness of the RSOPT-Isomorphic problem using the hardness result of Problem 3.

Problem 3 (RSD-Isomorphic). Given a path P, a set H of interactions, an alphabet Σ, an arity α, a delay δ, a seed $C_\sigma = [(P_\sigma, w_\sigma, H_\sigma)]$, find a ruleset $\mathcal{H}$ and a finite transcript w such that $C = [(P_\sigma P, w_\sigma w, H_\sigma \cup H)]$ is the unique terminal conformation of the OS $\Xi = (\Sigma, w, \mathcal{H}, \delta, \alpha, C_\sigma)$.

Figure 5 illustrates difference among the RSD-UniqConformation, the RSD-Isomorphic and the RSOPT-Isomorphic problems.

Lemma 1. *The RSD-Isomorphic problem is NP-hard when $\alpha, \delta \geq 2$ or $\delta \geq 3$.*

Theorem 1. *The RSOPT-Isomorphic problem is NP-hard when $\alpha, \delta \geq 2$ or $\delta \geq 3$.*

---

[1] In the original paper, the seed was defined by a single term σ.

Problem	Input	Output
RSD-UniqConformation	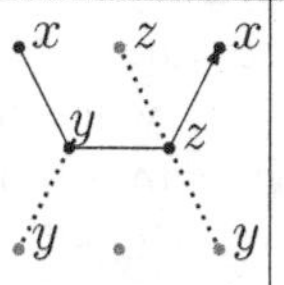 (P, w, H)	$\mathcal{H}$ that folds $[(P, w, H)]$
RSD-Isomorphic	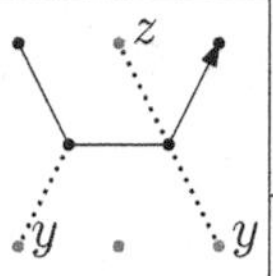	$(\mathcal{H}, w)$ that folds $[(P, w, H)]$
RSOPT-Isomorphic	(P, H)	$(\mathcal{H}, w)$ that folds $[(P, w, H)]$ while minimizing $|\mathcal{H}|$

Fig. 5. An example of the three problems. Seeds are colored in red. (Color figure online)

4   Ruleset Reduction by Bead Type Merging

Since the RSOPT-Isomorphic problem is NP-hard in general, we consider a poly-time heuristic for optimizing a ruleset efficiently. Because not all rules in a ruleset are useful, we start with removing useless rules. For a deterministic OS, it is sufficient to simulate the OS and find useless rules. The simulation takes $O(n \cdot 5^\delta)$ time, where n is the length of the transcript.

Corollary 1. *For a deterministic OS $\Xi = (\Sigma, w, \mathcal{H}, \delta, \alpha, C_\sigma = [(P_\sigma, w_\sigma, H_\sigma)])$, we can remove useless rules in $O(n \cdot 5^\delta)$ time, where $n = |w|$.*

For a nondeterministic OS, we show the hardness of the problem.

Theorem 2. *For a nondeterministic OS $\Xi = (\Sigma, w, \mathcal{H}, \delta, \alpha, C_\sigma = [(P_\sigma, w_\sigma, H_\sigma)])$ and a rule $r \in \mathcal{H}$, it is coNP-hard to determine whether or not r is useful.*

It is coNP-hard to identify and remove useless rules in general. Thus, we propose a method to reduce the ruleset size regardless of usefulness of rules. For two bead types a and b, suppose $(a, c) \in \mathcal{H}$ if and only if $(b, c) \in \mathcal{H}$ for all possible bead types c. If we merge beads a and b and replace all b's in the transcript and the seed to a's, it is straightforward to verify that the resulting OS is isomorphic to the original OS. We formally define the problem of finding a smallest ruleset based on the bead type merging.

Problem 4 (RSR-BTM-Isomorphic). Given a ruleset $\mathcal{H} \subseteq \Sigma \times \Sigma$ of an OS, find a minimum alphabet Σ' and a ruleset $\mathcal{H}' \subseteq \Sigma' \times \Sigma'$, where there exists a homomorphism $h : \Sigma \to \Sigma'$ such that for each pair of bead types $(x_1, x_2) \in \Sigma \times \Sigma$, $(x_1, x_2) \in \mathcal{H} \Leftrightarrow (h(x_1), h(x_2)) \in \mathcal{H}'$.

For the RSR-BTM-Isomorphic problem, we construct a binary string x_i for each bead type σ_i, where $x_i[j]$ is 1 if $(\sigma_i, \sigma_j) \in \mathcal{H}$ and 0 otherwise. It is straightforward that if $x_i = x_j$, then σ_i and σ_j can be successfully merged. We run radix sort for strings $x_1, x_2, \ldots, x_t$ where $t = |\sigma|$. After the sorting, any set of bead types corresponding to the same (consecutive) string can be successfully merged. Since the length of the strings is t, the radix sort requires $O(t^2)$ time using $O(t)$ space.

Theorem 3. *We can solve the RSR-BTM-Isomorphic problem in $O(t^2)$ time using $O(t)$ space, where $t = |\Sigma|$.*

5   Ruleset Reduction by Fuzzy Ruleset Construction

The bead type merging only uses information from the ruleset, not the whole OS. Note that we can remove or add some rules in the ruleset while maintaining an OS isomorphic. Thus, we propose another more efficient heuristic that finds a reduced ruleset from a set of rulesets for an isomorphic OS.

Given an alphabet Σ, we define a *fuzzy ruleset* to be a pair of a required ruleset $\mathcal{H}_P \subseteq \Sigma \times \Sigma$ and a forbidden ruleset $\mathcal{H}_N \subseteq \Sigma \times \Sigma$ such that $\mathcal{H}_P \cap \mathcal{H}_N = \emptyset$. Given an OS $\Xi = (\Sigma, w, \mathcal{H}, \delta, \alpha, C_\sigma)$, we say that a fuzzy ruleset $(\mathcal{H}_P, \mathcal{H}_N)$ is a *representative fuzzy ruleset* of the OS if $\Xi' = (\Sigma, w, \mathcal{H}', \delta, \alpha, C_\sigma)$ is isomorphic to Ξ for all $\mathcal{H}'$ satisfying the following conditions:

1. If $(a, b) \in \mathcal{H}_P$, then $(a, b) \in \mathcal{H}'$.
2. If $(a, b) \notin \mathcal{H}_N$, then $(a, b) \notin \mathcal{H}'$.

Namely, if a fuzzy ruleset $(\mathcal{H}_P, \mathcal{H}_N)$ is representative, then rules in $\mathcal{H}_P$ should be included in the ruleset, and rules in $\mathcal{H}_N$ should be excluded from the ruleset, which ensures that the system is isomorphic to the original system. These conditions are obtained by the property of the cotranscriptional folding. When we want to design an isomorphic system, we should keep the same location of the stabilized beads and the same interactions. While stabilizing a bead x, the bead and its $\delta - 1$ nascent beads choose a path P_x that maximizes the sum s_x of interactions. Then, for any alternative path P'_x where the bead is not stabilized at the target location, we can arbitrarily assign interactions for P'_x as long as the sum s'_x of interactions does not exceed s_x, as illustrated in Fig. 6.

We reduce the ruleset size in two phases: First, given an OS Ξ, we extract a representative fuzzy ruleset from Ξ. Second, we modify the ruleset graph reduction in Sect. 4 and reduce the size of the fuzzy ruleset. Then, using the fuzzy ruleset, we further reduce the ruleset size.

Problem 5 (FRS-Extraction).   Given an OS $\Xi = (\Sigma, w, \mathcal{H}, \delta, \alpha, C_\sigma = [(P_\sigma, w_\sigma, H_\sigma)])$, find a representative fuzzy ruleset $(\mathcal{H}_P, \mathcal{H}_N)$ minimizing $|\mathcal{H}_P| + |\mathcal{H}_N|$.

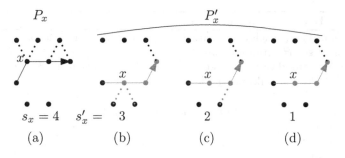

Fig. 6. When we stabilize the bead x, suppose the path P_x in (a) is the most stable one with the sum of interactions $s_x = 4$. Since the path P'_x in (b), (c) and (d) stabilizes x at the location different from P_x, we may assign arbitrary interactions for these paths as long as the sum s'_x of interactions does not exceed 4.

Theorem 4. *The FRS-Extraction problem is NP-hard when* $\alpha, \delta \geq 2$ *or* $\delta \geq 3$.

Next, we design a heuristic algorithm for the FRS-Extraction problem. Assume that an OS $\varXi$ folds the set $\{C_i = [(P_\sigma P_i, w_\sigma w, H_\sigma \cup H_i)]\}$ of t terminal conformations. We assume that $|w| = n$, and $|w_\sigma|$ and $|H_i|$'s are bounded to $O(n)$. We first propose an algorithm for one terminal conformation C_1, and then apply the algorithm for all terminal conformations. We take the following approach for the problem: First, for each point in P_σ or P_1, we assign a distinct bead type to retrieve w_σ and w. Second, we find conditions of the rules,, which are necessary and sufficient for an isomorphic OS. Third, we construct a representative fuzzy ruleset from these conditions. Let $P_1 = p_1 p_2 \cdots p_n$ and $P_\sigma = p_{n+1} p_{n+2} \cdots p_{n+m}$.

At first, let $\varSigma = \{\kappa_i \mid 1 \leq i \leq n + m\}$ and assume that κ_i is placed at p_i. We run Algorithm 1, which returns three conditions that are necessary and sufficient for an isomorphic OS. The required condition set P (the forbidden condition set N) includes the set of rules that should be included in (excluded from) the desired ruleset $\mathcal{H}$. Later, the construction of a representative fuzzy ruleset $(\mathcal{H}_P, \mathcal{H}_N)$ starts from (P, N). The last output is the conditional ruleset $\mathcal{H}_k = \{(K \in \varSigma \times \varSigma, s)\}$, which implies that the number of rules in $K \cap \mathcal{H}$ should not exceed s. The conditional ruleset has information of rules that are not explicitly shown in the most stable elongation but prevent the path from not following P_1. Figure 7 illustrates Algorithm 1.

Lemma 2. *Algorithm 1 runs in* $O(5^\delta \delta n)$ *time using* $O(5^\delta \delta n)$ *space.*

Since conditions in $\mathcal{H}_k$ are about the rules that are not explicitly shown in the most stable elongation, there is no necessary rule that should be added to P because of $\mathcal{H}_k$. Thus, we construct a representative fuzzy ruleset $(P, \mathcal{H}_N)$, where $\mathcal{H}_N = N \cup N_{add}$ and N_{add} satisfies conditions in $\mathcal{H}_k$. We prove that minimizing $|\mathcal{H}_N|$ is NP-complete.

Algorithm 1. ExtractConditionSets

Input: An arity α, a delay δ, a path P_σ for a seed, a path P_1 for a transcript
and a set H_1 of interactions

Output: A required condition set P, a forbidden condition set N and a
conditional ruleset $\mathcal{H}_k$

1 $\Sigma \leftarrow \{\kappa_i \mid 1 \leq i \leq n+m\}$.
2 place $\kappa_{n+1}, \kappa_{n+2}, \ldots, \kappa_{n+m}$ to $p_{n+1}, p_{n+2}, \ldots, p_{n+m}$ to form C_σ.
3 place κ_1 to p_1.
4 **for** $i \leftarrow 2$ **to** n **do**
5 $\quad$ place κ_i to p_i.
6 $\quad$ calculate the sum s_i of the interactions that led κ_i to the position p_i.
7 $\quad$ **for each** *annotated neighbors* p_j *of* p_i **do**
8 $\quad\quad$ **if** $\{p_i, p_j\} \in H_1$ **then** add (κ_i, κ_j) to P.
9 $\quad\quad$ **else** add (κ_i, κ_j) to N.
10 $\quad$ **for each** *unannotated path* $P' = p'_1 p'_2 \cdots p'_\delta$ *where* $p'_1 \neq p_i$ *is an unannotated*
$\quad\quad$ *neighbor of* p_{i-1} **do**
11 $\quad\quad$ $o_j \leftarrow 0, K \leftarrow \emptyset$
12 $\quad\quad$ **for** $j \leftarrow 1$ **to** δ **do**
13 $\quad\quad\quad$ **for each** *annotated neighbors* p_k *of* p'_j *where* p_k *has interactions less*
$\quad\quad\quad\quad$ *than* α **do**
14 $\quad\quad\quad\quad$ **if** $(\kappa_{i+j-1}, \kappa_k) \in P$ **then** $o_j \leftarrow o_j + 1$
15 $\quad\quad\quad\quad$ **else**
16 $\quad\quad\quad\quad\quad$ **if** $s_i = 1$ **then** add $(\kappa_{i+j-1}, \kappa_k)$ to N.
17 $\quad\quad\quad\quad\quad$ **else** add $(\kappa_{i+j-1}, \kappa_k)$ to K.
18 $\quad\quad$ **if** $s_i \neq 1$ **then** add $(K, s_i - o_j)$ to $\mathcal{H}_k$.

19 **return** $P, N, \mathcal{H}_k$

Lemma 3. *Given a set $\mathcal{H}_k \subseteq 2^{\Sigma \times \Sigma} \times \mathbb{N}$, let $N_{add} \subseteq \Sigma \times \Sigma$ be a set such that, for all $(K_i, s_i) \in \mathcal{H}_k$, $|K_i| - |K_i \cap N_{add}| < s_i$ holds. Then, it is NP-complete to find N_{add} with the minimum size.*

Since finding the minimum N_{add} is NP-complete, we use 3 heuristics to create N_{add}. We assume that a condition (K_i, s_i) is in $\mathcal{H}_k$.

1. While adding pairs to N to satisfy conditions in $\mathcal{H}_k$, we add as few pairs as possible, since more pairs in a negative condition set makes reduction harder.
2. We prefer (K_i, s_i) with the largest s_i, since we need to add more pairs to satisfy that condition.
3. For a pair $(\kappa_j, \kappa_k) \in K_i$, we prefer a pair with the most frequent appearances in all K_i's, since adding the pair (κ_j, κ_k) helps satisfying all these conditions.

Based on these heuristics, we run Algorithm 2.
Lemma 4. *Algorithm 2 runs in $O(5^\delta \delta n (\delta + \log n))$ time using $O(5^\delta \delta n)$ space.*

Once we have a representative fuzzy ruleset $(\mathcal{H}_P, \mathcal{H}_N)$, the next step is to construct a reduced ruleset that satisfies conditions of the fuzzy ruleset. We

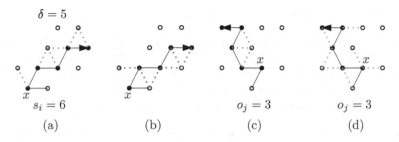

Fig. 7. An illustration of Algorithm 1. Beads in the elongation are represented by disks, and beads already stabilized are represented by circles. Suppose the delay of the system is 5 and the path in (a) is chosen to stabilize the first bead x, with the sum of interactions $s_i = 6$. Rules for blue interactions in (a) are added to P. In (b), interactions that should not exist are colored in red, and added to N. In (c), one path that is not chosen is illustrated. This path has the basic strength of $o_j = 3$ from rules in P. The set of all 7 possible interactions that are not from P are shown in (d), colored in brown. Rules for brown interactions are added to K, and $(K, 3)$ is added to $\mathcal{H}_k$, representing that we can add no more than 3 rules among rules for brown interactions to the ruleset. (Color figure online)

construct a *fuzzy ruleset graph* from $(\mathcal{H}_P, \mathcal{H}_N)$ by adding positive edges for rules in $\mathcal{H}_P$ and negative edges for rules in $\mathcal{H}_N$.

- $V = \Sigma$
- For each pair of molecules $(x_1, x_2) \in \Sigma \times \Sigma$,
  - add $(\{x_1, x_2\}, 1)$ to E if $(x_1, x_2) \in \mathcal{H}_P$,
  - add $(\{x_1, x_2\}, -1)$ to E if $(x_1, x_2) \in \mathcal{H}_N$.

Problem 6 (FRSR-BTM-Isomorphic). Given a representative fuzzy ruleset $(\mathcal{H}_P, \mathcal{H}_N)$ of an OS over an alphabet Σ, find a minimum alphabet Σ' and a ruleset $\mathcal{H}' \subseteq \Sigma' \times \Sigma'$, where there exists a homomorphism $h : \Sigma \to \Sigma'$ such that for every $(x_1, x_2) \in \Sigma \times \Sigma$, $((x_1, x_2) \in P \wedge (x_1, x_2) \notin N) \Leftrightarrow (h(x_1), h(x_2)) \in \mathcal{H}'$.

Algorithm 2. ExtractFuzzyRuleset

Input: a conditional ruleset $\mathcal{H}_k$
Output: A set N_{add}

1 **while** $\mathcal{H}_k \neq \emptyset$ **do**
2   **for each** (K_i, s_i) *with the largest* s_i **do**
3   count the number $occ_{(j,k)}$ of appearances of (κ_j, κ_k) in all K_i's.

4   **for each** (K_i, s_i) *with the largest* s_i **do**
5   **while** *the condition does not hold* **do**
6   find a pair (κ_j, κ_k) of bead types with the biggest $occ_{(j,k)}$.
7   add (κ_j, κ_k) to N_{add}.
8   delete (K_i, s_i) from $\mathcal{H}_k$.

9 **return** N_{add}

Lemma 5. *The FRSR-BTM-Isomorphic problem is NP-complete.*

Note that we reduce the vertex coloring problem to the FRSR-BTM-Isomorphic problem. We formally establish the function f from a fuzzy ruleset graph $G_r = (V, E_r)$ to a graph $G_c = (V, E_c)$ by the following rules: For all $(v_i, v_j) \in V^2$, $(v_i, v_j) \in E_c$ if and only if we cannot merge v_i and v_j. It requires $O(n^3)$ to construct $f(G_r)$ from G_r, when $n = |V|$. We establish the following lemma.

Lemma 6. *For a fuzzy ruleset graph $G_r = (V, E_r)$, let $G_c = f(G_r)$. Let v_1 and v_2 be two mergeble nodes in V. Let G'_r (G'_c) be the graph resulting from G_r (G_c) after merging v_1 and v_2. Then, $G'_c = f(G'_r)$.*

From Lemma 6, we know that any solution to the vertex coloring problem has its pair solution to the FRSR-BTM-Isomorphic problem. Therefore, we can use approximation algorithms for the vertex coloring problem to find approximate solutions for the FRSR-BTM-Isomorphic problem. One algorithm is Welsh-Powell algorithm [11]. Once all vertices v_i are ordered according to their degrees d_i, the algorithm runs in $O(n^2)$ time and gives at most $\max_i \min\{d_i + 1, i\}$ colors.

In summary, we first extract necessary and sufficient conditions of rules from the set of ruleset sizes by Algorithm 1. We accumulate P, N and $\mathcal{H}_c$ by running Algorithm 1 for $1 \leq i \leq t$, and then run Algorithm 2 to construct a representative fuzzy ruleset. We construct a fuzzy ruleset graph from the representative fuzzy ruleset, and use an approximation algorithm for the vertex coloring problem to find an approximate solution for the FRSR-BTM-Isomorphic problem. We establish the following theorem.

Theorem 5. *Using Algorithms 1 and 2, an approximation algorithm for the vertex coloring problem, we can approximately solve the RSOPT-Isomorphic problem in $O(5^\delta \delta n(\delta + \log n + t) + n^3)$ time using $O(5^\delta \delta n)$ space.*

Note that the bead type modification for a given ruleset in Sect. 4 is a special case of the FRSR-BTM-Isomorphic problem, where $\mathcal{H}_P \cup \mathcal{H}_N = \Sigma \times \Sigma$. Therefore, the method proposed in Sect. 5 is at least as efficient as bead type merging in the size of the reduced ruleset.

6   Conclusions

The oritatami system (OS) is a computational model inspired by RNA cotranscriptional folding, where an RNA transcript folds upon itself while being synthesized out of a gene. One element of the OS is the ruleset, which defines interactions between beads in the system, and it is crucial to reduce the ruleset size for implementing a simpler OS in experiments. We have first defined the concept of isomorphism of OSs. Then we have proved that it is NP-hard to find the smallest ruleset of an isomorphic OS in general. We proposed the bead type merging method and representative fuzzy ruleset construction to reduce the ruleset size.

There still are some open questions. For example, it is open to find approximate ratios of the proposed heuristic algorithms as well as to design an efficient algorithm that removes useless rules. We can also consider a ruleset optimization for a given path without considering the set of interactions, and a transcript optimization for a given ruleset.

References

1. Frieda, K.L., Block, S.M.: Direct observations of cotranscriptional folding in an adenine riboswitch. Science **338**(6105), 397–400 (2012)
2. Geary, C., Meunier, P., Schabanel, N., Seki, S.: Efficient universal computation by greedy molecular folding. CoRR, abs/1508.00510 (2015)
3. Geary, C., Meunier, P., Schabanel, N., Seki, S.: Programming biomolecules that fold greedily during transcription. In: Proceedings of the 41st International Symposium on Mathematical Foundations of Computer Science, pp. 43:1–43:14 (2016)
4. Geary, C., Rothemund, P.W.K., Andersen, E.S.: A single-stranded architecture for cotranscriptional folding of RNA nanostructures. Science **345**, 799–804 (2014)
5. Han, Y.-S., Kim, H., Ota, M., Seki, S.: Nondeterministic seedless oritatami systems and hardness of testing their equivalence. In: Rondelez, Y., Woods, D. (eds.) DNA 2016. LNCS, vol. 9818, pp. 19–34. Springer, Cham (2016). doi:10.1007/978-3-319-43994-5_2
6. Han, Y., Kim, H., Rogers, T.A., Seki, S.: Self-attraction removal from oritatami systems. In: Proceedings of the 19th International Conference on Descriptional Complexity of Formal Systems, pp. 164–176 (2017)
7. Lai, D., Proctor, J.R., Meyer, I.M.: On the importance of cotranscriptional RNA structure formation. RNA **19**, 1461–1473 (2013)
8. Ota, M., Seki, S.: Rule set design problems for oritatami system. Theor. Comput. Sci. **671**, 16–35 (2017)
9. Rivas, E., Eddy, S.R.: A dynamic programming algorithm for RNA structure prediction including pseudoknots. J. Mol. Biol. **285**(5), 2053–2068 (1999)
10. Rogers, J., Joyce, G.F.: A ribozyme that lacks cytidine. Nature **402**(6759), 323–325 (1999)
11. Rosen, K.H.: Discrete Mathematics and Its Applications. McGraw-Hill Education, New York (2006)
12. Watters, K.E., Strobel, E.J., Yu, A.M., Lis, J.T., Lucks, J.B.: Cotranscriptional folding of a riboswitch at nucleotide resolution. Nat. Struct. Mol. Biol. **23**(12), 1124–1131 (2016)
13. Xayaphoummine, A., Bucher, T., Isambert, H.: Kinefold web server for RNA/DNA folding path and structure prediction including pseudoknots and knots. Nucleic Acids Res. **33**, W605–W610 (2005)
14. Zuker, M.: Mfold web server for nucleic acid folding and hybridization prediction. Nucleic Acids Res. **31**(13), 3406–3415 (2003)
15. Zuker, M., Stiegler, P.: Optimal computer folding of large RNA sequences using thermodynamics and auxiliary information. Nucleic Acids Res. **9**(1), 133–148 (1981)

Unknotted Strand Routings of Triangulated Meshes

Abdulmelik Mohammed[1($\boxtimes$)] and Mustafa Hajij[2]

[1] Aalto University, Espoo, Finland
abdulmelik.mohammed@aalto.fi
[2] University of South Florida, Tampa, FL, USA
mhajij@usf.edu

Abstract. In molecular self-assembly such as DNA origami, a circular strand's topological routing determines the feasibility of a design to assemble to a target. In this regard, the Chinese-postman DNA scaffold routings of Benson et al. (2015) only ensure the unknottedness of the scaffold strand for triangulated topological spheres. In this paper, we present a cubic-time $\frac{5}{3}-$approximation algorithm to compute unknotted Chinese-postman scaffold routings on triangulated orientable surfaces of higher genus. Our algorithm guarantees every edge is routed at most twice, hence permitting low-packed designs suitable for physiological conditions.

Keywords: DNA origami · Knot theory · Graph theory · Chinese postman problem

1 Introduction

Since the pioneering work of Ned Seeman in 1982 [29], DNA has emerged as a versatile, programmable construction material at the nanoscale. Accordingly, DNA-based polyhedra [3,4,13,16,30,33], periodic- [36] and algorithmic [28] crystals, custom two- [2,27,34] and three-dimensional shapes [6,20] have since been demonstrated in the lab. With the introduction of the experimentally robust DNA origami technique [27], highly automated and well-abstracted derivative design methods [3,33] have become prominent, enabling design and synthesis of evermore complex three-dimensional geometries.

In the Chinese-postman-tour DNA origami design of triangulated topological spheres by Benson et al. [3], the long circular *scaffold* strand is first routed on the mesh skeleton and then held in place with hundreds of short *staple* strands. Henceforth, double helices, comprised half-and-half from the scaffold and staples, constitute the edges of the mesh, while the set of nearby strand transitions between edges form the vertices. However, the limitation of the method to topological spheres excludes simple but natural wireframes such as the nested cube synthesized by Veneziano et al. [33]. Evident with the view of the nested cube as a toroidal mesh, the class of higher-genus surfaces permits a much larger class of spatially embedded wireframes to be designed.

© Springer International Publishing AG 2017
R. Brijder and L. Qian (Eds.): DNA 23 2017, LNCS 10467, pp. 46–63, 2017.
DOI: 10.1007/978-3-319-66799-7_4

A fundamental topological constraint when employing a circular strand for assembly is that the strand routing must be unknotted. In a recent paper, Ellis-Monaghan et al. [8] have shown the scaffold routings of Benson et al. [3] can be knotted on higher-genus surfaces. In Fig. 1, we present another example of a knotted Chinese-postman-tour/Eulerian-tour routing on a triangulated torus. We leave it to the reader to verify that the routing corresponds to a trefoil knot on the torus. For higher-genus surfaces such as tori, a routing can also be knotted or unknotted depending on how the surface is embedded in real space. For instance, the Chinese-postman-tour/Eulerian-tour routing in Fig. 2 is knotted if the embedding of the torus is knotted (as in Fig. 3), but is unknotted if the embedding is standard. This can be verified by noting that the routing helically follows the meridional (horizontal) direction.

In this work, we examine the problem of finding unknotted Chinese-postman tours on the 1-skeleton of higher-genus triangulated surfaces. We present a cubic-time approximation algorithm to compute unknotted Chinese postman tours on such surfaces. Our algorithm further guarantees that the tour routes each edge at most twice, thus allowing low-packed helix bundle designs suitable for low-salt solutions [3, 33].

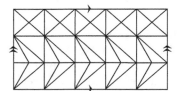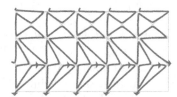

Fig. 1. Left: a planar representation of the 1-skeleton of a torus mesh. The torus is reconstructed, in a standard way, by glueing the horizontal boundaries together and likewise glueing the vertical boundaries together. Right: a knotted routing as a detached Chinese postman tour/Eulerian tour on the mesh skeleton. Note that since the boundaries of the planar representation are identified for glueing, only one copy of a boundary edge in the representation is visited by the Eulerian routing.

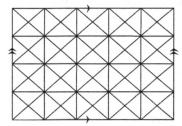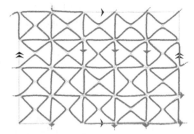

Fig. 2. Left: a planar representation of the 1-skeleton of a torus mesh. Right: a detached Chinese postman tour/Eulerian tour on the skeleton which is unknotted in a standard embedding of the torus, but is knotted in a knotted embedding of the torus. Note that a column of the mesh is a repeatable unit and thus this construction naturally leads to an infinite family of knotted and unknotted routings depending on the embedding of the torus in $\mathbb{R}^3$.

2   Preliminaries and Problem Definition

Assuming familiarity with basic topology [23] and graph theory [19], we only provide definitions to formally state our computational problem and prove our claims.

2.1   Triangulated Surfaces

A *surface* is a topological space that is Hausdorff, second countable, and locally homeomorphic to $\mathbb{R}^2$. We exclusively consider compact, connected, orientable surfaces, and for brevity refer to them as surfaces. Unless stated otherwise, a surface is assumed to be without boundary. The classification theorem of (compact, connected, orientable) surfaces states that any surface is homeomorphic to either the sphere or the connected sum of g tori, for $g \geq 1$. Surfaces without boundary can be classified up to homeomorphism via the *Euler characteristic*: $\chi(S) = 2 - 2g$, where g is the genus. Intuitively, the genus of a surface is the number of handles in that surface. For instance, a torus has genus one. A surface of genus zero and no boundary components is a *topological sphere*. A topological sphere with a single boundary component is called a *topological disk*. In practice surfaces are usually approximated by triangulated meshes (c.f. Fig. 3), or more formally by simplicial complexes.

Fig. 3. A triangulated surface (torus) embedded in $\mathbb{R}^3$ in a knotted manner.

Let k be a positive integer and $A = \{v_0, \cdots, v_k\}$ be a set of points in $\mathbb{R}^n$. We say that $v_0, \cdots, v_k$ are *affinely independent*, if $v_0 - v_1, \cdots, v_0 - v_k$ are linearly independent.[1] A *k-dimensional simplex*, or simply a *k-simplex*, is the set $\sigma^k = [v_0, \cdots, v_k] = \{\sum_{i=0}^k \lambda_i v_i \in \mathbb{R}^n : \sum_{i=0}^k \lambda_i = 1, \lambda \geq 0\}$. Note that a simplex is completely determined by its set of vertices. We often call a 0-simplex a *vertex*, a 1-simplex an *edge* and a 2-simplex a *face*. Given a simplex σ_A on a set A, any non-empty subset T of A is also affinely independent and determines a simplex σ_T called a *facet* of σ_A.

---

[1] By convention, if $k = 0$, then v_0 is affinely independent.

A *k-dimensional simplical complex* is a finite collection Σ of simplices of dimension at most k that satisfies the following conditions. First, if σ is in Σ, then all the facets of σ are also in Σ. Second, if $\sigma_1, \sigma_2 \in \Sigma$, $\sigma_1 \cap \sigma_2 \neq \emptyset$, then $\sigma_1 \cap \sigma_2$ is in Σ. Last, every point in Σ has a neighborhood that intersects at most finitely many simplices of Σ. The *i-skeleton* of a simplicial complex Σ is the union of the simplices of Σ of dimensions less than or equal to i. If Σ is a simplicial complex and $\sigma \in \Sigma$ then $star(\sigma) := \{\mu \in \Sigma$ *such that μ contains σ, or is a facet of a simplex which contains σ*$\}$. We denote by $|\Sigma|$ the set obtained by taking the union of all simplices of a simplicial complex Σ, and equipped with the relative subspace topology of the usual topology of $\mathbb{R}^n$.

In this paper we only need 2-dimensional simplicial complexes. If Σ is a 2-dimensional simplicial complex then we denote by $V(\Sigma)$, $E(\Sigma)$ and $F(\Sigma)$ the set of vertices, edges and triangles of Σ, respectively. A *simplicial surface* S is a simplicial complex consisting of a finite set of faces such that (1) Every vertex in Σ belongs to at least one face in $F(\Sigma)$ and (2) For every v in $V(\Sigma)$, $|star(v)|$ is homeomorphic to a 2-disk. If Σ is a simplicial surface and in addition there is a piecewise-linear embedding $\mathcal{F} : |\Sigma| \longrightarrow \mathbb{R}^3$ then we call $(\Sigma, \mathcal{F})$ a *triangulated surface*, or simply a *mesh*. Note that if Σ is a triangulated surface then $|\Sigma|$ is a topological surface embedded in $\mathbb{R}^3$. Figure 3 shows an example of a triangulated surface. The Euler characteristic of a mesh Σ is given by $\chi(\Sigma) = |V(\Sigma)| - |E(\Sigma)| + |F(\Sigma)|$.

2.2 Postman Tours and Knots

In this paper, we only consider finite, undirected, loopless graphs. By graph, we mean a simple graph, reserving the term multigraph for graphs which can have parallel edges. We denote a multigraph as $G = (V, E)$, and use $V(G)$ and $E(G)$ to refer to its vertices and edges, respectively. A *walk W of length $l \geq 0$* on a multigraph $G = (V, E)$ is a sequence of vertices and edges $(v_0, e_0, v_1, e_1, \cdots, e_{l-1}, v_l)$ such that $e_i = [v_{i-1}, v_i] \in E(G)$. We say W *visits* or *traces* an edge e if $e \in W$. A walk is said to be closed if $v_0 = v_l$. A *path* is a walk without repeated vertices and a *cycle* is like a path except that $v_0 = v_l$. A *trail* is a walk with distinct edges, and a trail is said to be *closed* if the walk is closed. A multigraph is said to be *connected* if there is a walk between any two vertices. A connected cycle-free graph is called a *tree* and more generally a cycle-free graph is called a *forest*.

A *postman tour* [7] is a closed walk which traces every edge at least once. The *length of a postman tour* is the length of the walk. A *Chinese postman tour* is a postman tour with minimum length. An *Eulerian tour* is a closed walk which visits every edge exactly once. A multigraph which admits an Eulerian tour is said to be *Eulerian*. A classical theorem of Euler [10] and Hierholzer [17] states that a multigraph is Eulerian if and only if it is connected and does not contain any odd-degree vertices. For an Eulerian multigraph, the notion of a Chinese postman tour and an Eulerian tour coincide (see the planar representations of Eulerian torus meshes and related routings in Figs. 1 and 2). In our work, we only work with postman tours on (simple) graphs. Moreover, we mostly view a postman tour on a graph as being an Eulerian tour on a related multigraph. In particular, the multiple traces of an edge by the postman tour are viewed as visits of independent copies of the edge in the Eulerian multigraph [7]. For further illustrations related to the concepts here, check the right most graph in Fig. 6, a Chinese postman tour of this graph depicted on the right in Fig. 7 and the related Eulerian multigraph shown on the left in Fig. 7.

An embedding of a multigraph G in $\mathbb{R}^3$ is a representation of G in $\mathbb{R}^3$ where the vertices of G are represented by points on $\mathbb{R}^3$ and the edges of G are represented by simple arcs on $\mathbb{R}^3$ such that: (1) no two arcs intersect at interior points to either of them, (2) the two vertices defining an edge e are associated with the endpoints of the arc associated with e, and (3) there is no arc which includes points that are associated with other vertices. Here we are interested in multigraphs that are embedded on meshes. In particular, we are interested in the 1-*skeleton* of a surface mesh M which corresponds to the embedded graph $G = (V, E)$, where $V = V(M)$ and $E = E(M)$.

A *knot* in $\mathbb{R}^3$ is a (piece-wise) linear embedding of the circle S^1 in $\mathbb{R}^3$. Knots are considered up to ambient isotopy, that is two knots are said to be ambient isotopic if we can continuously deform one to the other without tearing or self intersection. An *unknot* is a knot ambient isotopic to the standard circle on the plane. Equivalently, an *unknot* is a knot that bounds an embedded piece-wise linear disk in $\mathbb{R}^3$ [24]. For a formal treatment of knot theory see [24,26]. Note that the standard definition of a postman tour as presented above is purely combinatorial and does not yet specify a curve to be analyzed as a knot. In the next section, we introduce an interpretation of a postman tour as a knot which permits any postman tour as a candidate solution for the unknotted Chinese postman tour routing problem (c.f. Problem 1).

2.3   The Unknotted Chinese Postman Tour Problem

For a graph G embedded in $\mathbb{R}^3$, suppose the embedding of a postman tour T on G is the curve implied by the image of the vertex-edge sequence of the tour except that the repeated edges are mapped to parallel but non-overlapping curves which only meet at the endpoints. Consider the related Eulerian multigraph G'. Unless G' is a cycle, T repeats vertices, and hence the embedding of T either touches or intersects itself at vertices. For formal treatment of T as a knot, we need to make the embedding a simple curve while keeping the curve within the vicinity of G. For this purpose, we define the notion of a detachment of a postman tour embedding, which is simply a local unpinning of all the edge transitions at the vertices, as follows.

Let $T = (v_0, e_0, v_1, e_1, \cdots, e_{l-1}, v_l = v_0)$ be a postman tour on a graph G embedded in $\mathbb{R}^3$. Construct a new graph Δ with vertex set $\{\delta_0, \delta_1, \cdots, \delta_{l-1}\}$ and edge set $\{\alpha_0, \alpha_1, \cdots, \alpha_{l-1}\}$, where $\alpha_i = [\delta_i, \delta_{i+1}]$, for $0 \le i \le l - 2$, and $\alpha_{l-1} = [\delta_{l-1}, \delta_0]$. Clearly, Δ is a cycle and has an Eulerian tour $T_\Delta = (\delta_0, \alpha_0, \delta_1, \alpha_1, \cdots, \alpha_{l-1}, \delta_l = \delta_0)$. Observe that while T might repeat vertices, T_Δ does not. Suppose we embed Δ as follows: (1) Each δ_i is at most a small $\epsilon > 0$ distance, in $\mathbb{R}^3$, away from v_i, (2) Each edge α_i is embedded exactly like e_i, except at its ends where it is incident to δ_i and δ_{i+1} instead of v_i and v_{i+1}. Note that by construction, all the vertices of Δ are at distinct locations. We call the induced embedding on T_Δ a *detachment* of T.[2] A detachment of a postman tour is simple and is thus a knot. We say that a *detached postman tour is unknotted* if the detachment is an unknot. Recall Fig. 1 for an example of a knotted detachment of a Chinese postman tour on the 1-skeleton of a torus standardly embedded in $\mathbb{R}^3$. We now state our problem as follows.

Problem 1. **Unknotted Chinese postman tour problem (UCPT):** Given a triangulated oriented surface without a boundary and of genus $g \geq 1$, find a minimum length postman tour along its 1-skeleton which is detachable to an unknot.

An example input instance of **UCPT**, a genus one triangulated mesh, is shown on the left in Fig. 4. Next, we present a $\frac{5}{3}$-approximation algorithm for **UCPT**, that is a postman tour with length at most two-thirds greater than any unknotted Chinese postman tour, which moreover guarantees that no edge is traced more than twice. Although the notion of detachment is important for the consideration of all possible postman tours as solutions to **UCPT**, our algorithm outputs non-crossing postman tours (c.f. Sect. 3.3 and Fig. 7) where the detachments are clear without explicit construction.

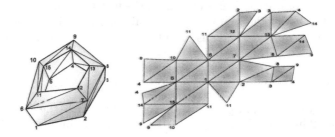

Fig. 4. Left: A toroidal mesh. Right: A polygonal schema of the mesh.

3   A Cubic Time Algorithm for Finding Unknotted Postman Tours

The main idea of our algorithm is to transfer the problem of finding an unknotted Chinese postman tour on an arbitrary surface mesh to the case of finding such a tour on the mesh's cutting to a topological disk. After such cutting, we can compute an unknotted Chinese postman tour on the disk, which then simply lifts to an unknotted approximate Chinese postman tour on the surface mesh. To be presented in detail in the subsequent sections, our algorithm, which we name as the **cut-and-route** algorithm, proceeds along the following steps:

---

[2] Detachments are also defined in the literature [11] for multigraphs without an associated embedding.

1. Cut the input surface mesh M to obtain a topological disk D where a re-gluing of partnered boundary edges reconstructs M.
2. Remove one instance of the partnered edges of D and extract an embedded subgraph H on D whose edges are in a one-to-one correspondence with M.
3. Find a non-crossing Chinese postman tour on H, which then maps to the desired unknotted approximate Chinese postman tour on M.

3.1   Cutting a Surface to a Disk

Surface cutting is a problem that has been extensively studied due to its importance in surface parameterization and texture mapping [9, 12]. For our algorithm, we adopt the basic algorithm of Dey and Schipper [5] for computing a polygonal schema. A *polygonal schema* [9] of a triangulated surface M is a topological disk that consists of all faces of M. A polygonal schema can be obtained by cutting a graph, called a *cut graph*, on the 1-skeleton of M. Dey and Schipper's cutting algorithm starts with topological disk D which consists of a single face f on the surface M and keeps expanding D by gluing faces to its boundary. We present it in Algorithm 1, named as Mesh2disk, for analysis within our cut-and-route algorithm.

Mesh2disk is simply a breadth first search (BFS) on the dual graph implied by the face-to-face adjacency list of M, and the BFS tree represents the connectivity of the faces in D. For the torus mesh in Fig. 4, the first few rounds of face addition are shown in Fig. 5 and the resulting polygonal schema is shown on the right in Fig. 4. Next, we prove three lemmas useful for the analysis of the cut-and-route algorithm.

Lemma 1. *Mesh2disk outputs a topological disk D.*

Proof. Let $(f_1, f_2, f_3, \cdots, f_{|F(M)|})$ be the order in which the faces of M are visited by Mesh2disk. Let D_j, for $j = 1$ to $|F(M)|$, be the simplicial complex after

---

Algorithm 1. Mesh2disk: Cutting a triangulated surface to a disk.

Input: A triangulated surface M given as a face-to-face adjacency list.
Output: A polygonal schema D of M.

1 $s \leftarrow$ the first face, $D \leftarrow$ an empty adjacency list, $Q \leftarrow$ empty queue of faces;
2 Enqueue s to Q;
3 **for** *all faces $f \in M$* **do** Mark f as not visited;
4 **while** *Q is not empty* **do**
5      $f \leftarrow$ dequeue from Q;
6   Mark f as visited;
7   **for** *g neighbor of f in M* **do**
8   **if** *g is not visited* **then**
9   Enqueue g to Q;
10   Append g to $D[f]$ and f to $D[g]$;
11 **return** D;

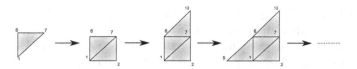

Fig. 5. Initial stages of Mesh2disk (Algorithm 1) constructing a polygonal schema of the torus mesh in Fig. 4.

faces f_1 through f_i have been appended. We prove D_j is a disk by induction on j. For the base case, D_1 consists of a single face f_1 and hence is a disk. Now assume, D_{j-1} is a disk. When f_j is added to D_j by gluing edge-wise to f_i for some $i < j$, $|V(D_j)| = |V(D_{j-1})| + 1$, $|E(D_j)| = |E(D_{j-1})| + 2$ and $|F(D_j)| = |F(D_{j-1})| + 1$ (c.f. Fig. 5.) Hence, $\chi(D_j) = \chi(D_{j-1})$. The number of boundary components remains the same. By the classification theorem of surfaces with boundary, D_j is a disk. □

Lemma 2. *For the input-output pair (M, D) of Mesh2disk, $|V(D)| = |F(M)| + 2$.*

Proof. Following the notation of Lemma 1, the claim follows by observing that D_j has one more vertex than D_{j-1}, for $j \geq 2$, and a straightforward induction on the number of faces of M. □

Lemma 3. *The algorithm Mesh2disk runs in $\mathcal{O}(|F(M)|)$ time.*

Proof. Mesh2disk is simply BFS as implied by the face-to-face adjacency list of M and thus takes $\mathcal{O}(|F(M)| + |E(M)|)$ time. By double counting of edges, $3|F(M)| = 2|E(M)|$. The claim follows. □

Each edge in the cut graph of a mesh determines exactly two boundary edges in the polygonal schema [5,9]. The boundary edges on D always come in pairs and if we glue all such pairs together we obtain the original mesh M. If e and e' are two boundary edges coming from the same edge in M, we say that e and e' are *partners*.

3.2   Removing Duplicate Edges on Disk Boundary

In principle, we can construct a Chinese postman tour on the polygonal schema D and map it back to an approximate Chinese postman tour on the input mesh M. However, an attempt to build a Chinese postman tour directly on D can repeat the cut graph edges on the mesh M three or four times since these edges appear twice as boundary edges of D. To guarantee that no edge of the mesh is traced more than twice, we instead find a Chinese postman tour on an embedded subgraph H of D with the following properties:

1. For any two partnered edges in D, exactly one of the two edges is in $E(H)$. Thus, the edge set of H has a one-to-one correspondence with the edge set of M.
2. H is a connected spanning subgraph of D, that is $V(H) = V(D)$.

After extracting such a spanning subgraph H, we can find a Chinese postman tour on H, which then maps to an approximate Chinese postman tour on M that visits the mesh edges at most twice. To extract H, we first identify two types of faces of D. A face in D is said to be a *peripheral face* of *type I* if it is bounded by only one boundary edge, and of *type II* if it is bounded by two boundary edges; see the boundary edges in the left-most image in Fig. 6. Our algorithm, named Declone and presented as Algorithm 2, runs through the peripheral faces, identifying partnered edges and removing the clones along the way. For the polygonal schema of the torus in Fig. 4, an intermediate output of Delcone is shown in the center in Fig. 6 while the final output is shown on the right.

Algorithm 2. Declone: Remove clone edges on a polygonal schema's boundary.

    Input: A polygonal schema D of a surface mesh M.
    Output: An embedded subgraph H of D with the two properties listed in
                Sect. 3.2.
1 **for** *all faces $f \in D$* **do**
2   Mark f as not processed;
3   **if** *f has a single boundary edge* **then** append f to F_I ;
4   **else if** *f has two boundary edges* **then** append f to F_{II} ;
5 $H \leftarrow D$;
6 **for** *all faces $f \in F_I$* **do**
7   **if** *f is not processed* **then**
8           $a \leftarrow$ a boundary edge of f;
9   Remove a from H;
10   Mark f as processed;
11         $g \leftarrow$ the face which contains a's partner;
12   **while** *g has two boundary edges* **do**
13            $b \leftarrow$ the boundary edge in g different from a's partner;
14   Remove b from H;
15   Mark g as processed;
16            $g \leftarrow$ the face which contains b's partner;
17            $a \leftarrow$ b's partner;
18   Mark g as processed;
19 **for** *all faces $f \in F_{II}$* **do**
20   **if** *f is not processed* **then**
21           $g \leftarrow$ f;
22           $a \leftarrow$ a boundary edge of g;
23   **while** *g is not processed* **do**
24            $b \leftarrow$ the other boundary edge of g;
25   Remove b from H;
26   Mark g as processed;
27            $g \leftarrow$ the face which contains b's partner;
28            $a \leftarrow$ b's partner;
29 **return** H

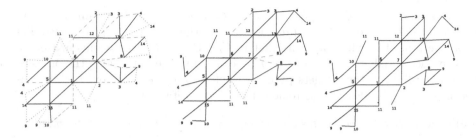

Fig. 6. Left, the 1-skeleton of the polygonal schema D of the torus mesh. Middle and right figures illustrate an intermediate output and the final output of Algorithm Declone, respectively. In the left and middle figures, the dashed edges highlight unprocessed type I peripheral faces while the dotted edges highlight unprocessed type II peripheral faces.

To prove Declone's output H satisfies the two properties listed in Sect. 3.2, we first define a new graph $X^\star$ which we refer to as the *cut-dual*.[3] The cut-dual is constructed by creating a vertex corresponding to each peripheral face in D, and adding an edge between two vertices u and v if and only if u's face contains an edge which has a partner in v's face. For clarity, we refer to the vertices in $X^\star$ through their corresponding faces in D. Note that $X^\star$ has a maximum degree two since a peripheral face has at most two boundary edges. Hence, $X^\star$ is a disjoint union of connected components, each of which is either a path or a cycle. Each path starts and ends with a type I face but is otherwise composed of type II faces. Analogously, each cycle is completely composed of type II faces. With $X^\star$ in mind, we now prove the following lemmas about Declone.

Lemma 4. *Let e and e' be two partner boundary edges in a polygonal schema D of a surface mesh M, then exactly one of the two edges e or e' is removed in H. Hence, there is a one-to-one correspondence between $E(H)$ and $E(M)$.*

Proof. To show that exactly one of the edges e and e' is removed in H, consider the peripheral faces f and g that contain the edges e and e', respectively. By construction of the cut-dual $X^\star$, f and g appear in the same path or the same cycle component of $X^\star$.

Suppose f and g appear in a path P. Let s be the type I face of P appended first to F_I and let t be the other type I face in P. Suppose we orient P from s to t. Declone (Line 6 to 19) processes the faces in P from s to t. Suppose, w.l.o.g, f appears before g in the (oriented) P. If $f = s$, then e gets removed in Line 9, but e' is retained, whether g is a type II face (in the while loop), or it is a type I face (outside the while loop.) If $f \neq s$, then it gets processed in the while loop. Once again, e gets removed (Line 14) and e' gets retained whether we stay in the while loop or exit it in the next iteration. Analogously, e' gets removed and e gets retained when g appears before f in P.

---

[3] $X^\star$ is the subgraph of the dual of M induced by the duals of the cut edges.

Now suppose f and g appear in a cycle C of $X^\star$. Since C contains no type I face, none of the faces in C are processed before the third for loop (Line 19). Let s be the face of C which appears first in F_{II}. Orient C from s outward based on the selected edge a in Line 22. If f precedes g in the path along the oriented C starting from s, then f gets processed before g and e is removed while e' is retained. If g precedes f, e' is removed while e is retained.   □

Lemma 5. *Declone computes a connected spanning subgraph H of the input D.*

Proof. Since no vertex is deleted in H, $V(H) = V(D)$. Hence, we only need to show that H is connected. Further noting that only a subset of the boundary edges of D get deleted, let us first analyze the state of the peripheral faces of D. In particular, we first show that for any peripheral face f in D, at most one boundary edge of f is deleted and the vertices of f remain connected in H.

Let u, v, w be the vertices of f and let $a = \{u, v\}, b = \{v, w\}, c = \{w, u\}$ be its edges. If f is a type I face, only one edge is a boundary edge and is potentially removed. If indeed the edge, w.l.o.g suppose a, is removed, its endpoints u and v remain connected through the path (u, c, w, b, v).

Now suppose f is a type II face, and w.l.o.g, assume that a and b are the boundary edges. Let g and g' be the two peripheral faces containing the partners of a and b, respectively. Reconsider the cut-dual $X^\star$ and the orientation of its paths and cycles (see proof of Lemma 4.) In the oriented path or cycle, the order is either (g, f, g') or (g', f, g). In the first case, b is removed while a is retained while in the second case, the reverse holds. If a gets removed, then u and v remain connected through the path (u, c, w, b, v) while if b is removed, v and w are connected through (v, a, u, c, w).

We can now show that H is a connected graph. Since D is connected, there is a walk W between any vertices u, v. For every boundary edge of D that appears in W but is deleted in H, replace it with one of the paths described above. This results in a new walk between u and v in H, and thus H is connected.   □

Lemma 6. *Declone runs in $\mathcal{O}(|F(M)|)$ time.*

Proof. Let p be the number of peripheral faces in D. Since the peripheral faces are a subset of the faces of D, $p \leq |F(D)| = |F(M)|$. The first for loop iterates $|F(D)| = |F(M)|$ times, each time consuming constant time. Note that checking face incidence takes constant time since each face is incident to three other faces. This also implies that the copy in Line 5 is linear in $|F(M)|$.

For the next two for loops, recall the proof in Lemma 4 and observe that each peripheral face gets processed once, either in a path or a cycle of $X^\star$. Hence, there are only p iterations of the total work done inside the while loops. The book-keeping, edge-removal and partner-checking can all be done in constant time, once again since each face is incident to three faces.   □

3.3   Finding Non-crossing Chinese Postman Tours on the Polygonal Schema

By Lemma 5, Declone outputs a connected spanning subgraph H of the polygonal schema D. Since H is generally a non-Eulerian graph, cut-and-route proceeds by adding a minimal set of duplicate edges which converts H to an Eulerian multigraph H' using Edmonds' Blossom algorithm [7,19]. For the H graph shown on the right in Fig. 6, the resulting Eulerian multigraph H' is shown on the left in Fig. 7. To keep H' in the polygonal schema, the duplicate edges are added in the interior side of the polygonal schema. In principle, we can then compute an Eulerian tour on H' with Hierholzer's algorithm [17,19], but such a tour can generally have crossings, which complicates the analysis of the unknottedness of the tour.

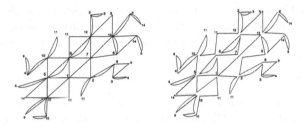

Fig. 7. Left, an Eulerian multigraph of H obtained by Edmonds algorithm. Right, a non-crossing Chinese postman tour on H, which then maps to an unknotted postman tour on the torus 1-skeleton.

To make the notion of crossing precise, note that a multigraph embedded on an orientable surface induces a local cyclic rotational order (fixed either clockwise or counterclockwise) of the incident edges of vertices. We say two pairs of consecutive edges in a closed-trail, all of which are incident to a common vertex, *cross* if the two pairs interleave with respect to the cyclic order of edges around the vertex. In other terms, a *crossing* is a quadruple of edges incident to a common vertex which can be grouped into two pairs according to their contiguity in the closed trail such that cyclically visiting edges around the vertex one alternates between the pairs. An Eulerian tour is said to be *non-crossing* if it does not contain any two crossing pairs of consecutive edges. Visually, a non-crossing Eulerian tour on a surface embedded multigraph can be drawn as a simple closed curve on the surface. In this sense non-crossing Eulerian tours detach to simple closed curves and are easier to analyze for unknottedness.

Abraham and Kotzig [1] as well as Grossman and Reingold [31] have shown that all Eulerian multigraphs embedded on a plane admit non-crossing Eulerian tours. More recently, Tsai and West [32] presented a technique to convert an arbitrary Eulerian tour on a multigraph embedded on a plane to a non-crossing one by a vertex-local re-splicing of the tour at crossing pairs.[4] Inspired by the proof of Grossman and Reingold [31], we introduce a linear-time algorithm in Lemma 7 to compute non-crossing Eulerian tours for multigraphs embedded on orientable surfaces. Briefly, the algorithm first computes an initial non-crossing closed-trail decomposition of the multigraph, and then iteratively resplices independent closed trails at each vertex, finally yielding a non-crossing Eulerian tour.

For a more precise description, we need the additional notion of a transition system of an Eulerian multigraph. A *transition system* [11] of an Eulerian multigraph is a set of partitions of edges, where each element of the set is a grouping (partitioning) of the incident edges of a vertex into (unordered) pairs. In a transition system, each edge is in two pairs corresponding to its' pairings at its' two end-points. There is a bijection between transition systems and closed-trail decomposition of Eulerian multigraphs [15]. In this setting, the pairs in the partitions of the transition system correspond to consecutive edges in the trails of the closed-trail decomposition. Given a transition system, we can compute the closed-trail decomposition in linear time by directionally following the pairings. The notion of non-crossing closed-trails extends naturally to closed-trail decompositions (and transition systems), in the sense that the closed trails are both self non-crossing and mutually non-crossing.

Lemma 7. *Let G be an Eulerian multigraph embedded on an orientable surface. There is an $\mathcal{O}(|V(G)| + |E(G)|)$ algorithm to compute a non-crossing Eulerian tour on G.*

Proof. As an input, assume that the embedded multigraph G is given as an adjacency list, describing for each vertex the list of the incident edges in the rotation order. The algorithm proceeds as follows. We first obtain an initial transition system by going through each vertex, and for each vertex, cycling around the incident edges in the rotation order and pairing the first with the second, the third with the fourth, etc. From this initial transition system, we follow the pairings to compute an initial closed-trail decomposition. Next, for each vertex, we again cycle through the incident edges of the vertex and every instance where the current edge b is not in the same closed trail as the previous edge a, we pair a with b and a's old mate a_p with b's old mate b_p.

To prove the algorithm computes a non-crossing Eulerian tour, it suffices to prove that all edges are in the same non-crossing closed trail after all the vertices have been processed. We prove all edge are in the same closed trail by showing that after a vertex has been processed, we obtain a non-crossing closed-trail decomposition where all the edges incident to that vertex are in the same closed trail. After all vertices have been processed, this local criterion suffices to ensure distal edges are in the same closed trail because G is connected. For an alternative proof of the sufficiency of the local criterion, observe that since G is Eulerian, it has some Eulerian tour T'. Now suppose an edge e is in some closed

---

[4] Although stated only for Eulerian multigraphs embedded on a plane, Tsai and West's proof also holds for Eulerian multigraphs embedded on any surface since the resplicing occurs locally at vertices.

trail T in the closed-trail decomposition. Starting from e, following the edges in T', we see that all the edges in T' are also in T since consecutive edges in T' are incident to a common vertex. Since T' is an Eulerian tour, all the edges in G are in T', and hence all edges in G are in T.

Hence, we just need to prove that after a vertex has been processed, we obtain a non-crossing closed-trail decomposition where all the local edges are in the same closed-trail. We prove this by induction on the incident edges of the vertex. In particular, we prove, for a local rotation index j (which runs from one up-to the degree of the vertex), the following two claims hold after the j-th edge has been processed: (1) The closed-trail decomposition is non-crossing, (2) The first j edges in the cyclic-order around the vertex are in the same closed-trail.

For the base case $j = 1$, (1) holds for the first processed vertex because in the initially computed transition system (and correspondingly closed-trail decomposition), the pairings are composed of neighboring edges and thus do not cross any other pairings. For the remaining vertices, (1) holds by induction on vertices. Claim (2) holds vacuously for $j = 1$.

Now suppose after the $(j - 1)$-th edge has been processed we have a non-crossing closed-trail decomposition where all local edges are in the same closed-trail. If the $(j - 1)$-th edge is in the same closed-trail as the j-th edge, nothing changes after the j-th edge is processed and both (1) and (2) still hold. Now suppose, j is not in the same closed-trail as $j - 1$. Let b be the j-th edge and let a be the $(j - 1)$-th edge, and let a_p and b_p be the mates of a and b before the re-pairing. To show that (1) holds, we only need to show that the new pairings $\{a, b\}$ and $\{a_p, b_p\}$ do not cross any of the unaltered pairings or each other. Indeed, the pairing $\{a, b\}$ is between neighboring edges and cannot cross any other pairing. To see that $\{a_p, b_p\}$ does not cross any other pairings, first note that b_p comes before a_p in the cyclic order around the vertex after a; that is, the cyclic order O is of the form $(\cdots, a, b, \cdots, b_p, \cdots, a_p, \cdots)$. This follows because $\{a, a_p\}$ does not cross $\{b, b_p\}$ by induction hypothesis. Now assume, for the sake of contradiction that $\{a_p, b_p\}$ crosses some other pairing $\{c, d\} \neq \{a, b\}$ at the vertex. Note that exactly one of c or d must be in between b_p and a_p in O since otherwise $\{c, d\}$ would not cross $\{a_p, b_p\}$. Without loss of generality, suppose c is the edge in between b_p and a_p in O. If d is between b and b_p, then $\{c, d\}$ crosses $\{b, b_p\}$ and if d is between a_p and a, then $\{c, d\}$ crosses $\{a, a_p\}$, in either case contradicting the induction hypothesis. Hence, $\{c, d\}$ does not cross $\{a_p, b_p\}$ and claim (1) holds. For claim (2), since a and b are now paired, we only need to show that the re-pairing does not leave the edges from 1 to $j - 1$ in different trails. This cannot be the case since breaking a closed trail at one pairing does not disconnect its' edges.

For the complexity, observe that, in both the computation of the initial decomposition and main processing, every vertex is processed once and every edge is checked at most twice. Hence, the $\mathcal{O}(|V(G)| + |E(G)|)$ time-complexity follows if we can show all the internal operations cost constant time. The initial pairing as well as re-pairing consumes constant time if we represent a transition system by maintaining, for each edge, the two mates of the edge at its two end-points. Checking whether two edges are in the same trail can be done in constant time if we associate with each edge, a pointer to their trail within the decomposition. That is, two pointers will point to the same trail if their corresponding edges are in the same trail. The pointers can be initialized in $\mathcal{O}(|E(G)|)$ time during the computation of the initial closed-trail decomposition. When re-pairing, updating the trail of a re-paired edge through its' pointer will simultaneously update the trails of all the edges in the same trail.   □

A non-crossing Eulerian tour of H', and correspondingly a non-crossing Chinese postman tour of H, for our running example is shown on the right in Fig. 7. Incorporating the algorithm presented in Lemma 7, cut-and-route outputs a non-crossing Chinese postman tour $\hat{T}$ on H which detaches to a simple closed curve on the polygonal schema D. Noting that, by Lemma 4, $\hat{T}$ is also a postman tour on the input mesh M, we now have the following theorems.

Theorem 1. *Cut-and-route outputs a postman tour which is detachable to the unknot on the input mesh M.*

Proof. Let C be the detached postman tour on M obtained by the cut-and-route algorithm. We prove that C is unknotted. By construction, C lies on the embedding of the polygonal schema D in $\mathbb{R}^3$. There is a homeomorphism h which maps the polygonal schema D, from its embedding in $\mathbb{R}^3$, to the standard disk on the plane. By homeo-morphism restriction, h maps C to a simple closed curve C' on the plane. By the Jordan-Schönfflies' theorem [26], C' bounds a disk D'. The inverse of D' under h is a disk on the embedding of D whose boundary is C. A simple closed curve that bounds an embedded disk is an unknot. Hence, C is unknotted.   □

Theorem 2. *Cut-and-route outputs an unknotted postman tour that visits any edge of the input mesh M at most twice. Moreover, it is a $\frac{5}{3}$-approximation algorithm for **UCPT**.*

Proof. Let $\hat{T}$ be the outputted non-crossing Chinese postman tour on H. The first claim follows because a Chinese postman tour visits every edge at most twice [7], and by Lemma 4, the edges of H are in one-to-one correspondence with the edges of M.

For the second claim, observe that the set of edges added to H to construct the Eulerian multigraph H' form a forest on a subset of $V(H)$. Indeed, if such a graph has a cycle, the edges in the cycle can be removed from H' while keeping it Eulerian. With the cycle removed, an Eulerian tour on H' yields a postman tour with less length than $\hat{T}$, thus contradicting the minimality of $\hat{T}$. By the relationship between the number of vertices and number of edges of a forest, there are at most $|V(H)| - 1$ extra edges in H'. By Lemmas 2 and 5, $|V(H)| = |V(D)| = |F(M)| + 2$. By double counting of edges, $|F(M)| = \frac{2}{3}|E(M)|$. Thus, at most $\frac{2}{3}|E(M)| + 1$ edges are repeated. Since any Chinese postman tour visits every edge, $|E(M)|$ is a lower-bound on the optimal unknotted Chinese postman tour. The approximation factor thus follows.   □

Theorem 3. *For an input mesh M, cut-and-route runs in $\mathcal{O}(|F(M)|^3)$ time.*

Proof. For the analysis, note the relations $|V(H)| = |V(D)| = |F(M)| + 2$, $|E(H)| = |E(M)|$ implied by Lemmas 2, 5 and 4. Also note that $|V(H'| = |V(H)|$ and $|E(H')| \leq 2 * |E(H)|$ from Edmonds' algorithm [7,19] and the fact that $2|E(M)| = 3|F(M)|$ by double counting. Mesh2disk and Declone were shown to run in $\mathcal{O}(|F(M)|)$ time in Lemmas 3 and 6. Edmonds' algorithm to convert H to the Eulerian counterpart H' runs in $\mathcal{O}(|V(H)|^3) = \mathcal{O}(|F(M)|^3)$ time [7,19]. It is easy to check (c.f. Supplementary methods in [2]) that we can compute, in no more than cubic time, the local rotation of edges at vertices for H (and in turn of H') from the polygonal schema description and the deleted edges of H. By Lemma 7, computing a non-crossing Eulerian tour on H', or equivalently finding a non-crossing Chinese postman tour on H, takes $\mathcal{O}(|V(H')|+|E(H')|) = \mathcal{O}(F(M))$ time. Hence, all the modules of cut-and-route run in no more than cubic-time with respect to the number of faces of the input mesh M.   □

4   Conclusions and Future Work

Eulerian tours have previously featured in experimental and theoretical considerations of DNA [3,8,18,25,35] and protein self-assembly [14,21]. Similarly, topological constraints have been implicitly considered in previous works [3,6]. Here, we formally investigated an unknottedness constraint of a circular strand's routing on triangulated higher-genus surfaces, mostly within the design-framework of Benson et al. [3]. We presented a cubic-time algorithm to compute unknotted approximate Chinese postman tours on such surfaces.

There are numerous theoretical questions available within the prescribed theory of unknotted Chinese postman tours. In the specified setting, the complexity of finding unknotted Chinese postman tours (**UCPT** defined in Sect. 2.3) on surface meshes remains. Simultaneously, further improvements to the approximation with respect to the approximation factor and run-time can also be pursued. More generally, **UCPT** on straight-line graph embeddings in $\mathbb{R}^3$ can be studied with an aim to design arbitrary non-manifold wireframe structures.

On the experimental side, it remains to be seen whether the current approach is viable, especially with the implied generality. First, unknottedness, while necessary, is likely insufficient for knotted surface embeddings such as the one depicted in Fig. 3. As evident in such instances, an unknotted routing can still have a self-threading of the strand through loops. Although it has been previously shown [22] that such self-threadings are attainable through careful design of the folding pathway, such designs may be unlikely to fold purely from thermodynamic optimization.

References

1. Abrham, J., Kotzig, A.: Construction of planar Eulerian multigraphs. In: Proceedings of Tenth Southeastern Conference on Combinatorics, Graph Theory, and Computing, pp. 123–130 (1979)

2. Benson, E., Mohammed, A., Bosco, A., Teixeira, A.I., Orponen, P., Högberg, B.: Computer-aided production of scaffolded DNA nanostructures from flat sheet meshes. Angew. Chem. Int. Ed. **55**(31), 8869–8872 (2016)
3. Benson, E., Mohammed, A., Gardell, J., Masich, S., Czeizler, E., Orponen, P., Högberg, B.: DNA rendering of polyhedral meshes at the nanoscale. Nature **523**(7561), 441–444 (2015)
4. Chen, J., Seeman, N.C.: Synthesis from DNA of a molecule with the connectivity of a cube. Nature **350**(6319), 631 (1991)
5. Dey, T.K., Schipper, H.: A new technique to compute polygonal schema for 2-manifolds with application to null-homotopy detection. Discrete Comput. Geom. **14**(1), 93–110 (1995)
6. Douglas, S.M., Dietz, H., Liedl, T., Högberg, B., Graf, F., Shih, W.M.: Self-assembly of DNA into nanoscale three-dimensional shapes. Nature **459**(7245), 414–418 (2009)
7. Edmonds, J., Johnson, E.L.: Matching, Euler tours and the Chinese postman. Math. Program. **5**(1), 88–124 (1973)
8. Ellis-Monaghan, J.A., Pangborn, G., Seeman, N.C., Blakeley, S., Disher, C., Falcigno, M., Healy, B., Morse, A., Singh, B., Westland, M.: Design tools for reporter strands and DNA origami scaffold strands. Theoret. Comput. Sci. **671**, 69–78 (2016)
9. Erickson, J., Har-Peled, S.: Optimally cutting a surface into a disk. Discrete Comput. Geom. **31**(1), 37–59 (2004)
10. Euler, L.: Solutio problematis ad geometriam situs pertinentis. Commentarii Academiae Scientiarum Petropolitanae **8**, 128–140 (1741)
11. Fleischner, H.: Eulerian Graphs and Related Topics, vol. 1. Elsevier, Amsterdam (1990)
12. Floater, M.S.: Parametrization and smooth approximation of surface triangulations. Comput. Aided Geom. Des. **14**(3), 231–250 (1997)
13. Goodman, R.P., Schaap, I.A., Tardin, C.F., Erben, C.M., Berry, R.M., Schmidt, C.F., Turberfield, A.J.: Rapid chiral assembly of rigid DNA building blocks for molecular nanofabrication. Science **310**(5754), 1661–1665 (2005)
14. Gradišar, H., Božič, S., Doles, T., Vengust, D., Hafner-Bratkovič, I., Mertelj, A., Webb, B., Šali, A., Klavžar, S., Jerala, R.: Design of a single-chain polypeptide tetrahedron assembled from coiled-coil segments. Nature Chem. Biol. **9**(6), 362–366 (2013)
15. Gross, J., Yellen, J.: Handbook of Graph Theory. Discrete Mathematics and Its Applications. CRC Press, Boca Raton (2004)
16. He, Y., Ye, T., Su, M., Zhang, C., Ribbe, A.E., Jiang, W., Mao, C.: Hierarchical self-assembly of DNA into symmetric supramolecular polyhedra. Nature **452**(7184), 198–201 (2008)
17. Hierholzer, C., Wiener, C.: Über die Möglichkeit, einen Linienzug ohne Wiederholung und ohne Unterbrechung zu umfahren. Mathematische Annalen **6**(1), 30–32 (1873)
18. Jonoska, N., Seeman, N.C., Wu, G.: On existence of reporter strands in DNA-based graph structures. Theoret. Comput. Sci. **410**(15), 1448–1460 (2009)
19. Jungnickel, D., Schade, T.: Graphs, Networks and Algorithms. Springer, New York (2008)
20. Ke, Y., Ong, L.L., Shih, W.M., Yin, P.: Three-dimensional structures self-assembled from DNA bricks. Science **338**(6111), 1177–1183 (2012)

21. Klavzar, S., Rus, J.: Stable traces as a model for self-assembly of polypeptide nanoscale polyhedrons. MATCH Commun. Math. Comput. Chem. **70**, 317–330 (2013)
22. Kočar, V., Schreck, J.S., Čeru, S., Gradišar, H., Bašić, N., Pisanski, T., Doye, J.P., Jerala, R.: Design principles for rapid folding of knotted DNA nanostructures. Nature Commun. **7**, 1–18 (2016)
23. Lee, J.: Introduction to Topological Manifolds, vol. 940. Springer Science & Business Media, New York (2010)
24. Lickorish, W.R.: An Introduction to Knot Theory, vol. 175. Springer Science & Business Media, New York (2012)
25. Morse, A., Adkisson, W., Greene, J., Perry, D., Smith, B., Ellis-Monaghan, J., Pangborn, G.: DNA origami and unknotted A-trails in torus graphs. arXiv preprint arXiv:1703.03799 (2017)
26. Rolfsen, D.: Knots and Links, vol. 346. American Mathematical Society, Providence (1976)
27. Rothemund, P.W.: Folding DNA to create nanoscale shapes and patterns. Nature **440**(7082), 297–302 (2006)
28. Rothemund, P.W., Papadakis, N., Winfree, E.: Algorithmic self-assembly of DNA Sierpinski triangles. PLoS Biol. **2**(12), e424 (2004)
29. Seeman, N.C.: Nucleic acid junctions and lattices. J. Theoret. Biol. **99**(2), 237–247 (1982)
30. Shih, W.M., Quispe, J.D., Joyce, G.F.: A 1.7-kilobase single-stranded DNA that folds into a nanoscale octahedron. Nature **427**(6975), 618–621 (2004)
31. Singmaster, D., Grossman, J.W.: E2897. Am. Math. Mon. **90**(4), 287–288 (1983)
32. Tsai, M.T., West, D.B.: A new proof of 3-colorability of Eulerian triangulations. Ars Mathematica Contemporanea **4**(1), 73–77 (2011)
33. Veneziano, R., Ratanalert, S., Zhang, K., Zhang, F., Yan, H., Chiu, W., Bathe, M.: Designer nanoscale DNA assemblies programmed from the top down. Science **352**(6293), 1534–1534 (2016)
34. Wei, B., Dai, M., Yin, P.: Complex shapes self-assembled from single-stranded DNA tiles. Nature **485**(7400), 623–626 (2012)
35. Wu, G., Jonoska, N., Seeman, N.C.: Construction of a DNA nano-object directly demonstrates computation. Biosystems **98**(2), 80–84 (2009)
36. Zheng, J., Birktoft, J.J., Chen, Y., Wang, T., Sha, R., Constantinou, P.E., Ginell, S.L., Mao, C., Seeman, N.C.: From molecular to macroscopic via the rational design of a self-assembled 3D DNA crystal. Nature **461**(7260), 74–77 (2009)

The Design Space of Strand Displacement Cascades with Toehold-Size Clamps

Boya Wang[1($\boxtimes$)], Chris Thachuk[2], Andrew D. Ellington[1],
and David Soloveichik[1($\boxtimes$)]

[1] University of Texas at Austin, Austin, USA
{bywang,david.soloveichik}@utexas.edu, andy.ellington@mail.utexas.edu
[2] California Institute of Technology, Pasadena, USA
thachuk@caltech.edu

Abstract. DNA strand displacement cascades have proven to be a uniquely flexible and programmable primitive for constructing molecular logic circuits, smart structures and devices, and for systems with complex autonomously generated dynamics. Limiting their utility, however, strand displacement systems are susceptible to the spurious release of output even in the absence of the proper combination of inputs—so-called leak. A common mechanism for reducing leak involves clamping the ends of helices to prevent fraying, and thereby kinetically blocking the initiation of undesired displacement. Since a clamp must act as the incumbent toehold for toehold exchange, clamps cannot be stronger than a toehold. In this paper we systematize the properties of the simplest of strand displacement cascades (a translator) with toehold-size clamps. Surprisingly, depending on a few basic parameters, we find a rich and diverse landscape for desired and undesired properties and trade-offs between them. Initial experiments demonstrate a significant reduction of leak.

1 Introduction

DNA strand displacement is a powerful mechanism for molecular information processing and dynamics [11]. A strand displacement reaction is the process where two strands hybridize with each other and displace a pre-hybridized strand. The displaced strand could then serve, in turn, as the displacing strand for downstream strand displacement events. Through concatenation of strand displacement reactions, a variety of programmable behaviors have been experimentally achieved, such as performing logical computation [5], engineering molecular mechanical devices [4], and implementing chemical reaction networks [1]. In strand displacement cascades, single-stranded DNA typically fulfills the role

B. Wang—Supported by NSF grants CCF-1618895 and CCF-1652824.
C. Thachuk—Supported by NSF grant CCF-1317694.
A.D. Ellington—Supported by NSF grant DBI-0939454, international funding agency ERASynBio 1541244, and by the Welch Foundation F-1654.
D. Soloveichik—Supported by NSF grants CCF-1618895 and CCF-1652824.

© Springer International Publishing AG 2017
R. Brijder and L. Qian (Eds.): DNA 23 2017, LNCS 10467, pp. 64–81, 2017.
DOI: 10.1007/978-3-319-66799-7_5

of *signals* that carry information, while pre-hybridized DNA complexes drive their interaction (and are consequently called the *fuels*).

Although the DNA strand displacement mechanism has shown great promise for programming molecular systems, the current scale of these systems remains limited. The main obstacle is *leak*, which occurs when undesired reactions get spontaneously triggered in the absence of triggering strand. Since leak results from a spurious interaction of the fuel complexes, fuels are necessarily kept at low concentration to reduce the leak reaction, which limits the general speed of the cascade. Instead of seconds, complex cascades often take hours [5].

To combat leak, a number of approaches have been tested such as introducing Watson-Crick mismatches [2,3], or adding a threshold species that can consume the leaked signal at a faster rate than it propagates to downstream components [5,8]. It is understood that leak occurs as a result of fraying at the end of a double helix which exposes a nucleation point for spurious displacement. Adding 1 to 3 nucleotides as the *clamp* domains [5,8,9] has proven useful in reducing undesired leak, since leak can only occur after the entire clamp and one or more additional nucleotides that are adjacent to the clamp fray. Clamps, in some form, are now commonly used in the majority of strand displacement systems.

Since longer clamps should better prevent fraying, and the probability of spontaneously opening the end of double-stranded helix decreases exponentially with the length of the clamp, we want to make the clamps as long as possible. However, with clamp domains, the intended strand displacement reaction must be a toehold exchange reaction, which limits the size of the clamp to that of a toehold [12]. In this paper, we extend the size of clamp to its maximum— toehold size—and generalize a design principle for strand displacement systems: Every fuel has a toehold-size clamp, and every reaction is a toehold exchange reaction. We consider the simplest kind of strand displacement cascade—a cascade of translators (which are logically equivalent to repeater gates). In a single translator, a signal strand serves as the input and through a series of strand displacement reactions, an output signal strand is produced whose sequence is independent of the input strand. Chaining single translators allows us to build systems of translators. Such translator chains have been used in a molecular automata system that can selectively target cellular surface [6]. Translators can also be composed to perform logic *OR* computation through having two translators convert two different input signals to the same output signal. A catalytic system can be constructed if a translator chain's output is the same as its input. Although translators are the simplest kind of strand displacement module, they can already exhibit complex and useful behavior.

Our simple formulation allows rigorous formal arguments about leak reduction and other desired properties that affect the intended reaction pathway. We prove that although every reaction is reversible, the system completion level does not decrease arbitrarily with the depth of the cascade, which allows long cascades to be constructed. We show that by adjusting two parameters which define the length of the double-stranded region (N) in a fuel and the minimal distance between two fuels $(shift)$, a variety of schemes can be achieved, each with unique properties. We prove a tradeoff theorem which says that no scheme satisfies all the properties we could want. Thus understanding the taxonomy of schemes is necessary to make proper design choices.

To analyze properties of interest for the various schemes, we use a thermodynamic argument which assumes that "enthalpy" and "entropy" are the dominating factors deciding whether a configuration is favorable or not. More specifically, we assume the main contribution of enthalpy is the number of bound toehold-size domains and that of entropy is the number of separate complexes.[1] Thus when comparing two configurations, if the number of bound toehold-size domains is the same while one configuration has n fewer separate complexes than the other, then this configuration is considered unfavorable as it incurs n units of entropic penalty relative to the other. Similarly, all else being equal, a configuration with n fewer bound toehold-size domains than another configuration is unfavorable and has a relative enthalpic penalty of n units.[2]

Recently, a leak reduction method relying on increasing a redundancy parameter N_r has been proposed where leak requires binding of N_r separate fuels ("NLD scheme" in [10]).[3] However, in the NLD scheme, a leaked upstream signal can start a cascade which gains one unit of enthalpy for every downstream strand displacement step. Thus to ensure the "leakless" property, the NLD system needs to have enough entropic penalty to compensate for this enthalpic driving force. In contrast, since every reaction is a toehold exchange reaction in our toehold-size clamp design, a leaked upstream signal cannot be driven by the enthalpy of forming new bonds downstream. Unlike the leak reduction method of [10], which solely relies on the entropic penalty, our design has an additional enthalpic penalty for leak, which scales as $N/shift$. Leak reduction based on the additional enthalpic penalty could be preferable especially at high concentrations, where the entropic penalty to leak is smaller. Note that high concentration regimes are of particular interest because they result in faster kinetics.

We also show that if the clamps in the NLD scheme are extended to toehold size, the NLD scheme can be categorized into one class of the toehold-size clamp design with the parameter N representing the length of double-stranded region

---

[1] Although our use of the words enthalpy and entropy are meant to evoke the respective physical chemistry concepts, the mapping is not 1–1. We note especially that the contribution of forming additional base pairs to the free energy has both substantial enthalpic and entropic parts (which can be physically distinguished based on their temperature dependence).

[2] Roughly speaking, "one unit of enthalpic penalty" corresponds to an average of $l \cdot 1.5$ kcal/mol, where l is the length of the domain (typically 5–10 nucleotides for a toehold). "One unit of entropic penalty" at concentration C M corresponds to $\Delta G^{\circ}_{\mathrm{assoc}} + RT \ln(1/C) \approx 1.96 + 0.6 \ln(1/C)$ kcal/mol [7]. With these numbers, at roughly 650 nM concentration, binding an additional $l = 7$ domain is equal to one unit of entropy. At low concentrations the entropic penalty becomes dominant, while the enthalpic penalty prevails at high concentrations.

[3] Our length parameter N is related to the redundancy parameter of [10], but whereas we count the number of short domains, they count the long domains.

and *shift* representing the length of one long domain. According to the taxonomy in this paper, the extended *NLD* scheme for any redundancy N_r has the property that toehold occlusion [5,9] and spurious strand displacement—(partial) displacement of a strand on a fuel by a spurious invader—cannot be avoided. In this sense, the toehold-size clamp design principle has a broader design space, allowing for more flexibility in balancing desired and undesired properties.

We conclude with an experimental demonstration of the toehold-size clamp design with one set of parameters. Leak reduction compares positively both in terms of the kinetic leak rate and the maximum amount of leak ever generated with the previously proposed *NLD* schemes. (Although the absolute leak rate is smaller for our scheme, the lower completion level due to toehold exchange reaction reversibility results in an overall smaller "signal to noise ratio" compared with the *NLD* schemes.)

2   Design Space of Toehold-Size Clamp Translators

2.1   System Description

We first introduce the conventions used in this paper. We use the domain level abstraction for DNA strands. A domain represents a concatenation of DNA bases treated as a unit, which can hybridize to or dissociate from a complementary domain. Unlike the traditional representation that divides domains into two classes where long domain indicates irreversible binding and short (toehold) domain indicates reversible binding, here all the domains have equal length of a toehold. As a result, if two strands are only held by a single domain, they can dissociate (see Fig. 1a). We assume that all domains are orthogonal (no cross-talk). Note that this is a strong assumption because the size of a domain is restricted to that of a toehold and there are a limited number of distinct toeholds that could be designed (see Discussion). The desired pathway consists entirely of toehold exchange strand displacement reactions. Additionally, to capture leak, we consider blunt-end strand displacement, which is not preceded by toehold binding but rather is mediated by fraying at the ends of helices. We assume that fraying cannot open a whole domain. The unique domains can be aligned in a row, and their identity is represented as their horizontal position (i.e., numbering on top of Fig. 1b). Domains aligned vertically have the same or complementary sequence.

By a *domain instance*, we mean a particular domain on a particular complex. In contrast, when we refer to a (domain) *position*, we mean all domain instances that have the same or complementary sequences, and are drawn vertically aligned in our figures.

Domain instances can be either *single-stranded* or *double-stranded*. Single-stranded domain instances are subdivided into *toehold* and *overhang* types. Note that toehold domain instances initiate the toehold mediated strand displacement. Double-stranded domain instances are subdivided into *left flank*, *clamp*, and *unused*. Specifically, *clamp*s are the domain instances located at the right end of a double-stranded helix, *left flank*s are the domain instances located at the

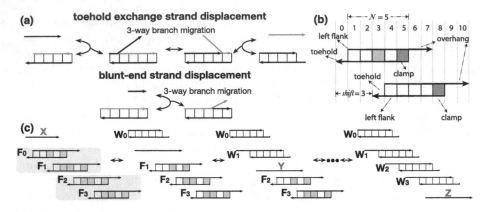

Fig. 1. (a) Fundamental reaction steps we consider. (b) The conventions of the toehold-size clamp design. (c) Desired reaction pathway of a 2-translator cascade ($X \rightarrow Y \rightarrow Z$). In the presence of the input signal strand, after 4 elementary translation steps, the signal strand X is translated to signal strand Z.

left end of a double-stranded helix (see Fig. 1b). The remaining domain instances are *unused*. The name "clamp" comes from historical use, as structurally similar domains were added to previous schemes to "clamp-down" the ends of helices to reduce leak. Note that without the clamp domain on the second fuel in Fig. 1b, the overhang of the first fuel can initiate blunt-end strand displacement.

The coloring of double-stranded domain instances refers to whether or not there are toehold (orange) or clamp (gray) domain instances at the same position. In particular, the color orange indicates that there is a toehold domain instances at the same position If a domain position overlaps with both toehold and clamp domain instances, it is colored in both orange and gray (e.g., see Fig. 2). If a domain position does not have toehold or clamp domain instances, it is colored in white.

2.2 Translator Design

A translator, composed of different fuels where each fuel is responsible for an elementary translation step, can translate an input signal strand to an independent output signal strand. When the input signal strand is present, it reacts with the first fuel displacing the top strand, which then serves as the input to trigger the downstream fuel (Fig. 1c). Note that the figure shows two translators, of two fuels each (since Y is sequence independent from X, and Z is sequence independent from Y).

To design a translator system with toehold-size clamps, two parameters are necessary and sufficient. We use the parameter N to represent the number of the double-stranded domains in a fuel and the parameter *shift* to represent the minimal distance between two single-stranded toeholds (see Fig. 1b). Since a toehold-size clamp is in every fuel, *shift* should be between 1 and N. We illustrate the diversity of schemes for $N = 6$ for all values of *shift* in Fig. 2.

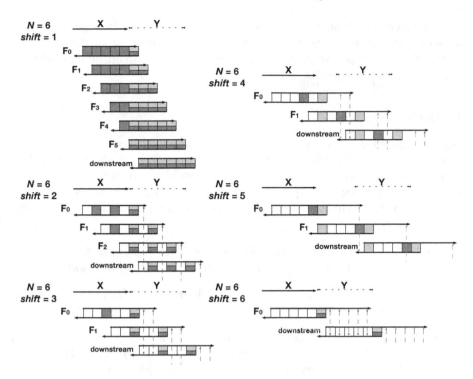

Fig. 2. The example schemes of translators with $N = 6$ and all values of *shift*. Every domain instance is colored according to the domain types of the domain instances at that position. The dashed red lines represent possible spurious strand displacement events between different fuels. The dashed blue lines represent possible toehold occlusion events when a toehold domain instance and an overhang domain instance exist in the same position. (Color figure online)

Once the parameters N and *shift* are determined, the translator design and the types of each domain instance can be assigned. Using 0 indexing, we start from the 0th fuel which is responsible for the first elementary translation step (e.g. Fig. 1c). The toehold domain instance of the 0th fuel is located at position 0. Since the length of the double-stranded domains in a fuel is N, the clamp instance in this fuel lies at position N. To ensure that every desired reaction is a toehold exchange strand displacement reaction (i.e., there is a clamp), the number of overhang domain instances in a fuel should be $shift - 1$ (we write $overhang = shift - 1$).

Generalizing these rules, the domain types of each domain instance in every fuel can be determined. The position of the ith fuel (which is responsible for the $i + 1$ elementary translation step) is shifted to the right by $shift \cdot i$. Thus the toehold domain instance in the ith fuel lies at the position $shift \cdot i$; the left flank domain instance lies at the position $shift \cdot i + 1$; and the clamp domain instance lies at the position $shift \cdot i + N$. How many fuels do we need to compose a single translator? Recall that single translator consists of as many fuels as are necessary to generate an output signal that is sequence independent of the input. Thus the number of fuels per translator is $\lfloor \frac{N+overhang}{shift} \rfloor$, which can be equivalently written as $\lfloor \frac{N+shift-1}{shift} \rfloor = \lfloor \frac{N-1}{shift} \rfloor + 1$.

2.3 Useful Lemmas

In subsequent sections, we prove properties of schemes based on their para-metrization of N and *shift*. Many of our arguments rely on showing whether regularly spaced positions with certain domain instances (such as toeholds) can intersect with other regularly spaced positions or intervals with different domain instances (such as overhangs). To simplify those arguments we first establish the following claims.

Lemma 1. *Let p, q, and r be natural numbers with $p > 0$ and $r - q > 0$. $\forall i \in \mathbb{N}, \exists j \in \mathbb{N}$ such that $j \cdot p + q = i \cdot p + r$ if and only if $r - q$ is a multiple of p.*

Proof. Fix any $i \in \mathbb{N}$. Suppose $j \cdot p + q = i \cdot p + r$, for some $j \in \mathbb{N}$. Then $j = i + \frac{r-q}{p}$ and $r - q$ must be a multiple of p. $\qquad\square$

Lemma 2. *Let p, q, and r be natural numbers with $p > 1$ and $r - q > 0$. $\forall i \in \mathbb{N}, \exists j \in \mathbb{N}$ such that $j \cdot p + q$ is contained in the interval $[i \cdot p + r, (i+1) \cdot p + r - 2]$ if and only if $r - q - 1 > 0$ and $r - q - 1$ is not a multiple of p.*

Proof. Call $k \cdot p + q$ a *valid position*, for any $k \in \mathbb{N}$. Fix any $i \in \mathbb{N}$. First we consider the case when $r - q - 1 = 0$ (and thus $r = q + 1$). Consider any interval $[i \cdot p + r, (i+1) \cdot p + r - 2] = [i \cdot p + q + 1, (i+1) \cdot p + q - 1]$. Suppose by contradiction that a valid position $j \cdot p + q$ intersects the interval. Then we have the following: (i) $i \cdot p + q + 1 \leq j \cdot p + q$ which simplifies to $j \geq i + \frac{1}{p}$ and thus $j \geq i + 1$ (since $j \in \mathbb{N}$). (ii) $j \cdot p + q \leq (i+1) \cdot p + q - 1$ which simplifies to $j \leq i + 1 - \frac{1}{p} < i + 1$. Contradiction.

Finally, consider two cases when $(r - q - 1) > 0$: (1) $(r - q - 1)$ is a multiple of p. Then $i \cdot p + q + (r - q - 1) = i \cdot p + r - 1$ is a valid position and the next valid position occurs at $(i + 1) \cdot p + r - 1$. Thus there is no valid position in the interval $[i \cdot p + r, (i + 1) \cdot p + r - 2]$. (2) $(r - q - 1)$ is not a multiple of p. Let δ be the remainder; $1 \leq \delta \leq p - 1$. Thus the smallest valid position larger than $i \cdot p + q + (r - q - 1)$ occurs at $i \cdot p + r - 1 + \delta$. Since $1 \leq \delta \leq p - 1$, we know that it falls in the interval $[i \cdot p + r, (i + 1) \cdot p + r - 2]$. $\qquad\square$

3   Thermodynamic Properties

An effective translator cascade ideally has a high signal-to-leak ratio, even at thermodynamic equilibrium. We next show that by varying scheme parameters N and *shift*, we vary the thermodynamic barrier to leak. We go on to show that while every translator cascade scheme is reversible, there is a lower bound on fraction of output signal when input is present, regardless of its depth.

3.1   Thermodynamic Barrier to Leak in Translator Cascades

Suppose there is a hypothetical experiment implementing a single copy of every fuel in a translator cascade of arbitrary depth, coupled with a single copy of a downstream reporter. For any scheme, a downstream reporter is identical to a fuel of that scheme but contains no overhang. Leak occurs when the top and bottom strands of the reporter dissociate into separate complexes—in typical experimental settings, this would increase a fluorescence signal used as a proxy to measure the produced output. To give a barrier to leak based on N and $shift$, we first develop some useful lemmas.

Lemma 3. *If a position contains an overhang domain instance, then it does not contain a clamp domain instance.*

Proof. If $shift = 1$ there are no overhang domain instances and we are done. Suppose $shift > 1$. The ith clamp domain instance lies at position $i \cdot shift + N$. The jth overhang domain instance lies at positions between $j \cdot shift + N + 1$ and $(j + 1) \cdot shift + N - 1$. Let $r = N + 1$, $p = shift$ and $q = N$. By Lemma 2, the position that has a clamp domain instance cannot intersect the position that has an overhang domain instance since $r - q - 1 = 0$.   □

Lemma 4. *Let c_i and c_{i+1} be the positions of the clamp domain instances for neighboring fuels i and $i + 1$, respectively. Then positions in the (possibly empty) interval $[c_i + 1, c_{i+1} - 1]$ contain overhang domain instances.*

Proof. $c_i = i \cdot shift + N$ and $c_{i+1} = (i + 1) \cdot shift + N$. The overhang domain instances for fuel i lie at positions between $i \cdot shift + N + 1$ and $(i + 1) \cdot shift + N - 1$, establishing the claim. (Note that if the interval is empty, then fuels do not have overhangs and the claim in trivially true.)   □

For the purposes of the next argument, it is convenient to refer to domain instances as either a *top* domain instance (if it occurs on a top strand), or a *bottom* domain instance (if it occurs on a bottom strand). We will refer to a double-stranded domain instance as a top domain instance that has a *bond* to a bottom domain instance. A *configuration* is a matching between top and bottom domain instances, where each match is one bond. Our arguments are based solely on counting the maximum number of possible bonds, given certain constraints. The barrier to leak implied by our result, even in the presence of pseudoknots, is entirely enthalpic in nature since it assumes no entropic penalty for joining two complexes into one.

Theorem 1. *Given a translator cascade of arbitrary depth, a downstream reporter, and no input signal, if M is its maximum possible number of bonds then any configuration having the two strands of the reporter in distinct complexes can have at most $M - \lceil \frac{N}{shift} \rceil$ bonds.*

Proof. Let M be the maximum possible number of bonds, of any configuration, for the translator cascade with downstream reporter in the absence of input.

Suppose the reporter complex is in the ith layer; then its toehold domain instance is at position $i \cdot shift$ and the remainder of its domain instances lie in positions between $p = i \cdot shift + 1$ and $q = i \cdot shift + N$.

Let L be any maximally bound configuration of top and bottom domain instances in $[0, p - 1]$, and let M_L equal the number bonds in L. Let R be the intended configuration (in the absence of input) of top and bottom domain instances in $[p, q]$: all bottom domain instances of fuel j have a bond to a top domain instance, also from fuel j, if their position is in $[p, q]$ (and similarly for the reporter). Let M_R equal the number of bonds in R. Since there are no toehold domain instances in $[p, q]$, then that interval contains at least as many top as bottom domain instances. Thus R is a maximally bound configuration of top and bottom domain instances in $[p, q]$, and $M_L + M_R = M$.

Let R' be a maximally bound configuration of top and bottom domain instances in $[p, q]$ subject to no top domain instance in the reporter complex being bound to a bottom domain instance in the reporter complex. Let $M_{R'}$ be equal to the number of bonds in R'. For each position $j \in [p, q]$, there are two possibilities: (i) There is an excess of top domain instances, and thus one of those can bind the reporter bottom domain instance, keeping the total number of bonds in position j unchanged. (ii) There is an equal number of top and bottom domain instances, and thus position j now has one fewer bond. Therefore, the difference $M_R - M'_R$ can be determined by counting the number of positions in $[p, q]$ with an equal number of top and bottom domain instances. Let p' be the maximal position containing a clamp domain instance where $p' < p$. Note that p' must exist since in any translator design there is a clamp domain instance at position N, and since $p > N$ as otherwise the reporter would have domain instances in common with the input signal. By Lemmas 3 and 4 every position in $[p', q]$ has either overhang domain instances or clamp domain instances, but not both. The same is true for positions in $[p, q]$ and none of those positions have instances of a toehold domain. Thus, the number of positions in $[p, q]$ with an equal number of top and bottom domains is exactly the number of clamp domain instances in that interval. There is a clamp domain instance at position q, and every position $q - k \cdot shift \geq p$, for $k \in \mathbb{N}$. Since $q - p + 1 = N$, then there are $\lceil \frac{N}{shift} \rceil$ positions that have clamp domain instances in $[p, q]$. Therefore $M_R - M_{R'} = \lceil \frac{N}{shift} \rceil$. By (i) and (ii) above, every position in $[p, q]$ has at least one unbound top domain instance. Let R'' be a reconfiguration of R', in the obvious way of swapping bonds, such that the reporter top strand forms its own complex. Since $M_{R''} = M_{R'}$, with $M_{R''}$ being equal to the number of bonds in R'', then $M - M_{R''} = \lceil \frac{N}{shift} \rceil$ establishing the claim. $\square$

This theorem implies that in the absence of input there is an enthalpic barrier of $\lceil \frac{N}{shift} \rceil$ bonds to separate the reporter strands from a maximally bound state. In contrast, when the input is present, the signal can be propagated all the way until the reporter, where separating the reporter strands incurs the loss of only 1 bond (breaking the bonding of the top and bottom clamp domains on the reporter, which have no other binding partners). Thus by increasing N we can enlarge the enthalpic barrier to leak without increasing the enthalpic barrier to correct output.

3.2 Asymptotic Completion Level of Translator Cascades

With a cascade of effectively irreversible strand displacement reactions (not relying on toehold exchange), it is safe to assume that most of the input signal should propagate through to the end. However, with a cascade of reversible reactions such as those we necessarily obtain with toehold size clamps, it might seem that the signal will decrease with the length of the cascade if the signal "spreads out" across the layers. Does this mean that translators with toehold size clamps cannot be composed into long cascades? In this section we prove a lower-bound on the amount of final signal output by a chain of translators that is *independent* of the length of the chain, which shows that long cascades are indeed feasible.

To analyze a system with a cascade of translators, we simplify each translator reaction to be a bimolecular reversible toehold exchange reaction $X + F \rightleftharpoons Y + W$, where X is the input signal, F is the fuel, Y is the output signal and W is the waste species. Assuming that the two toeholds in a toehold exchange reaction (i.e., toehold and clamp domain instances in our nomenclature) have the same thermodynamic binding strength, the net reaction of a translator has $\Delta G^o \approx 0$ and the equilibrium constant of each reaction can be treated as 1. Thus for a single translator, if the initial concentration for the reactants are $[X]_0 = \alpha$, $[F]_0 = 1$, at chemical equilibrium, the concentration of output strand Y will be $\frac{\alpha}{\alpha+1}$.

We then ask how much output signal will be produced if multiple translators are cascaded together. Suppose we can have the n-layer reaction system, where each reaction represents a translator reaction:

$$X_1 + F_1 \rightleftharpoons X_2 + W_1$$

$$\cdots$$

$$X_i + F_i \rightleftharpoons X_{i+1} + W_i$$

$$\cdots$$

$$X_n + F_n \rightleftharpoons X_{n+1} + W_n$$

The system starts with all the fuels (F_i, $i = 1$, 2, ..., n) at concentration 1 and X_1 at α. If there is relatively little signal compared with fuel (we set $\alpha < \frac{1}{2}$), the reactions are driven forward by the imbalance between F_i and W_i. By conservation of mass, $F_i + W_i = 1$. Since $W_i \leqslant \alpha$ (we can't produce more waste than there was input), we get $F_i = 1 - W_i \geqslant 1 - \alpha$. Thus, at chemical equilibrium, for each reaction we have:

$$\frac{X_{i+1}}{X_i} = \frac{F_i}{W_i} \geqslant \frac{1 - \alpha}{\alpha}.$$

Letting $\beta = \frac{1-\alpha}{\alpha}$, we obtain a lower bound for X_{i+1}: $X_{i+1} \geqslant \beta X_i$.

Since the total concentration of all signal strands is conserved, we have:

$$X_1 + X_2 + \cdots + X_i + \cdots + X_{n+1} = \alpha$$

$$(\beta^{-n} + \beta^{-(n-1)} + \cdots + \beta^{-1} + 1) \cdot X_{n+1} \geqslant \alpha$$

Since $\displaystyle\sum_{i=-n}^{0} \beta^i = \beta^{-n}\frac{1 - \beta^{n+1}}{1 - \beta}$, the above equation can be simplified as

$$X_{n+1} \cdot \beta^{-n}\frac{1 - \beta^{n+1}}{1 - \beta} \geqslant \alpha$$

Thus the concentration of X_{n+1} is

$$X_{n+1} \geqslant \frac{\alpha\beta^n(1 - \beta)}{1 - \beta^{n+1}} = \frac{\alpha(\frac{1}{\alpha} - 2)}{\frac{1}{\alpha} - 1 - \frac{1}{(\frac{1}{\alpha}-1)^n}} \geqslant \frac{\alpha(\frac{1}{\alpha} - 2)}{\frac{1}{\alpha} - 1} = \alpha\frac{1 - 2\alpha}{1 - \alpha}.$$

This result indicates that increasing the number of reaction layers does not affect the lower bound of the equilibrium concentration of the output signal. Therefore, concatenating the translators composed of toehold exchange reactions can always generate at least a constant fraction of signal independent of the number of layers.

4   Kinetic Properties

Beyond thermodynamic properties, the kinetic properties of translator schemes can vary depending on the choice of N and *shift*. In this section, we show that schemes are susceptible to undesirable properties such as toehold-occlusion, spurious strand displacement, or reconfiguration of fuels, to varying degrees. As we will see, certain schemes preclude some of these phenomena entirely.

4.1   Toehold Occlusion

In strand displacement systems, reaction kinetics can be controlled by the strength of a toehold [12]. Stronger toeholds can enable faster reaction kinetics; however, if the toehold strength is too strong, toehold dissociation can become a rate limiting step. This is problematic when overhangs of fuel can bind to toeholds of other fuel, since fuel is typically present in high concentration. This creates so-called *toehold occlusion* [5,9], which can significantly slow down the intended reaction kinetics in the presence of input signal.

Theorem 2. *Toehold occlusion is not possible in a translator scheme if and only if N is a multiple of* shift.

Proof. Toeholds are occluded when overhangs can bind to them. The toehold domain instance of fuel i lies at position $i \cdot shift$. Overhang domain instances of fuel i lie at positions between $i \cdot shift + N + 1$ and $i \cdot shift + N + shift - 1$. The claim follows by Lemma 2 ($p = shift$, $q = 0$ and $r = N + 1$). □

4.2   Spurious Strand Displacement

Spurious strand displacement events, even if they do not lead to leak of output signal or dissociation of strands of any kind, are unproductive reactions that can slow down the intended kinetics of the system. Spurious displacement occurs when any proper prefix or suffix of a fuel's double-stranded helix is displaced by a spurious invader. A spurious invader of a fuel is any complex not equal to its intended input strand. We partition our analysis into two categories: (i) spurious displacement in the absence of input, and (ii) spurious displacement in its presence.

Spurious Displacement in the Absence of Input. Spurious displacement between fuels can become increasingly problematic with respect to the intended kinetics of a cascade—since fuels involved in spurious displacement can be unavailable for their intended reaction—as the concentration of the system is increased.

We find it convenient to refer to specific *top* or *bottom* domain instances, as in Sect. 3.1. We begin by looking at spurious displacement of bottom domain instances which can only occur in positions containing toehold domain instances as these are the only positions with bottom domains in excess.

Lemma 5. *In the absence of input, spurious displacement of bottom domain instances in a translator scheme is possible (i) in left flank domain instances if and only if* shift $= 1$ *(i.e. the toehold domain instance can invade the left flank domain instance, see Fig. 2), and (ii) in clamp domain instances if and only if N is a multiple of* shift.

Proof. (i) Left flank domain instances are offset of toehold domain instances by 1. By Lemma 1, setting $p = shift$, $q = 0$ and $r = 1$, it follows that $r - q = 1$ is a multiple of p, and thus $shift = 1$, if and only if a domain position overlaps with both a toehold domain instance and a left flank domain instance. (ii) Clamp domain instances are offset of toehold domain instances by N. By Lemma 1, setting $p = shift$, $q = 0$ and $r = N$, it follows that $N = r - q$ is a multiple of $shift = p$ if and only if a domain position overlaps with both a toehold domain instance and a clamp domain instance. $\square$

Now consider spurious displacement of top domain instances which can only occur in positions containing overhang domain instances as these are the only positions with top domain instances in excess.

Lemma 6. *In the absence of input, spurious displacement of top domain instances in a translator scheme is possible if and only if $N - 1$ is not a multiple of* shift.

Proof. By construction, when $shift = 1$ there are no overhang domain instances and therefore spurious displacement of top domain instances is not possible. Assume $shift > 1$.

We first establish that it is not possible to spuriously displace top domain instances of clamp instances, because the positions of clamp domain instances cannot intersect that of overhang domain instances, by Lemma 3.

Thus any displacement of top domain instances of a fuel must be a proper prefix of its helix and therefore must include a left flank domain instance. Left flank domain instances are offset by 1, relative to toehold domain instances. Setting $p = shift$, $q = 1$ and $r = N + 1$ it follows from Lemma 2 that the positions of left flank domain instances can intersect that of overhang domain instances if and only if $r - q - 1 = N - 1$ is not a multiple of $p = shift$.   □

Spurious Displacement in the Presence of Input. A second type of spurious displacement is when a free signal strand (including the input), can act as a spurious invader of a fuel other than its designed target. In this case, particularly when the input concentration is significantly lower than fuel as is typical, signal strands can become involved in numerous unproductive reactions thus slowing (possibly significantly) signal propagation through every layer of the cascade.

Lemma 7. *Spurious displacement between signal strands and fuels is not possible in a translator scheme if and only if* shift $\geq N - 1$.

Proof. Domain instances of signal strand i lie at positions between $i \cdot shift$ and $i \cdot shift + N - 1$. Signal strand i is a spurious invader if it can displace any domain instances on some fuel $j > i$. Signal strand i cannot displace the clamp domain instance on fuel i, at position $i \cdot shift + N$, and therefore cannot displace the clamp domain instance of fuel j, at position $j \cdot shift + N > i \cdot shift + N$. Suppose signal strand i is a spurious invader of fuel j; it must invade a prefix of fuel j's double-stranded domain instances (its helix) which necessarily includes its left flank domain instance at position $j \cdot shift + 1$. It follows that $j \cdot shift + 1 \leq i \cdot shift + N - 1$, and thus $shift \leq \frac{N-2}{j-i} \leq N - 2$ since $j > i$. Finally, suppose signal strand i is not a spurious invader of any fuel $j > i$; then it cannot displace the left flank domain instance of fuel j, so $j \cdot shift + 1 > i \cdot shift + N - 1$ which implies $shift > \frac{N-2}{j-i} = N - 2$ when $j = i + 1$.   □

Spurious Displacement with or Without Input. By Lemmas 5, 6, and 7 we have the following.

Theorem 3. *Spurious displacement is not possible in a translator scheme if and only if* shift $= N - 1$.

4.3   Reconfiguration

Spurious strand displacements can be more complicated when multiple species are involved. Some may result in the formation of complex multi-stranded structures. In this section, we show contrasting examples of possible reconfiguration after spurious strand displacement. The first example requires a bimolecular reaction to undo, while unimolecular reconfiguration is sufficient for the second. (To the first approximation, even at the "high" concentrations used here, bimolecular reactions are relatively slower than unimolecular reactions.)

Consider the extreme case when $shift = 1$ in the absence of input signal. By Lemma 5, we have shown that the left flank bottom domain instance can be displaced by a toehold domain instance on the next fuel. Since $shift = 1$, the ith left flank domain instance appears at position $shift \cdot i + 1 = i + 1$, for all $i \in \mathbb{N}$. Therefore, multiple spurious displacement events could result in all of the bottom domain instances of one fuel being displaced by toehold domain instances of other fuels. This results in a free (unbound) bottom strand. To restore the original configuration, a bimolecular reaction pathway is needed. Although the multiple blunt-end displacement events to cause this reconfiguration are unlikely, once formed, it requires a slow bimolecular reaction to undo. Figure 3a demonstrates this pathway.

In other cases, spurious strand displacement seems unable to cause any major reconfiguration problem. For example, when $N = 5$ and $shift = 3$ any reconfiguration can be undone via fast unimolecular steps. An example of a structure that can form via spurious displacement in this scheme is shown in Fig. 3b.

These two examples suggest that there is likely a separation between schemes with harmful spurious displacement and those in which spurious displacement occurs but can be quickly undone. Formally differentiating the two cases is an area for further research.

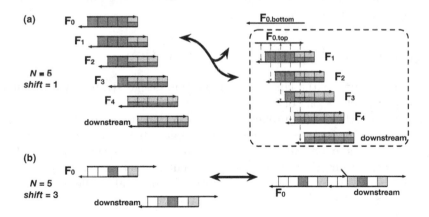

Fig. 3. Examples of configurations reachable with spurious strand displacement.

5 Trade-Offs Between the Properties

We have shown that different choices of N and $shift$ yield translator schemes with varying thermodynamic and kinetic properties. By Theorems 2 and 3 we have the following fundamental kinetic trade-off.

Corollary 1. *There is no translator scheme that avoids both toehold occlusion and spurious displacement.*

Thus, every translator scheme has some undesirable kinetic property. There is a quantitative trade-off for the thermodynamic property of enthalpic barrier to leak, given by Theorem 1, based on the ratio of N and *shift*. Schemes can also have a trade-off between unfavorable kinetic and thermodynamic properties. By Theorem 3, only schemes with $shift = N - 1$ can avoid spurious strand displacement. However, by Theorem 1, these schemes only have a constant enthalpic barrier to leak.

Corollary 2. *A large enthalpic barrier to leak is incompatible with avoiding spurious displacement.*

In fact, schemes with the largest enthalpic barrier to leak also have the most potential spurious interactions between fuels and signal strands. As an example of this, compare the spurious interactions caused by the X input signal with fuels, besides the initial fuel, as *shift* increases in Fig. 2.

In summary, there is no best translator scheme with respect to all thermodynamic and kinetic properties studied here. Instead, the entire taxonomy we develop informs the choice of translator scheme, and one should be chosen based on the expected conditions of its planned use. For example, high concentration conditions may be best served by a scheme with no toehold occlusion and a balance between its enthalpic barrier to leak and its potential number of spurious displacement reactions.

6   Preliminary Experimental Verification

To experimentally test the kinetic leak reduction design strategy with toehold-size clamps, we chose one of the parameter pairs $N = 5$ and $shift = 3$. This combination has the desired property that (1) in the absence of the input signal strand, leak requires at least two units of enthalpic penalty (breaking two bonds) compared with the maximally bound state, and (2) even if spurious strand displacement can occur, there exists a unimolecular reaction pathway that can reverse the spurious interactions and restore the original configuration of the system (see Sect. 4.3).

6.1   Leak Reduction with a Single Translator

We compare with the previously proposed leak reduction method (*NLD* scheme) based exclusively on an entropic penalty [10]. More specifically, we choose two *NLD* redundancy parameter values $N_r = 1$ and $N_r = 2$. For $N_r = 1$, the scheme is the typical leaky translator (named *SLD* scheme). For $N_r = 2$, the scheme is the previously described "leakless" translators (named *DLD* scheme).

In our experiments, every fuel is kept at 5 μM, which is 50–100 times larger than the typical concentration used in strand displacement systems. Figure 4a compares the kinetic behaviors of *SLD*, *DLD*, and the toehold-size clamp design, all in the absence of input signal. As expected, the leak rate in the *SLD* scheme (measured at the first 20 min) is 30 times higher than that of the *DLD* scheme. However, the *DLD* leak rate is still roughly 0.03 nM/min throughout the 10 h. In contrast, after quickly generating 10 nM initial leak (which is hypothesized to be caused by misfolded fuel structure or synthesis error of DNA strands), the toehold-size clamp design does not show gradual leak at our experimental setting.

In addition to kinetic measurements, we tested how much leak each design has at thermodynamic equilibrium, which sets an upper bound of the total leak for an isolated translator. To achieve thermodynamic equilibrium, fuels and reporter are slowly annealed together. Figure 4b compares the total leak amount of these designs. The toehold-size clamp design shows the least amount of leak even at thermodynamic equilibrium.

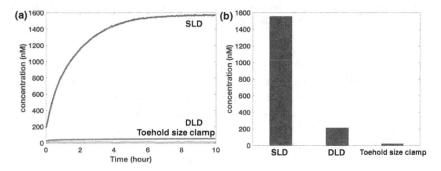

Fig. 4. Comparison of leak between the *SLD*, *DLD* and toehold-size clamp design ($N = 5$, *shift* $= 3$) from (a) a kinetics and (b) a thermodynamic equilibrium perspective. Since leak is measured, the systems do not contain the input signal strands. To measure leak at thermodynamic equilibrium, fuels and reporter are slowly annealed. The concentrations for fuels are 5 μM. Reporter in the toehold-size clamp design is at 5 μM. Reporters in the *SLD* and *DLD* designs are at 6 μM. The reaction temperature is 37 °C. See the sequences and methods in the full version of this paper.

6.2   Leak Reduction with Translator Cascade

Beyond a single translator, we wanted to know (1) how much leak (without input signal) and (2) desired output signal (with input signal) a translator cascade can generate with increasing number of translators.

Figure 5a shows that in the time period of the experiment, in the absence of input signal strand, the translator cascades of 1 to 6 fuels (1 to 3 translators) all show no apparent leak. In the presence of input signal, the completion level decreases with the number of layers. However, as more layers are added, the completion level does not decrease linearly, and indeed seems to approach an asymptote, a behavior consistent with the theoretical prediction of Sect. 3.2.

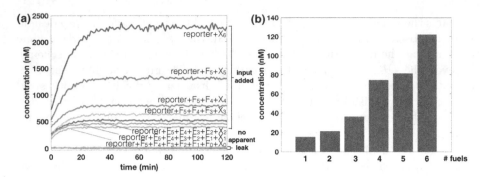

Fig. 5. Kinetics and thermodynamic equilibrium of translator cascades with the toehold-size clamp design ($N = 5$, $shift = 3$). (a) Kinetic behaviour in the presence and absence of input signal, for cascades of different length. (b) The total amount of leak in the absence of input signal at thermodynamic equilibrium, for cascades of different length. The concentrations of the reporter and the fuels are around 5 μM. The concentration of each input is 2.5 μM. The reaction temperature is 37 °C. See the sequences and methods in the full version of this paper.

Figure 5b studies the leak of translator cascades of varying depth at thermodynamic equilibrium. The leak at equilibrium increases as the number of fuels increases. However, even if there are 6 fuels (3 translators), the total leak concentration is still less than 3% of the fuel concentration.

These results suggest that the absolute leak concentration of the toehold-size clamp design is significantly less than the previously proposed *DLD* design. Nonetheless, the relative positive signal to background noise ratio of the toehold-size clamp design is smaller than of the *DLD* design because of the significant smaller completion level due to the reversibility of all displacement steps. Our results suggest that the toehold-size clamp design could be preferable, especially when a system requires absolutely smaller leak, such as when concatenating the translators with downstream catalytic or auto-catalytic systems.

7   Discussion

In this work, we study schemes for constructing strand displacement systems which utilize toehold size clamps to decrease leak. The full diversity of such schemes for translators is accessible by varying two parameters N and *shift*. We provide rigorous guarantees on the enthalpic barrier to leak as a function of these parameters. We further prove that certain parameter values result in other desirable properties like no spurious displacement, and no toehold occlusion. We prove a tradeoff theorem which says that no scheme satisfies all desired properties; consequently, understanding the properties of the full assortment of schemes helps to make the proper design choices. Since no single scheme can be judged to be "best", and tradeoffs are inherent, future work will also experimentally compare different parameter sets in different experimental regimes.

In contrast to previously reported methods for arbitrarily decreasing leak which rely on entropic barrier arguments, we describe how the enthalpic barrier to leak can be raised arbitrarily. The enthalpic barrier argument is particularly germane for the high concentration regime where the entropy penalty for joining complexes is smaller.

Our argument relies on the strong assumption that all domains are orthogonal. In reality, given the limited size of a domain (that of a toehold), as the number of distinct domains increases, it is not possible to make all the domains orthogonal. Nonetheless we note that in certain cases having the same sequence in multiple domain positions seems to pose no problem. (For example, in schemes where N is not a multiple of *shift*, all the clamp domain instances could have the same sequences.) Future work could further explore how to assign the same domains without undesired interactions and how the number of orthogonal domains needed scales with the length of double-stranded region N.

References

1. Chen, Y.-J., Dalchau, N., Srinivas, N., Phillips, A., Cardelli, L., Soloveichik, D., Seelig, G.: Programmable chemical controllers made from DNA. Nature Nanotechnol. **8**(10), 755–762 (2013)
2. Jiang, Y.S., Bhadra, S., Li, B., Ellington, A.D.: Mismatches improve the performance of strand-displacement nucleic acid circuits. Angew. Chem. **126**(7), 1876–1879 (2014)
3. Olson, X., Kotani, S., Padilla, J.E., Hallstrom, N., Goltry, S., Lee, J., Yurke, B., Hughes, W.L., Graugnard, E.: Availability: a metric for nucleic acid strand displacement systems. ACS Synth. Biol. **6**(1), 84–93 (2017)
4. Omabegho, T., Sha, R., Seeman, N.C.: A bipedal DNA Brownian motor with coordinated legs. Science **324**(5923), 67–71 (2009)
5. Qian, L., Winfree, E.: Scaling up digital circuit computation with DNA strand displacement cascades. Science **332**(6034), 1196–1201 (2011)
6. Rudchenko, M., Taylor, S., Pallavi, P., Dechkovskaia, A., Khan, S., Butler Jr., V.P., Rudchenko, S., Stojanovic, M.N.: Autonomous molecular cascades for evaluation of cell surfaces. Nature Nanotechnol. **8**(8), 580–586 (2013)
7. SantaLucia, Jr., J., Hicks, D.: The thermodynamics of DNA structural motifs. Ann. Rev. Biophys. Biomol. Struct. **33**, 415–440 (2004)
8. Seelig, G., Soloveichik, D., Zhang, D.Y., Winfree, E.: Enzyme-free nucleic acid logic circuits. Science **314**(5805), 1585–1588 (2006)
9. Srinivas, N.: Programming chemical kinetics: engineering dynamic reaction networks with DNA strand displacement. Ph.D. thesis, California Institute of Technology (2015)
10. Thachuk, C., Winfree, E., Soloveichik, D.: Leakless DNA strand displacement systems. In: Phillips, A., Yin, P. (eds.) DNA 2015. LNCS, vol. 9211, pp. 133–153. Springer, Cham (2015). doi:10.1007/978-3-319-21999-8_9
11. Zhang, D.Y., Seelig, G.: Dynamic DNA nanotechnology using strand-displacement reactions. Nature Chem. **3**(2), 103–113 (2011)
12. Zhang, D.Y., Winfree, E.: Control of DNA strand displacement kinetics using toehold exchange. J. Am. Chem. Soc. **131**(47), 17303–17314 (2009)

A Stochastic Molecular Scheme for an Artificial Cell to Infer Its Environment from Partial Observations

Muppirala Viswa Virinchi, Abhishek Behera, and Manoj Gopalkrishnan[(✉)]

India Institute of Technology Bombay, Mumbai, India
axlevisu@gmail.com, abhishek.enlightened@gmail.com,
manoj.gopalkrishnan@gmail.com

Abstract. The notion of entropy is shared between statistics and thermodynamics, and is fundamental to both disciplines. This makes statistical problems particularly suitable for reaction network implementations. In this paper we show how to perform a statistical operation known as Information Projection or E projection with stochastic mass-action kinetics. Our scheme encodes desired conditional distributions as the equilibrium distributions of reaction systems. To our knowledge this is a first scheme to exploit the inherent stochasticity of reaction networks for information processing. We apply this to the problem of an artificial cell trying to infer its environment from partial observations.

1 Introduction

Biological cells function in environments of high complexity. Transmembrane receptors allow a cell to sample the state of its environment, following which biochemical reaction networks integrate this information, and compute decision rules which allow the cell to respond in sophisticated ways. One challenge is that receptors may be imperfectly specific, activated by multiple ligands with various propensities. What algorithmic and statistical ideas are needed to deal with this challenge, and how would these ideas be implemented with reaction networks? We begin to address these questions. The two questions do not decouple because the attractiveness of algorithmic and statistical ideas towards these challenges is tied in with their ease of implementation with reaction networks. We are interested in statistical algorithms that fully exploit the native dynamics and stochasticity of reaction networks. To fix ideas, consider an example.

Example 1. Consider an artificial cell with two types of transmembrane receptors R_1 and R_2 in an environment with three ligand species L_1, L_2, and L_3 (Fig. 1). Receptor R_1 has equal sensitivity to ligands L_1 and L_3, and no sensitivity to L_2. Receptor R_2 has equal sensitivity to ligands L_2 and L_3, and no sensitivity to L_1. This information can be summarized in a **sensitivity matrix** of nonnegative rational numbers

© Springer International Publishing AG 2017
R. Brijder and L. Qian (Eds.): DNA 23 2017, LNCS 10467, pp. 82–97, 2017.
DOI: 10.1007/978-3-319-66799-7_6

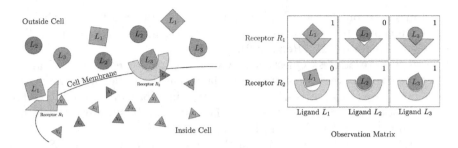

Fig. 1. An artificial cell with two transmembrane receptors R_1 and R_2 and extracellular ligands L_1, L_2, L_3. R_1 has equal sensitivity to both L_1 and L_3. R_2 has equal sensitivity to both L_2 and L_3. All sensitivities are nonnegative rational numbers.

$$\mathcal{S} = \begin{array}{c} \\ R_1 \\ R_2 \end{array}\begin{array}{c} L_1\ L_2\ L_3 \\ \left(\begin{array}{ccc} 1 & 0 & 1 \\ 0 & 1 & 1 \end{array}\right) \end{array}$$

The question of interest is how to design a cytoplasmic chemical reaction network to estimate the numbers l_1, l_2, l_3 of the ligands from receptor activation information. We assume that a prior probability distribution over ligand states $(l_1, l_2, l_3) \in \mathbb{Z}_{\geq 0}^3$ is given. We further assume that this prior probability distribution is a product of Poisson distributions specified by given Poisson rate parameters $q_1, q_2, q_3 \in \mathbb{R}_{>0}$ respectively. Lemma 2 provides intuition for the product-Poisson assumption. The following questions concern us.

1. Given information on the exact numbers r_1 and r_2 of activation events of receptors R_1 and R_2, obtain samples over populations (l_1, l_2, l_3) of the ligand species according to the Bayesian posterior distribution $\Pr[(l_1, l_2, l_3) \mid (r_1, r_2, \text{Poisson}(q_1, q_2, q_3))]$.
2. Given information on the average numbers $\langle r_1 \rangle$ and $\langle r_2 \rangle$ of activation events of receptors R_1 and R_2, obtain samples over populations (l_1, l_2, l_3) of the ligand species according to the Bayesian posterior distribution $\Pr[(l_1, l_2, l_3) \mid (\langle r_1 \rangle, \langle r_2 \rangle, \text{Poisson}(q_1, q_2, q_3))]$.

We investigate these questions for arbitrary numbers of receptors and ligands, arbitrary sensitivity matrices $\mathcal{S}$, and arbitrary product-Poisson rate parameters q, and make the following new contributions:

- In Sect. 3, we precisely state our question in the general setting. In Sect. 4, we illustrate our main ideas on Example 1.
- In Sect. 5.1, we describe a reaction network scheme Proj that takes as input a sensitivity matrix $\mathcal{S}$ and outputs a prime chemical reaction network. Our proposed reaction networks have the following merits that make them promising candidates for molecular implementation. Implementing the reactions requires only thermodynamic control and not kinetic control because the reaction rate

constants need only be specified upto the equilibrium constant for the reactions (Remark 1). Our scheme avoids catalysis, and so is robust to "leak reaction" situations [21] (Remark 2).

- In Sect. 5.2, we address Question 1. We show that for each fixed S and q, when the chemical reaction system is initialized as prescribed according to the numbers r_i of activation events of receptors, and allowed to evolve according to stochastic mass-action kinetics, then the system evolves towards the desired Bayesian posterior distribution (Theorem 3).
- In Sect. 5.3, we address Question 2. We show that for each fixed S and q, when the chemical reaction system is initialized as prescribed according to the average numbers $\langle r_i \rangle$ of activation events of receptors, and allowed to evolve according to deterministic mass-action kinetics, then the distribution of unit-volume aliquots of the system evolves towards the desired Bayesian posterior distribution (Theorem 6).
- In Sect. 6, we compare our scheme with other reaction network schemes that process information. Exploiting inherent stochasticity and free energy minimization appear to be the two key new ideas in our scheme.
- In Sect. 7, we discuss limitations and directions for future work, including a reaction scheme for the expectation-maximization algorithm, a commonly used algorithm in machine learning.

2   Background

2.1   Probability and Statistics

For $n \in \mathbb{Z}_{>0}$, following [15], $D : \mathbb{R}_{\geq 0}^n \times \mathbb{R}_{\geq 0}^n \to \mathbb{R}$ is the function

$$D(x \,\|\, y) := \sum_{i=1}^{n} x_i \log\left(\frac{x_i}{y_i}\right) - x_i + y_i$$

with the convention $0 \log 0 = 0$ and for $p > 0$, $p \log 0 = -\infty$. If x, y are probability distributions then $\sum_{i=1}^{n} -x_i + y_i = 0$ and D is the same as relative entropy $\sum_{i=1}^{n} x_i \log\left(\frac{x_i}{y_i}\right)$. When the index i takes values over a countably infinite set, we define D by the same formal sum as above, and understand it to be well-defined whenever the infinite sum converges in $[0, \infty]$. For $x \in \mathbb{R}_{>0}^k$, by Poisson(x) we mean $\Pr[n_1, n_2, \ldots, n_k \mid x] = \prod_{i=1}^{k} e^{-x_i} \frac{x_i^{n_i}}{n_i!}$. The following lemma is well-known and easy to show.

Lemma 1. $D(\text{Poisson}(x) \,\|\, \text{Poisson}(y)) = D(x \,\|\, y)$ for all $x, y \in \mathbb{R}_{>0}^k$.

The Exponential-Projection or **E-Projection** [15] (or Information-Projection or I-Projection [6]) of a probability distribution q onto a set of distributions P is $p^* = \arg\min_{p \in P} D(p \,\|\, q)$. The Mixture-Projection or **M-Projection** (or reverse I-projection) of a probability distribution p onto a set of distributions Q is $q^* = \arg\min_{q \in Q} D(p \,\|\, q)$.

2.2   Reaction Networks

We recall notation, definitions, and results from reaction network theory [1,10–12,14]. For $x, y \in \mathbb{R}^k$, by x^y we mean $\prod_{i=1}^{k} x_i^{y_i}$, and by e^x we mean $\prod_{i=1}^{k} e^{x_i}$. For $m \in \mathbb{Z}_{\geq 0}^k$, by $m!$ we mean $\prod_{i=1}^{k} m_i!$.

Fix a finite set S of **species**. By a **reaction** we mean a formal chemical equation

$$\sum_{i \in S} y_i X_i \to \sum_{i \in S} y_i' X_i$$

where the numbers $y_i, y_i' \in \mathbb{Z}_{\geq 0}$ are the **stoichiometric coefficients**. This reaction is also written as $y \to y'$ where $y, y' \in \mathbb{Z}_{\geq 0}^S$. A **reaction network** is a pair $(S, \mathcal{R})$ where S is finite, and $\mathcal{R}$ is a finite set of reactions. It is **reversible** iff for every reaction $y \to y' \in \mathcal{R}$, the reaction $y' \to y \in \mathcal{R}$. Fix $n, n' \in \mathbb{Z}_{\geq 0}^S$. We say that $n \mapsto_{\mathcal{R}} n'$, read n **maps to** n' iff there exists a reaction $y \to y' \in \mathcal{R}$ with $y_i \leq n_i$ for all $i \in S$ and $n' = n + y' - y$. We say that $n \Rightarrow_{\mathcal{R}} n'$, or in words that n' is $\mathcal{R}$-**reachable** from n, iff there exist a nonnegative integer $k \in \mathbb{Z}_{\geq 0}$ and $n(1), n(2), \ldots, n(k) \in \mathbb{Z}_{\geq 0}^S$ such that $n(1) = n$ and $n(k) = n'$ and for $i - 1$ to $k - 1$, we have $n(i) \mapsto_{\mathcal{R}} n(i + 1)$. A reaction network $(S, \mathcal{R})$ is **weakly reversible** iff for every reaction $y \to y' \in \mathcal{R}$, we have $y' \Rightarrow y$. Trivially, every reversible reaction network is weakly reversible. The **reachability class** of $n_0 \in \mathbb{Z}_{\geq 0}^S$ is the set $\Gamma(n_0) = \{n \mid n_0 \Rightarrow_{\mathcal{R}} n\}$. The **stoichiometric subspace** $H_{\mathcal{R}}$ is the real span of the vectors $\{y' - y \mid y \to y' \in \mathcal{R}\}$. The **conservation class** containing $x_0 \in \mathbb{R}_{\geq 0}^S$ is the set $C(x_0) = (x_0 + H_{\mathcal{R}}) \cap \mathbb{R}_{\geq 0}^S$.

Fix a weakly reversible reaction network $(S, \mathcal{R})$. Let $x = (x_i)_{i \in S}$. The associated ideal $I_{(S,\mathcal{R})} \subseteq \mathbb{C}[x]$ is the ideal generated by the binomials $\{x^y - x^{y'} \mid y \to y' \in \mathcal{R}\}$. A reaction network is **prime** iff its associated ideal is a prime ideal, i.e., for all $f, g \in \mathbb{C}[x]$, if $fg \in I$ then either $f \in I$ or $g \in I$.

A **reaction system** is a triple $(S, \mathcal{R}, k)$ where $(S, \mathcal{R})$ is a reaction network and $k : \mathcal{R} \to \mathbb{R}_{>0}$ is called the **rate function**. It is **detailed balanced** iff it is reversible and there exists a point $q \in \mathbb{R}_{>0}^S$ such that for every reaction $y \to y' \in \mathcal{R}$:

$$k_{y \to y'} \, q^y \, (y' - y) = k_{y' \to y} \, q^{y'} \, (y - y')$$

A point $q \in \mathbb{R}_{>0}^S$ that satisfies the above condition is called a **point of detailed balance**.

Fix a reaction system $(S, \mathcal{R}, k)$. Then **stochastic mass action** describes a continuous-time Markov chain on the state space $\mathbb{Z}_{\geq 0}^S$. A state $n = (n_i)_{i \in S} \in \mathbb{Z}_{\geq 0}^S$ of this Markov chain represents a vector of molecular counts, i.e., each n_i is the number of molecules of species i in the population. Transitions go from $n \to n + y' - y$ for each $n \in \mathbb{Z}_{\geq 0}^S$ and each $y \to y' \in \mathcal{R}$, with transition rates

$$\lambda(n \to n + y' - y) = k_{y \to y'} \frac{n!}{(n - y)!}$$

The following theorem states that the stationary distributions of detailed-balanced reaction networks are obtained from products of Poisson distributions. It is well-known, see for example [23] for a proof.

Theorem 1. *If $(S, \mathcal{R}, k)$ is detailed balanced with q a point of detailed balance then the corresponding stochastic mass action Markov chain admits on each reachability class $\Gamma \subset \mathbb{Z}_{\geq 0}^S$ a unique stationary distribution*

$$\pi_\Gamma(n) \propto \begin{cases} e^{-q}\frac{q^n}{n!} & \text{for } n \in \Gamma \\ 0 & \text{otherwise} \end{cases}$$

Deterministic mass action describes a system of ordinary differential equations in *concentration* variables $\{x_i(t) \mid i \in S\}$:

$$\dot{x}(t) = \sum_{y \to y' \in \mathcal{R}} k_{y \to y'}\, x(t)^y\, (y' - y) \tag{1}$$

Note that every detailed balance point is a fixed point to Eq. 1. For detailed balanced reaction systems, every fixed point is also detailed balanced. Moreover, every conservation class $C(x_0)$ has a unique detailed balance point x^* in the positive orthant. Further if the reaction network is prime then x^* is a "global attractor," i.e., all trajectories starting in $C(x_0) \cap \mathbb{R}_{>0}^S$ asymptotically reach x^*. (Recently Craciun [5] has proved the global attractor theorem for all detailed-balanced reaction systems with a much more involved proof. We do not need Craciun's theorem, the special case which holds for prime detailed-balanced reaction systems and is much easier to prove, suffices for our purposes.) The following Global Attractor Theorem for Prime Detailed Balanced Reaction Systems follows from [12, Corollary 4.3, Theorem 5.2]. See [13, Theorem 3] for another restatement of this theorem.

Theorem 2. *Let $(S, \mathcal{R}, k)$ be a prime, detailed balanced reaction system with point of detailed balance q. Fix a point $x_0 \in \mathbb{R}_{>0}^S$. Then there exists a point of detailed balance x^* in $C(x_0) \cap \mathbb{R}_{>0}^S$ such that for every trajectory $x(t)$ to Eq. 1 with initial conditions $x(0) \in C(x_0)$, the limit $\lim_{t \to \infty} x(t)$ exists and equals x^*. Further $D(x(t) \| q)$ is strictly decreasing along non-stationary trajectories and attains its unique minimum value in $C(x_0)$ at x^*.*

3   Problem Statement

We argue in the next lemma that a product of Poisson distributions is not an unreasonable form to use as a prior on ligand populations. The ideas are familiar from statistical mechanics as well as stochastic processes. We recall them in a chemical context.

Lemma 2. *Consider a well-mixed vessel of infinite volume with n species X_1, $X_2, \ldots, X_n$ at concentrations $x_1, x_2, \ldots, x_n$ respectively. Assume that the solution* *is sufficiently dilute, and that molecule volumes are vanishingly small. A unit volume aliquot is taken. Then the probability of finding the population in the aliquot in state $(m_1, m_2, \ldots, m_n) \in \mathbb{Z}_{\geq 0}$ is given by the product-Poisson distribution* $\prod_{i=1}^{n} \frac{e^{-x_i} x_i^{m_i}}{m_i!}$.

Proof. We will first do the analysis for a finite volume V and then let $V \to \infty$.

Consider a container of finite volume V, which contains species $X_1, X_2, \ldots, X_n$ at concentrations $x_1, x_2 \ldots, x_n$. Consider a unit volume aliquot within this particular container. The probability of finding a particular molecule from the vessel within the unit volume aliquot is $\frac{1}{V}$. The number of molecules of species X_i in the vessel is $V x_i$ for $i = 1 \ldots n$. Hence the probability of finding m_i molecules of species X_i in the aliquot is given by the binomial coefficient

$$\binom{V x_i}{m_i} \left(\frac{1}{V} \right)^{m_i} \left(1 - \frac{1}{V} \right)^{V x_i - m_i} .$$

We assume that the solution is sufficiently dilute, and that molecular sizes are vanishingly small, so that the probability of finding one molecule in the aliquot is independent of the probability of finding a different molecule in the aliquot. This assumption leads to:

$$\Pr(m_1, m_2, \ldots, m_n \mid x_1, x_2, \ldots, x_n) = \prod_{i=1}^{n} \binom{V x_i}{m_i} \left(\frac{1}{V} \right)^{m_i} \left(1 - \frac{1}{V} \right)^{V x_i - m_i}$$

The RHS follows because for all $i \in \{1, 2, \ldots, n\}$: the limit

$$\lim_{V \to \infty} \frac{V x_i (V x_i - 1) \ldots (V x_i - m_i + 1)}{V^{m_i} m_i!} \left[(1 - 1/V)^V \right]^{x_i - m_i/V} = e^{-x_i} x_i^{m_i} / m_i!$$

Fix positive integers $n_R, n_L \in \mathbb{Z}_{\geq 0}$ with $n_R \leq n_L$ denoting the number of receptor species and the number of ligand species respectively. Fix $q = (q_1, q_2, \ldots, q_{n_L}) \in \mathbb{R}_{>0}^{n_L}$ denoting Poisson rate parameters for the product-Poisson distribution Poisson(q) which we consider as a prior over ligand numbers. Following [24], we fix an $n_R \times n_L$ **sensitivity matrix** $\mathcal{S}$ with entries s_{ij} in the nonnegative rational numbers $\mathbb{Q}_{\geq 0}$. The entry s_{ij} denotes the sensitivity of the i'th receptor R_i to the j'th ligand L_j. The intuition is that when ligand j encounters receptor i, the conditional probability that the receptor activates is proportional to s_{ij}. So a high-sensitivity ligand will trigger a receptor more often than a low-sensitivity ligand with the same concentration, with the number of times they trigger the receptor in proportion to their corresponding entries in the sensitivity matrix.

Our results in this paper will hold for a subclass of sensitivity matrices which we term tidy. A sensitivity matrix $\mathcal{S} = (s_{ij})_{n_R \times n_L}$ is **tidy** iff for each receptor R_i there exists a **message vector** $m_i \in \mathbb{Z}_{\geq 0}^{n_L}$ such that $\mathcal{S} m_i = e_i$ where $e_i \in \mathbb{R}^{n_R}$ is the unit vector with a 1 in the row corresponding to the i'th receptor. Every time receptor R_i is activated, it will trigger a cascade leading to the synthesis inside the cell of m_{ij} molecules of species X_j for $j = 1$ to n_L. The intuition is that for $j = 1$ to n_L, the "signal" species X_j is the cell's internal representation of the ligand L_j. Message vectors ensure that the difference in the numbers of X_j's and the numbers of L_j's lies in $\ker \mathcal{S}$, so that the numbers of X_j's are a feasible state of the world, consistent with the observations, though not the most likely such feasible state.

Note that there could be multiple message vector sets $\{m_i\}_{i=1 \text{ to } n_R}$, so the cell need not choose the "correct" one. The task of figuring out the most likely state of the environment will be left to the intracellular reaction network between the $X_1, X_2, \ldots, X_{n_L}$ molecules. The following questions concern us.

1 Given information on the exact numbers $r = (r_1, r_2, \ldots, r_{n_R}) \in \mathbb{Z}_{\geq 0}^{n_R}$ of receptor activation events, obtain samples over populations $l = (l_1, l_2, \ldots, l_{n_L}) \in \mathbb{Z}_{\geq 0}^{n_L}$ of the ligand species according to the Bayesian posterior distribution $\Pr[l \mid (r, \mathrm{Poisson}(q_1, q_2, \ldots, q_{n_L}))]$.

2 Given information on the average numbers $\langle r \rangle = (\langle r_1 \rangle, \langle r_2 \rangle, \ldots, \langle r_{n_R} \rangle) \in \mathbb{R}_{>0}^{n_R}$ of receptor activation events (averaged over the surface of the cell, or time, or both), obtain samples over populations $l = (l_1, l_2, \ldots, l_{n_L}) \in \mathbb{Z}_{\geq 0}^{n_L}$ of the ligand species according to the Bayesian posterior distribution $\Pr[l \mid (\langle r \rangle, \mathrm{Poisson}(q_1, q_2, \ldots, q_{n_L}))]$.

4 An Example

Before moving to the general solution, we illustrate our main ideas with an example.

Example 2 (continues from p. 1). Consider the sensitivity matrix

$$S = \begin{array}{c} \\ R_1 \\ R_2 \end{array}\begin{pmatrix} \overset{L_1}{1} & \overset{L_2}{0} & \overset{L_3}{1} \\ 0 & 1 & 1 \end{pmatrix}$$

and the point $q = (q_1, q_2, q_3) \in \mathbb{R}_{>0}^3$ from Example 1. To describe the reactions, we compute a basis B for the right kernel of S. In this case, the vector $(1, 1, -1)^T$ is a basis for the right kernel. (To be precise, we will view the right kernel as a free group in the integer lattice, and take a basis for this free group. This ensures not only that each basis vector has integer coordinates, but also that the corresponding reaction network is prime, which we use crucially in our proofs.)

We describe a chemical reaction system $(\mathrm{Proj}(S, B), k_q)$ as follows. There is one chemical species X_i corresponding to each ligand L_i, so that the species are $X_1, X_2,$ and X_3. Each basis vector is written as a reversible reaction, with negative numbers representing stoichiometric coefficients on one side of the chemical equation, and positive numbers representing stoichiometric coefficients on the other side. So the vector $(1, 1, -1)^T$ describes the reversible pair of reactions $X_1 + X_2 \rightleftharpoons X_3$.

The rates k_q of the reactions need to be set so that q is a point of detailed balance. For this example, calling the forward rate $k_1 \in \mathbb{R}_{>0}$ and the backward rate $k_2 \in \mathbb{R}_{>0}$, the balance condition is $k_1 q_1 q_2 = k_2 q_3$ so that $k_1/k_2 = \frac{q_3}{q_1 q_2}$. One choice satisfying this condition is $k_1 = q_3$ and $k_2 = q_1 q_2$. Note that our scheme requires only the ratio of the rates to be specified (Remark 1).

Solution to Question 1: Given $r = (r_1, r_2) \in \mathbb{Z}_{\geq 0}^2$ interpreted as $(r_1, r_2)^T = \mathcal{S}(l_1, l_2, l_3)^T$, we want to draw samples from the conditional distribution $\Pr[(l_1, l_2, l_3) \mid (r_1, r_2, \text{Poisson}(q_1, q_2, q_3))]$. The statistical solution is to multiply the Bayesian prior $\text{Poisson}(q_1, q_2, q_3)$ by the likelihood $\Pr[(r_1, r_2) \mid (l_1, l_2, l_3, \text{Poisson}(q_1, q_2, q_3))]$, and normalize so probabilities add up to 1. The likelihood is the characteristic function of the set

$$L = \{l = (l_1, l_2, l_3) \in \mathbb{Z}_{\geq 0}^3 \mid \mathcal{S}l^T = r^T\}.$$

Note that $\mathcal{S}$ is tidy with message vectors $m_1 = (1, 0, 0)^T$ and $m_2 = (0, 1, 0)^T$. The reaction system $(\text{Proj}(\mathcal{S}, B), k_q)$ which is $X_1 + X_2 \underset{q_1 q_2}{\overset{q_3}{\rightleftharpoons}} X_3$ here, is initialized at $n(0) = (r_1, r_2, 0) = \sum_i r_i m_i$, and allowed to evolve according to stochastic mass-action kinetics with master equation:

$$
\begin{aligned}
\dot{p}(n, t) = {} & p(n_1 - 1, n_2 - 1, n_3 + 1, t)\left(\frac{q_1 q_2}{q_3}(n_3 + 1) - n_1 n_2\right) \\
& + p(n_1 + 1, n_2 + 1, n_3 - 1, t)\left((n_1 + 1)(n_2 + 1) - \frac{q_1 q_2}{q_3}n_3\right)
\end{aligned}
$$

where $p(n, t)$ is the probability that the system is in state n at time t. We claim that the steady-state distribution is the required Bayesian posterior. First note that this reaction system has a detailed balanced point q, so it admits $\text{Poisson}(q)$ as a steady-state distribution. Since $n(0) \in L$, it is enough to show that L forms an irreducible component of the Markov chain. Together we conclude that the steady-state distribution will be a restriction of $\text{Poisson}(q_1, q_2, q_3)$ to the set L.

To obtain that L forms an irreducible component of the Markov chain, we will crucially use the fact that we chose a basis of the free group to generate our reactions, and not just a basis of the real vector space. This will allow us to prove that the corresponding reaction network is prime, and hence that L forms an irreducible component. Note, for example, that if we had chosen the vector $(2, 2, -2)^T$ in the kernel instead of $(1, 1, -1)^T$, that would have given us the reaction $2X_1 + 2X_2 \rightleftharpoons 2X_3$ in which case L does not form an irreducible component of the Markov chain since each reaction conserves parity of molecular counts.

Solution to Question 2: Given $\langle r \rangle = (\langle r_1 \rangle, \langle r_2 \rangle) \in \mathbb{R}_{\geq 0}^2$ activation events of receptors R_1 and R_2, with $\langle r \rangle$ interpreted as empirical average of $\mathcal{S}(l_1, l_2, l_3)^T$ over a large number of samples of (l_1, l_2, l_3), we want to draw samples from the conditional distribution $\Pr[(l_1, l_2, l_3) \mid (\langle r_1 \rangle, \langle r_2 \rangle, \text{Poisson}(q_1, q_2, q_3))]$. Note that we are conditioning over an event whose probability tends to 0 unless $\mathcal{S}q^T = \langle r \rangle^T$, so the conditional distribution needs to be defined using the notion of regular conditional distribution [8]. As the number of samples goes to infinity, by the conditional limit theorem [8, Theorem 7.3.8, Corollary 7.3.5], this conditional distribution converges to Poisson(x^*) where $x^* = (x_1^*, x_2^*, x_3^*) \in \mathbb{R}_{\geq 0}^3$ minimizes $D(x \,\|\, q)$ among all x satisfying $S x^T = \langle r \rangle^T$. Because these results are stated in the reference in much greater generality, to show that these results actually apply to our case will need some technical work which is the content of Sect. 5.3.

To compute x^*, we allow $(\mathrm{Proj}(\mathcal{S}, B), k_q) = X_1 + X_2 \underset{q_1 q_2}{\overset{q_3}{\rightleftharpoons}} X_3$ to evolve according to deterministic mass-action kinetics starting from $x(0) = (\langle r_1 \rangle, \langle r_2 \rangle, 0) = \sum_i \langle r_i \rangle m_i$.

$$\begin{pmatrix} \dot{x}_1(t) \\ \dot{x}_2(t) \\ \dot{x}_3(t) \end{pmatrix} = \left(x_1(t) x_2(t) - \frac{q_1 q_2}{q_3} x_3(t) \right) \begin{pmatrix} -1 \\ -1 \\ 1 \end{pmatrix}$$

Then the equilibrium concentration is the desired x^* by Theorem 2. The required sample can be drawn by sampling a unit aliquot, as in Lemma 2.

Our scheme suggests that the reactions are carried out in infinite volume, which seems impractical. In practise, infinite volume need not be necessary because the chemical dynamics of even molecular numbers as small as 50 molecules are often described fairly accurately by the infinite-volume limit. Further, our scheme suggests an infinite number of samples for this to work correctly, which also looks impractical. However, the rate of convergence is exponentially fast, so the scheme can be expected to work quite accurately even with a moderate number of samples. Analysis beyond the scope of the current paper is needed to explore the tradeoffs in volume and number of samples (also see Sect. 7).

5   Main

5.1   A Reaction Scheme

In this subsection, we present a reaction scheme Proj (short for projection) that takes as input a matrix $\mathcal{S}$ with rational entries, and a basis B for the free group $\mathbb{Z}^{n_L} \cap \ker \mathcal{S}$ and outputs a reversible reaction network $\mathrm{Proj}(\mathcal{S}, B)$ that is prime. The same scheme, appropriately initialized, serves to perform M-projection (as we showed in [13]) and E-projection, as we show here.

Definition 1. Fix a matrix $\mathcal{S} = (s_{ij})_{m \times n}$ with rational entries $s_{ij} \in \mathbb{Q}$, and a basis B for the free group $\mathbb{Z}^n \cap \ker \mathcal{S}$. The reaction network $\mathrm{Proj}(\mathcal{S}, B)$ is described by species $X_1, X_2, \ldots, X_n$ and for each $b \in B$ the reversible reaction: $\sum_{j : b_j > 0} b_j X_j \rightleftharpoons \sum_{j : b_j < 0} -b_j X_j$

Remark 1. Exquisitely setting the specific rates of individual reactions to desired values requires a detailed understanding of molecular dynamics, and is forbiddingly difficult with current molecular technology. When we set rates, we will only require that a given point remains a point of detailed balance. This is equivalent to specifying the equilibrium constants of all the reactions. This is an equilibrium thermodynamics condition, hence much less forbidding.

Lemma 3. *Fix a matrix $\mathcal{S} = (s_{ij})_{m \times n}$ with rational entries $s_{ij} \in \mathbb{Q}$, and a basis B for the free group $\mathbb{Z}^n \cap \ker \mathcal{S}$. Then the reaction network $\mathrm{Proj}(\mathcal{S}, B)$ is prime.*

Proof. [18, Corollary 1.15] establishes this when $\mathcal{S}$ is a matrix of integers. Scaling the rational entries to make them all integers makes no difference to the kernel.

Remark 2. From [12, Theorem 5.2], prime reaction networks are free of catalysis. Catalysts require care to implement. Ideally a catalyst should act as a switch, so that its absence completely shuts off the catalyzed reaction. In practice, there is always a "leak reaction" [21] even in the absence of the catalyst species. Care needs to be taken that the timescales of the leak are much slower than the timescales of the catalyzed reaction to get an acceptable approximation to the final answer. It is therefore notable that our scheme is able to perform a nontrivial computation even though it admits an implementation wholly free of catalysis.

Example 3. Consider the reaction $2X \rightleftharpoons 0$. On the state space $\mathbb{Z}_{\geq 0}$, this reaction will preserve the parity of the initial number n_0 of X. This is a case where the intersection of a conservation class $C(n_0)$ with the state space does not equal the reachability class $\Gamma(n_0)$. It turns out that these "non-benign" situations only happen when the reaction network is not prime. We will use this property when answering Questions 1 and 2, so we establish it now.

Definition 2. A weakly-reversible reaction network $(S, \mathcal{R})$ is **benign** iff for all $n_0 \in \mathbb{Z}_{\geq 0}^S$, the reachability class $\Gamma(n_0) = C(n_0) \cap \mathbb{Z}_{\geq 0}^S$ where $C(n_0)$ is the conservation class of n_0.

Lemma 4. *Every prime reaction network is benign.*

Proof. Let $(S, \mathcal{R})$ be a prime reaction network. This means that the associated ideal $(x^y - x^{y'})_{y \to y' \in \mathcal{R}}$ is prime. We define the **associated lattice** as

$$
\mathcal{L} = \left\{ \sum_{y \to y' \in \mathcal{R}} a_{y \to y'}(y' - y) \mid a_{y \to y'} \in \mathbb{Z} \text{ for all } y \to y' \in \mathcal{R} \right\}.
$$

Note from [18] that $\mathcal{L}$ is **saturated**, i.e., if $k \in \mathbb{Z}$ and $v \in \mathbb{Z}^S$ are such that $kv \in \mathcal{L}$ then $v \in \mathcal{L}$

Suppose $n_0, n_0' \in \mathbb{Z}_{\geq 0}^S$ such that $n_0' \in C(n_0)$ but n_0' is not reachable from n_0. The condition $n_0' \in C(n_0)$ means that there is a rational combination

$$
n_0' - n_0 = \sum_{y \to y' \in \mathcal{R}} b_{y \to y'}(y' - y)
$$

This shows that for some sufficiently large integer M, the quantity $M(n_0' - n_0) \in \mathcal{L}$. Since $\mathcal{L}$ is saturated, $n_0' - n_0 \in \mathcal{L}$. Hence there is an integer combination

$$
n_0' - n_0 = \sum_{y \to y' \in \mathcal{R}} c_{y \to y'}(y' - y).
$$

Since $(S, \mathcal{R})$ is weakly-reversible, there is a path $y' \Rightarrow_{\mathcal{R}} y$ for every $y \to y' \in \mathcal{R}$, and therefore there is a combination over nonnegative integers. This implies that $n_0 \Rightarrow_{\mathcal{R}} n'_0$. Hence the network is benign.

5.2   Solution to Question 1

In this section we solve Question 1 using the reaction network $\mathrm{Proj}(S, B)$.

Fix an $n_L \times n_R$ tidy sensitivity matrix $S = (s_{ij})_{n_R \times n_L}$ with non-negative rational entries $s_{ij} \in \mathbb{Q}_{\geq 0}$, and message vectors $\{m_i \in \mathbb{Z}_{\geq 0}^{n_L}\}_{i=1,2,\dots,n_R}$, Poisson rate parameter vector $q \in \mathbb{R}_{\geq 0}^{n_L}$, and number $r \in \mathbb{Z}_{\geq 0}^{n_R}$ of receptor activation events observed. Fix a basis B for the free group $\mathbb{Z}^{n_L} \cap \ker S$. Let k_q be a function of rate constants for the reaction network $\mathrm{Proj}(S, B)$ such that q is a point of detailed balance of the reaction system $(\mathrm{Proj}(S, B), k_q)$. For example, the choice $k_q(y \to y') = q^{y'}$ satisfies this requirement.

Theorem 3. *Consider Stochastic Mass Action for the reaction system* $(\mathrm{Proj}(S, B), k_q)$ *from the initial state* $n(0) = \sum_{i=1}^{n_R} r_i m_i$. *Then the Bayesian Posterior* $\Pr[l \mid (r, \mathrm{Poisson}(q))]$ *is the stationary distribution of this Markov chain.*

Proof. Let $L = \left\{ l \in \mathbb{Z}_{\geq 0}^{n_L} \mid Sl^T = r^T \right\}$. The prior is $\mathrm{Poisson}(q)$ and the likelihood is $\Pr[r \mid l] = \Pr[Sl^T = r^T]$ which is the characteristic function on L. Therefore from Bayes' Theorem,

$$\Pr[l \mid (r, \mathrm{Poisson}(q))] \propto \begin{cases} e^{-q}\dfrac{q^l}{l!} & \text{for } l \in L \\ 0 & \text{otherwise} \end{cases}$$

Since the reaction network $\mathrm{Proj}(S, B)$ is prime, by Lemmas 3 and 4, $\mathrm{Proj}(S, B)$ is benign. By construction $n(0) \in L$, and so L is the reachability class $\Gamma(n(0))$. Applying Theorem 1 to $L = \Gamma(n_0)$

$$\pi_L(l) \propto \begin{cases} e^{-q}\dfrac{q^l}{l!} & \text{for } l \in L \\ 0 & \text{otherwise} \end{cases}$$

which is exactly the Bayesian Posterior $\Pr[l \mid (r, \mathrm{Poisson}(q))]$.

In the following theorem, we show that our reaction scheme has computed an E-Projection.

Theorem 4. *Let* $\mathcal{P} := \{Probability\ measure\ P\ on\ \mathbb{Z}_{\geq 0}^{n_L} \mid P(l) = 0\ for\ all\ l \notin L\}$. *Then* $\Pr[l \mid (r, \mathrm{Poisson}(q))]$ *is the E-Projection of* $\mathrm{Poisson}(q)$ *on* $\mathcal{P}$.

Proof. Let $P^* = \arg\min_{P \in \mathcal{P}} D(P \,\|\, \mathrm{Poisson}(q))$ be the E-projection of $\mathrm{Poisson}(q)$ onto $\mathcal{P}$. To minimize $D(P \,\|\, \mathrm{Poisson}(q))$ with constraints $\sum_{l \in L} P(l) = 1$ and $P(l) = 0$ for $l \notin L$, write the Lagrangian

$$F(P, \lambda, \mu) = D(P \,\|\, \mathrm{Poisson}(q)(l)) + \lambda \left(\sum_{l \in L} P(l) - 1 \right) + \sum_{l \notin L} \mu_l P(l)$$

At P^*, $\frac{\partial F}{\partial P(l)} = 0$ for all $l \in \mathbb{Z}_{\geq 0}^{n_L}$. That is, $\log\left(\frac{P^*(l)}{\text{Poisson}(q)}\right) + 1 + \lambda = 0$ if $l \in L$ and $P^*(l) = 0$ if $l \notin L$. That is,

$$P^*(l) \propto \begin{cases} \text{Poisson}(q)(l) & \text{for } x \in L \\ 0 & \text{otherwise} \end{cases}$$

which is the Bayesian Posterior $\Pr[l \mid (r, \text{Poisson}(q))]$

5.3 Solution to Question 2

In this subsection we solve Question 2 using the reaction network $\text{Proj}(\mathcal{S}, B)$. We first characterize the Bayesian Posterior $\Pr[l \mid (\langle r \rangle, \text{Poisson}(q))]$ as an E-projection using a conditional limit theorem.

Definition 3. Fix $\langle r \rangle \in \mathbb{R}_{>0}^{n_R}$. Then $\mathcal{P}_{\langle r \rangle}$ is the set of those probability measures on $\mathbb{Z}_{\geq 0}^{n_L}$ such that if Y is a random variable distributed according to $P \in \mathcal{P}_{\langle r \rangle}$ then the expected value $\mathcal{S}\langle Y \rangle_P^T = \langle r \rangle^T$.

Theorem 5. *Fix $\langle r \rangle \in \mathbb{R}_{>0}^{n_R}$. Then $\Pr[l \mid (\langle r \rangle, \text{Poisson}(q))]$ is a Poisson distribution, as well as the E-Projection $\arg\min_{P \in \mathcal{P}_{\langle r \rangle}} D(P \,\|\, \text{Poisson}(q))$ of $\text{Poisson}(q)$ on $\mathcal{P}_{\langle r \rangle}$.*

Proof. We apply the Gibbs Conditioning Principle ([9, Theorem 7.3.8]) n_R times with a sequence of energy functions $U_1, \ldots, U_{n_R}$ which iteratively set the expected values of the n_R rows of $\mathcal{S}$ to the corresponding values from $\langle r \rangle$. The intuition is that this is a formal way of doing Lagrange optimization.

To show that this result can be applied, we choose the space Σ as $\mathbb{R}^{n_L}$, the initial distribution $\mu = \mu_0$ as $\text{Poisson}(q)$ on $\mathbb{Z}_{\geq 0}^{n_L}$ and 0 everywhere else, and for $i = 1$ to n_R, we define the function $U_i : \Sigma \to [0, \infty)$ by $U_i(n) = \frac{(\mathcal{S}n)_i}{\langle r_i \rangle}$. The sequence of Gibbs distributions are then defined by $\frac{d\mu_{i+1}}{d\mu_i} = \frac{e^{-\beta_i U_i(n)}}{Z_{\beta_i}}$ where Z_{β_i} is the normalizing constant. It is easily checked that each of these is a Poisson distribution since the U_i's are linear functions. Since $\langle r \rangle \in \mathbb{R}_{>0}^{n_R}$, there is nonzero probability under μ_{i-1} that $(\mathcal{S}x)_i < \langle r_i \rangle$ for all i. Hence for $i = 1$ to n_R it follows that $\mu_{i-1}(\{x \mid U_i(x) < 1\}) > 0$. The other condition $\mu_{i-1}(\{x \mid U(x) > 1\}) > 0$ is true since under a Poisson distribution, $(\mathcal{S}x)_i$ can take arbitrarily large integer values with nonzero probability. Since the μ_i are all Poisson, $\beta_\infty = -\infty$ since Poisson distributions converge for arbitrarily small non-negative values of rate parameters. Hence the assumptions of [9, Lemma 7.3.6] are satisfied and we get to apply [9, Theorem 7.3.8] sequentially n_R times and conclude that the empirical distribution on the space $\mathbb{Z}_{\geq 0}^{n_L}$ converges weakly to a Poisson distribution $\mu_{n_R} = \text{Poisson}(p^*) \in \mathcal{P}_{\langle r \rangle}$, which is also the E-projection $\arg\min_{P \in \mathcal{P}_{\langle r \rangle}} D(P \,\|\, \text{Poisson}(q))$.

Now fix an $n_L \times n_R$ tidy sensitivity matrix $\mathcal{S} = (s_{ij})_{n_R \times n_L}$ with non-negative rational entries $s_{ij} \in \mathbb{Q}_{\geq 0}$, and message vectors $\{m_i \in \mathbb{Z}_{\geq 0}^{n_L}\}_{i=1,2,\ldots,n_R}$, Poisson rate parameter vector $q \in \mathbb{R}_{\geq 0}^{n_L}$, and average number $\langle r \rangle \in \mathbb{R}_{>0}^{n_R}$ of receptor activation events observed. Fix a basis B for the free group $\ker \mathcal{S} \cap \mathbb{Z}^{n_L}$. Let k_q be a function of rate constants for the reaction network $\mathrm{Proj}(\mathcal{S}, B)$ such that q is a point of detailed balance of the reaction system $(\mathrm{Proj}(\mathcal{S}, B), k_q)$. For example, the choice $k_q(y \rightarrow y') = q^{y'}$ satisfies this requirement.

Theorem 6. *Consider the solution $x(t)$ to the Deterministic Mass Action ODEs for the reaction system $(\mathrm{Proj}(\mathcal{S}, B), k_q)$ from the initial concentration $x(0) = \sum_{i=1}^{n_R} \langle r_i \rangle m_i$. Let $x^* = \lim_{t \rightarrow \infty} x(t)$. Then x^* is well-defined, and the Bayesian Posterior $\Pr[l \mid (r, \mathrm{Poisson}(q))]$ equals $\mathrm{Poisson}(x^*)$. That is, one obtains samples from the Bayesian Posterior by measuring the state of a unit volume aliquot of the system at equilibrium.*

Proof. Note that $\mathrm{Poisson}(x(0)) \in \mathcal{P}_{\langle r \rangle}$. Further the reaction vectors span the kernel of $\mathcal{S}$ so we have $x \in C(x(0)) \cap \mathbb{R}_{>0}^{n_L}$ iff $\mathrm{Poisson}(x) \in \mathcal{P}_{\langle r \rangle}$. By Theorem 5, the distribution $\Pr[l \mid (r, \mathrm{Poisson}(q))]$ equals $\mathrm{Poisson}(y)$ for some $y \in \mathbb{R}_{>0}^{n_L}$. Further, it is an E-projection so that, among all Poisson distributions in $\mathcal{P}_{\langle r \rangle}$, the relative entropy $D(\mathrm{Poisson}(y) \,\|\, q)$ is minimum. By Lemma 1, the E-projection of $\{\mathrm{Poisson}(x) \mid x \in C(x(0)) \cap \mathbb{R}_{>0}^{n_L}\}$ to $\mathrm{Poisson}(q)$ is the Poisson distribution of the E-projection of $C(x(0)) \cap \mathbb{R}_{>0}^{n_L}$ to q.

By Lemma 3, the reaction network $\mathrm{Proj}(\mathcal{S}, B)$ is prime. Further the reaction system $(\mathrm{Proj}(\mathcal{S}, B), k_q)$ is detailed balanced with q a point of detailed balance, by assumption. Hence by Theorem 2, the limit x^* is well-defined and is the E-projection of $C(x(0)) \cap \mathbb{R}_{>0}^{n_L}$ to q. Together we have $\Pr[l \mid (r, \mathrm{Poisson}(q))] = \mathrm{Poisson}(x^*)$. We can sample from a unit aliquot at equilibrium due to Lemma 2.

6   Related Work

Various schemes have been proposed to perform information processing with reaction networks, for example, [21,22] which shows how Boolean circuits and perceptrons can be built, [20] which shows how to implement linear input/output systems, [7] exploiting analogies with electronic circuits, [2] for computing algebraic functions, etc. Some of these schemes have even been successfully implemented in vitro.

Each of these schemes has been inspired by analogy with some existing model of computation. However, reaction networks as a computing platform has some unique opportunities and challenges. It is an inherently distributed and stochastic platform. Noise manifests as leaks in catalyzed reactions. We can tune equilibrium thermodynamic parameters, but kinetic-level control is very difficult. In addition, one needs to keep in mind the tasks that reaction networks are called upon to solve in biology, or might be called upon to solve in technological applications. Keeping these factors in mind, there is value in considering a scheme which attempts to uncover the class of problems that is suggested by the mathematical structure of reaction network dynamics.

In trying to uncover such a class of problems, we have looked to the ideas of Maximum Entropy or MaxEnt [16] which form a natural bridge between Machine

Learning and Reaction Networks. The systematic foundations of statistics based on the minimization of KL-divergence (equivalently, free energy) go back to the pioneering work of Kullback [17]. The conceptual, technical, and computational advantages of this approach have been brought out by subsequent workers [4,6,15]. This work has also been put forward as a mathematical justification of Jaynes' MaxEnt principle. Our hope is that those parts of statistics and machine learning that can be expressed in terms of minimization of free energy should naturally suggest reaction network algorithms for their computation.

The link between statistics/machine learning and reaction networks has been explored before by Napp and Adams [19]. They propose a deterministic mass-action based reaction network scheme to compute single-variable marginals from a joint distribution given as a factor graph, drawing on "message-passing" schemes. Our work is in the same spirit of finding more connections between machine learning and reaction networks, but the nature of the problem we are trying to solve is different. We are trying to estimate a full distribution from partial observations. In doing so, we exploit the inherent stochasticity of reaction networks to represent correlations and do Bayesian inference.

One previous work which has engaged with stochasticity in reaction networks is by Cardelli et al. [3]. They give a reaction scheme that takes an arbitrary finite probability distribution and encodes it in the stationary distribution of a reaction system. In comparison, we are taking samples from a marginal distribution and encoding the full distribution in terms of the stationary distribution. Thus our scheme allows us to do conditioning and inference.

In Gopalkrishnan [13], one of the present authors has proposed a molecular scheme to do Maximum Likelihood Estimation in Log-Linear models. The reaction networks employed in that work are essentially identical to the reaction networks employed in this work, modulo some minor technical differences. In that paper, the reaction networks were used to obtain M-projections (or reverse I-projections), and thereby to solve for Maximum Likelihood Estimators. In this paper, we obtain E-projections, and sample from conditional distributions. The results in that paper were purely at the level of deterministic mass-action kinetics. The results in this paper obtain at the level of stochastic behavior.

7 Discussions

We have shown that reaction networks are particularly well-adapted to perform E-projections. In a previous paper [13], one of the authors has shown how to perform M-projections with reaction networks. Intuitively, an E-projection corresponds to a "rationalist" who interprets observations in light of previous beliefs, and an M-projection corresponds to an "empiricist" who forms new beliefs in light of observations.

Not surprisingly, these two complementary operations keep appearing as blocks in various statistical algorithms. Our two schemes should be viewed together as building blocks for implementing more sophisticated statistical algorithms. For example, the **EM algorithm** works by alternating E and M projections [15]. If our two reaction networks are coupled so that the point q is obtained by the scheme in [13], and the initialization of the scheme in this paper is used to perturb the conservation class for the M-projection correctly, then an "interior point" version of the EM algorithm may be possible, though perhaps not with detailed balance but in a "driven" manner reminiscent of futile cycles.

We have illustrated how E-projections might apply to the situation of an artificial cell trying to infer its environment from partial observations. We are acutely aware that our illustration is far from complete. A more sophisticated algorithm would work in an "online" fashion, adjusting its estimates on the fly to each new receptor activation event. This certainly appears within the scope of the kind of schemes we have outlined, but more careful design and analysis is necessary before formal theorems in this direction can be shown. Also we think it likely that the schemes that will prove most effective will work neither purely in the regime of the first scheme, nor purely in the regime of the second scheme, but somewhere in between. How long a time window they average over, and how large a volume is optimal, and how these choices tradeoff between sensitivity and reliability, these are questions for further analysis.

One glaring gap in our narrative is that we require the internal species X_i to be as numerous as the outside ligands L_i. A much more efficient encoding of ligand population vectors should be possible, drawing on ideas from graphical models, so that the number of representing species need only be a logarithm of the number of ligands being represented. Moreover it may be possible to perform E and M projections directly on these graphical model representations.

Our constructions and results of Sect. 5.1 were carried out for arbitrary matrices with rational number entries. We only used the assumption of "tidy" matrices to set initial conditions in Theorems 3 and 6. If some other method of setting initial conditions correctly is available, for example by performing matrix inversions with a reaction network, then the technical condition of tidy matrices can be dropped. In defence of the assumption that our sensitivity matrices are tidy, it is not inconceivable that through evolution a biological cell would have evolved its receptors so that the sensitivity matrix allows for simple meaningful messages to be transmitted inside the cell. For example, if for each receptor, there exists at least one ligand, that attaches to that receptor and no other, then the sensitivity matrix is tidy, and all the messages are unit vectors.

Note that the mathematics does not require the restriction of the sensitivities s_{ij} to **nonnegative** rational numbers. We could have admitted negative numbers, and all our results would go through.

Acknowledgements. Work of Abhishek Behera was supported in part by Bharti Centre for Communication in IIT Bombay.

References

1. Anderson, D.F., Craciun, G., Kurtz, T.G.: Product-form stationary distributions for deficiency zero chemical reaction networks. Bull. Math. Biol. **72**(8), 1947–1970 (2010)

2. Buisman, H.J., ten Eikelder, H.M.M., Hilbers, P.A.J., Liekens, A.M.L., Liekens, A.M.L.: Computing algebraic functions with biochemical reaction networks. Artif. Life **15**(1), 5–19 (2009)
3. Cardelli, L., Kwiatkowska, M.Z., Laurenti, L.: Programming discrete distributions with chemical reaction networks. CoRR, abs/1601.02578 (2016)
4. Cencov, N.N.: Statistical Decision Rules and Optimal Inference. Translations of Mathematical Monographs. American Mathematical Society, New York (2000)
5. Craciun, G., Toric differential inclusions, a proof of the global attractor conjecture. arXiv preprint arXiv:1501.02860 (2015)
6. Csiszár, I., Shields, P.C., et al.: Information theory and statistics: a tutorial. Found. Trends® Commun. Inf. Theor. **1**(4), 417–528 (2004)
7. Daniel, R., Rubens, J.R., Sarpeshkar, R., Lu, T.K.: Synthetic analog computation in living cells. Nature **497**(7451), 619–623 (2013)
8. Dembo, A., Zeitouni, O.: Large Deviations Techniques and Applications. Stochastic Modelling and Applied Probability, vol. 38. Springer, Heidelberg (2010)
9. Dupuis, P., Ellis, R.S.: A Weak Convergence Approach to the Theory of Large Deviations, vol. 902. Wiley, New York (2011)
10. Feinberg, M.: On chemical kinetics of a certain class. Arch. Rational Mech. Anal. **46**, 1–41 (1972)
11. Feinberg, M.: Lectures on chemical reaction networks (1979). http://www.che.eng. ohio-state.edu/FEINBERG/LecturesOnReactionNetworks/
12. Gopalkrishnan, M.: Catalysis in reaction networks. Bull. Math. Biol. **73**(12), 2962–2982 (2011)
13. Gopalkrishnan, M.: A scheme for molecular computation of maximum likelihood estimators for log-linear models. In: Rondelez, Y., Woods, D. (eds.) DNA 2016. LNCS, vol. 9818, pp. 3–18. Springer, Cham (2016). doi:10.1007/978-3-319-43994-5_1
14. Horn, F.J.M.: Necessary and sufficient conditions for complex balancing in chemical kinetics. Arch. Rational Mech. Anal. **49**(3), 172–186 (1972)
15. Amari, S.: Information Geometry and its Applications, 7th edn. Springer, Osaka (2016)
16. Jaynes, E.T.: Information theory and statistical mechanics. Phys. Rev. **106**(4), 620 (1957)
17. Kullback, S.: Information Theory and Statistics. Courier Corporation, New York (1997)
18. Miller, E.: Theory and applications of lattice point methods for binomial ideals. In: Combinatorial Aspects of Commutative Algebra and Algebraic Geometry, pp. 99–154. Springer, Heidelberg (2011)
19. Napp, N.E., Adams, R.P.: Message passing inference with chemical reaction networks. In: Advances in Neural Information Processing Systems, pp. 2247–2255 (2013)
20. Oishi, K., Klavins, E.: Biomolecular implementation of linear I/O systems. Syst. Biol. IET **5**(4), 252–260 (2011)
21. Qian, L., Winfree, E.: A simple DNA gate motif for synthesizing large-scale circuits. J. R. Soc. Interface **8**(62), 1281–1297 (2011)
22. Qian, L., Winfree, E.: Scaling up digital circuit computation with DNA strand displacement cascades. Science **332**(6034), 1196–1201 (2011)
23. Whittle, P.: Systems in Stochastic Equilibrium. Wiley, New York (1986)
24. Zwicker, D., Murugan, A., Brenner, M.P.: Receptor arrays optimized for natural odor statistics. In: Proceedings of the National Academy of Sciences, p. 201600357 (2016)

Complexities for High-Temperature Two-Handed Tile Self-assembly

Robert Schweller, Andrew Winslow$^{(\boxtimes)}$, and Tim Wylie

University of Texas - Rio Grande Valley, Edinburg, TX 78539, USA
{robert.schweller,andrew.winslow,timothy.wylie}@utrgv.edu

Abstract. Tile self-assembly is a formal model of computation capturing DNA-based nanoscale systems. Here we consider the popular *two-handed tile self-assembly model* or *2HAM*. Each 2HAM system includes a *temperature* parameter, which determines the threshold of bonding strength required for two assemblies to attach. Unlike most prior study, we consider general temperatures not limited to small, constant values. We obtain two results. First, we prove that the computational complexity of determining whether a given tile system uniquely assembles a given assembly is coNP-complete, confirming a conjecture of Cannon et al. (2013). Second, we prove that larger temperature values decrease the minimum number of tile types needed to assemble some shapes. In particular, for any temperature $\tau \in \{3, \dots\}$, we give a class of shapes of size n such that the ratio of the minimum number of tiles needed to assemble these shapes at temperature τ and any temperature less than τ is $\Omega(n^{1/(2\tau+2)})$.

1 Introduction

This work considers problems in a variation of *DNA tile self-assembly*, an approach for precise control of nanoscale structures that uses DNA base-pair interactions between four-sided DNA molecules first introduced by Seeman [23] and formalized by Winfree [27] as the mathematical *abstract Tile Assembly Model* or *aTAM*.

The wide range of complex and useful behaviors of the aTAM has since been established, including the model's ability to execute any algorithm [27] and assemble desired shapes using few tile types [2,20,25]. Since then, dozens of tile assembly models sharing traits with the aTAM have been studied, even giving rise to a structural complexity theory for tile assembly models [28].

Two-Handed Assembly. One of the most popular models of tile self-assembly is the *two-handed tile assembly model (2HAM)* [1,4,11,19], also referred to the *hierarchical* [5,9] or *polyomino* [15] model. The 2HAM differs from the original aTAM in its lack of a "seed": in the aTAM, assembly is limited to single-tile

---

This research was supported in part by National Science Foundation Grant CCF-1555626.

R. Brijder and L. Qian (Eds.): DNA 23 2017, LNCS 10467, pp. 98–109, 2017.
DOI: 10.1007/978-3-319-66799-7_7

addition to a growing seed assembly, while in the 2HAM, assembly may occur by attachment of *any* two assemblies via bonds of sufficient strength. The difficulty of experimentally enforcing seeded assembly [21] motivates the study of the 2HAM.

Temperature. A recurring question in many model variations, including the 2HAM, is the importance of *temperature*: the threshold of bonding strength needed for attachment between assemblies. A long-standing open problem in tile assembly concerns the capabilities of systems at the lowest temperature, where one bond suffices for attachment [12, 16–18]. Dynamically varied temperature has also been studied as a mechanism for guiding assembly [14, 26].

In the aTAM, systems at higher temperatures exhibit a greater range of *dynamics*: behaviors that occur during the assembly process [6], and these additional behaviors can be used to reduce the *tile complexity* of some shapes: the number of tile types needed to assemble the shape [24]. On the other hand, if scaling (replacement of each tile by a square block of tiles) is permitted, then these additional dynamics (and corresponding reductions in tile complexity) can be recreated or *simulated* by lower temperature systems [10]. In contrast, higher temperature 2HAM systems exhibit additional dynamics that cannot be simulated by lower temperature 2HAM systems [8].

Our Results. This work considers whether the additional dynamics in higher temperature 2HAM systems confer additional capabilities. We prove two results, one complexity theoretic and the other combinatorial, that give positive evidence.

The first result (Sect. 3) affirms a conjecture from 2013 [4] regarding the complexity of *verifying* that a system yields a unique specified terminal assembly. The proof critically uses high-temperature dynamics to demonstrate that such verification is coNP-hard.

The second result (Sect. 4) proves that for some shapes, higher temperatures yield more efficient assembly. Specifically, the ratio between the tile complexities of some shapes at temperature τ and any lower temperature is polynomial in the shape size. Seki and Ukuno [24] achieved a similar result in the aTAM, but for only a constant additive gap in tile complexity.

2 Definitions

Here we give a presentation of the two-handed tile assembly model (2HAM) and associated definitions used throughout the paper.

2.1 Tiles and Assemblies

Tiles. A tile is an axis-aligned unit square centered at a point in $\mathbb{Z}^2$, where each edge is labeled by a *glue* selected from a glue set Π. A *strength function* str : $\Pi \rightarrow \mathbb{N}$ denotes the *strength* of each glue. Two tiles that are equal up to translation have the same *type*.

Assemblies. A *positioned shape* is any subset of $\mathbb{Z}^2$. A *positioned assembly* is a set of tiles at unique coordinates in $\mathbb{Z}^2$, and the positioned shape of a positioned assembly A is the set of coordinates of those tiles.

For a given positioned assembly Υ, define the *bond graph* G_Υ to be the weighted grid graph in which each tile of Υ is a vertex and the weight of an edge between tiles is the strength of the matching coincident glues or 0.[1] A positioned assembly C is said to be τ-*stable* for positive integer τ provided the bond graph G_C has minimum edge cut at least τ.

For a positioned assembly A and integer vector $\boldsymbol{v} = (v_1, v_2)$, let $A_{\boldsymbol{v}}$ denote the assembly obtained by translating each tile in A by vector $\boldsymbol{v}$. An *assembly* is a set of all translations $A_{\boldsymbol{v}}$ of a positioned assembly A. A *shape* is the set of all integer translations for some subset of $\mathbb{Z}^2$, and the shape of an assembly A is the shape consisting of the set of all the positioned shapes of all positioned assemblies in A. The *size* of either an assembly or shape X, denoted as $|X|$, refers to the number of elements of any positioned element of X.

Combinable Assemblies. Informally, two assemblies are τ-*combinable* provided they may attach to form a τ-stable assembly. Formally, two assemblies A and B are τ-*combinable* into an assembly C provided there exists $A' \in A$ and $B' \in B$ such that $A' \bigcup B'$ is a τ-stable element of C.

2.2   Two-Handed Tile Assembly Model (2HAM)

A *two-handed tile assembly system (2HAM system)* is an ordered pair (T, τ) where T is a set of single tile assemblies, called the *tile set*, and $\tau \in \mathbb{N}$ is the *temperature*. Assembly proceeds by repeated combination of assembly pairs to form new τ-stable assemblies, starting with single-tile assemblies. The *producible assemblies* are those constructed in this way. Formally:

Definition 1 (2HAM producibility). *For a given 2HAM system $\Gamma = (T, \tau)$, the set of* producible assemblies *of Γ, denoted PROD$_\Gamma$, is defined recursively:*

- *(Base) $T \subseteq$ PROD$_\Gamma$*
- *(Recursion) For any $A, B \in$ PROD$_\Gamma$ such that A and B are τ-combinable into C, then $C \in$ PROD$_\Gamma$.*

For a system $\Gamma = (T, \tau)$, we say $A \to_1^\Gamma B$ for assemblies A and B if A is τ-combinable with some producible assembly to form B, or if $A = B$. Intuitively this means that A may grow into assembly B through one or fewer combination reactions. We define the relation $\to^\Gamma$ to be the transitive closure of $\to_1^\Gamma$, i.e., $A \to^\Gamma B$ means that A may grow into B threw a sequence of combination reactions.

---

[1] Note that only matching glues have positive strength. The more general model of "flexible glues" where non-matching glue pairs may also have positive strength has been considered [7].

Definition 2 (Terminal assemblies). *A producible assembly A of a 2HAM system $\Gamma = (T, \tau)$ is terminal provided A is not τ-combinable with any producible assembly of Γ.*

Definition 3 (Unique assembly). *A 2HAM system uniquely assembles an assembly A if for all $B \in \text{PROD}_\Gamma$, $B \rightarrow^\Gamma A$.*

3   Unique Assembly Verification in the 2HAM Is coNP-Complete

Definition 4 (Unique assembly verification (UAV) problem). *Given a 2HAM system Γ and assembly A, does Γ uniquely assemble A?*

Adleman et al. [3] proved that the UAV problem in the aTAM is in P. Cannon et al. [4] first considered the UAV problem in the 2HAM. They proved that the problem is in coNP and conjectured that the problem is coNP-hard, suggested by their proof of the same result for an extension of the model to three dimensions (with cubic tiles). Here we confirm their conjecture, using high temperature to overcome previous planarity "barriers".

Theorem 1. *The UAV problem is coNP-complete.*

The reduction is from a problem involving *grid graphs*: graphs whose vertices are a subset of $\mathbb{Z}^2$ and two vertices are connected by an (undirected) edge if they have distance 1. Itai, Papadimitriou, and Szwarcfiter [13] proved that the following problem is NP-hard:

Definition 5 (Hamiltonian cycle problem in grid graphs). *Given a grid graph G, does G contain a Hamiltonian cycle?*

We reduce from the complement of this problem.

Lemma 1. *The UAV problem in the 2HAM is coNP-hard.*

Proof. Consider a grid graph $G = (V, E)$. From G we construct a tile system Γ_G and an assembly A_G such that Γ_G uniquely assembles A_G if an only if G has no Hamiltonian cycle. Without loss of generality, assume the leftmost and rightmost vertices of G have x-values 0 and n, and the bottommost and topmost vertices have y-values 0 and m, respectively. Construct a tile set T_G from G as described in Fig. 1 to yield the system $\Gamma_G = (T_G, \tau = |V|)$.

The system Γ_G has a terminal assembly A_G consisting of a $2(n+1) \times 2(m+1)$ block of blue tiles connected to a $2(n-1) \times 2(m-1)$ block of red tiles, as shown in Fig. 2. We also claim that this is the unique terminal assembly of Γ_G if and only if G has no Hamiltonian cycle.

Correctness: G has cycle $\Rightarrow$ No unique terminal assembly. Suppose G has a Hamiltonian cycle. Then there exists a producible assembly C_{inner} of red 3×3 blocks

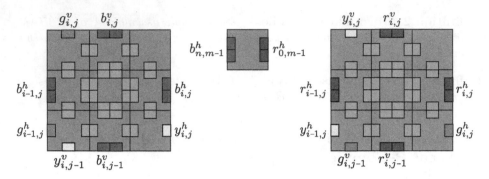

Fig. 1. This tileset consists of a collection of 3×3 blocks and a single connector tile. The center tile of each 3×3 block has bond strength of τ with its four neighbors. Each corner tile bonds to its horizontal and vertical neighbors with $\lceil \tau/2 \rceil$, $\lfloor \tau/2 \rfloor$ strength, respectively. A blue block is constructed for every location in $\{0, \ldots, n\} \times \{0, \ldots, m\}$, and a red block is constructed for locations in $\{1, \ldots, n-1\} \times \{1, \ldots, m-1\}$. Red and blue glues have strength τ, while green and yellow glues have strength 1 or 0 as determined by the grid graph: $g_{i,j}^h$ and $y_{i,j}^h$ have strength 1 if (i,j) and $(i, j-1)$ are vertices in G and strength 0 otherwise. The glues $g_{i,j}^v$ and $y_{i,j}^v$ have strength 1 when (i,j) and $(i-1,j)$ are vertices in G and strength 0 otherwise. (Color figure online)

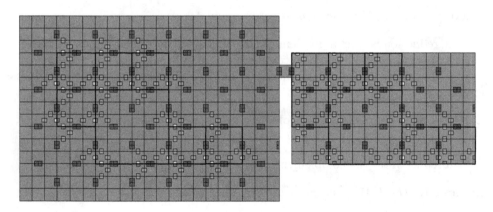

Fig. 2. For a given grid graph, the following assembly is the unique terminal assembly if and only if no Hamiltonian cycle exists. (Color figure online)

corresponding to the interior of the cycle. By design, C_{inner} has exactly $\tau = |V|$ yellow and green glues exposed. Similarly, there exists a producible assembly C_{outer} of blue 3×3 blocks corresponding to the exterior of the cycle with $\tau = |V|$ yellow and green glues in the same relative locations as those of C_{inner}. At temperature τ, C_{inner} and C_{outer} attach to form a large assembly that is *not* a subassembly of the previously described terminal assembly. See Fig. 3 for such a pair combinable C_{inner} and C_{outer} and the grid graph they correspond to.

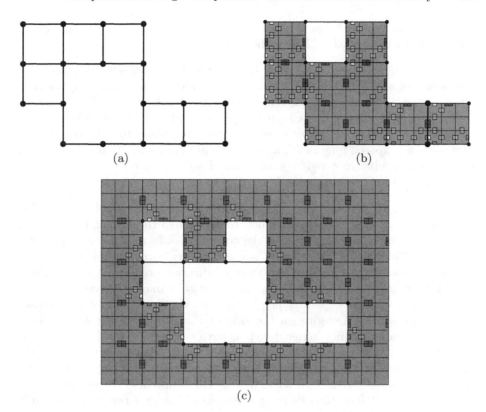

Fig. 3. (a) The input graph for the example reduction. If a Hamiltonian cycle exists, then the producible assemblies consisting of C_{outer} (c), the exterior of the cycle, and C_{inner} (b), the interior of the cycle, are combinable with exactly τ strength.

Correctness: No unique terminal assembly $\Rightarrow$ G has cycle. Now suppose that $\Gamma_G = (T_G, \tau = |V|)$ does not have a unique assembly, i.e., there exists some producible assembly X that cannot assemble further into A_G (equivalently, is not a subassembly of A_G. The existence of X implies that there exists some producible assembly R, consisting of red blocks, that is attachable to a producible assembly of blue blocks by way of τ or more green and yellow glues. We first use R to construct a second "cleaned-up" producible assembly R' that is also attachable to a producible assembly of blue blocks. We then show that R' implies a Hamiltonian cycle determined by its shape.

Consider modifying the assembly R in the following way. First, if R contains the center tile for any 3×3 block of red tiles, add all missing tiles of the corresponding block. Second, for all other blocks, remove all tiles of this block from R. Call the resulting assembly of completed 3×3 blocks R'.

The assembly R' has the following properties. First, R' is producible. In particular, the removal of tiles from R as specified cannot disconnect the assembly.

Second, R' may attach to a producible blue assembly by way of yellow and green glues summing to at least τ. This is the case because:

- No tile removed from R to yield R' is a corner tile of a block, since the presence of any corner tile of a block in a producible assembly implies that the assembly also contains the center tile of the block (and so R' contains all tiles from such a block if R contained any corner tile of the block).
- Any exposed green or yellow glue of R used to attach to a blue assembly remains an exposed glue in R', as any such glue is adjacent to a 3×3 block containing a blue (not red) center tile and so cannot be "covered" by the addition of tiles to R to create R'.

We now use R', an assembly that is producible and combinable with a blue assembly through yellow and green glues, to generate a Hamiltonian cycle in G. Consider the polyomino consisting of the collection of faces of G corresponding to each 3×3 block in R'. Starting at some arbitrary corner of this polyomino, walk its perimeter to generate a sequence of distinct points $p_0, \ldots p_{r-1}$. Each consecutive pair are adjacent in G, but points may or may not be in V.

For each consecutive pair $p_i, p_{(i+1) \bmod r} \in V$, the assembly R' exposes exactly 1 green or yellow glue on the side of the corresponding 3×3 block. On the other hand, for any consecutive pair with either point not in V, no green or yellow glues are exposed. Then since no location repeats and the total number of green and yellow glues must sum to at least τ (for attachment to a blue assembly), the sequence must be a length-V permutation of V where consecutive points are adjacent (and thus share an edge), implying that this permutation is a Hamiltonian cycle of G.   □

4   Tile Complexity Gaps Between Temperatures

The *tile complexity* of a shape S at temperature τ is the minimum number of tile types in a 2HAM system at temperature τ that uniquely assembles S. The *tile complexity gap* of a shape S between two temperatures τ_1 and τ_2 is the ratio of the minimum number of tile types in 2HAM systems at temperatures τ_1 and τ_2 that uniquely assemble S. Here we give, for any distinct pair of temperatures, an infinite family of shapes with large tile complexity gap at these temperatures.

We start by describing the construction for the special case of $\tau_1 = 2$ and $\tau_2 = 3$, shown in Fig. 5. The shapes each consist of a base rectangle and all gadgets of the form shown in Fig. 4 for some integer $m \geq 3$.

The gadget has three horizontal *sections* with m locations where the vertical bar connects with the base. At most m tiles are used for the left and right vertical column ($2m$ for the center column), and the height difference between any two of the three horizontal sections is at most $2m - 1$. There are m^6 different gadgets, since each section has m^2 possible column locations.

Theorem 2. *There exists a shape of size n with a tile complexity gap of $\Omega((\frac{n}{\log n})^{1/7})$ between $\tau = 2$ and $\tau = 3$.*

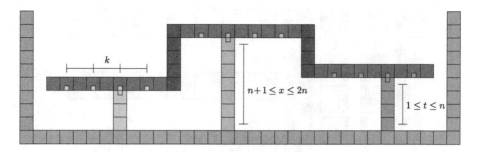

Fig. 4. A single gadget on the shape with m possible glue locations and m possible heights for each vertical bar. Note the spacing between horizontal glues ensures that the "hat" can not attach to the shape shifted because of the walls.

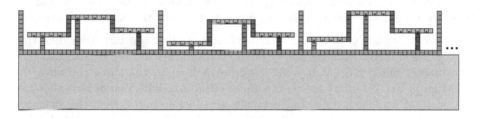

Fig. 5. The terminal assembly shape which consists of a rectangle used to seed all possible gadgets to attach to the top.

Proof. Since there are $\Theta(m^6)$ possible gadgets, and each gadget is $\Theta(m)$ tiles in width, the width of the shape is $\Theta(m^7)$ tiles. The base rectangle of the shape is a $\Theta(\log m) \times \Theta(m^7)$ rectangle requiring $\Theta(\log m)$ tile types. Thus, the shape contains $\Theta(m^7 \log m)$ total tiles.

The remainder of the proof is dedicated to proving that (1) the shape can be assembled at $\tau = 3$ using $O(m)$ tile types, and (2) requires $\Omega(m^2)$ tile types at $\tau = 2$. Thus the tile complexity gap is $\Omega(m)$. Since the size of S is $n = \Theta(m^7 \log m)$, $\Omega(m) = \Omega((\frac{n}{\log n})^{1/7})$.

The Tile Complexity at $\tau = 3$ is $O(m)$. All hat assembly can be assembled using the same $O(m)$ tile types at $\tau = 3$ as follows. Each of the three horizontal sections is built deterministically with m strength-1 glues exposed on the south side. The two vertical columns connecting the sections are assembled nondeterministically and may have any length from 2 to $2m$. This means every possible configuration is built ($4m^2$ hats). The three columns are seeded from the base and expose a strength-1 glue matching their respective section.

The hat can only attach if all three glues can match (columns and hat sections). The tiles for constructing the hat piece are shown in Fig. 6. Note the spacing tiles in between each horizontal glue tile to ensure that the hat attaches without shifting left or right (because of the enclosing walls). Such gadgets can be assembled from $3(2m+1)$ tile types for the horizontal sections, and $2(4m-2)$

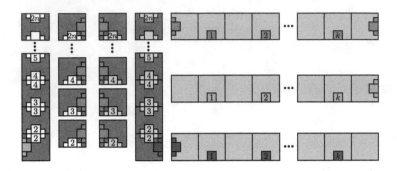

Fig. 6. Building the "hat" for the gadget nondeterministically. The single blocks represent a strength 1 bond and the three blocks a strength-3 bond.

tile types for the vertical connecting strips. The columns from the base use $4m$ tiles. Since we use these same tiles for every gadget, the tile complexity is $\Theta(m)$.

The Tile Complexity at $\tau = 2$ is $\Omega(m^2)$. Assembling the hat using few tiles at $\tau = 2$ is difficult because only 2 of the 3 columns can ever be necessary for attachment. Since the hat is built before attaching to the three columns, the situation in Fig. 7(a) may occur, or similar situations with one of the other two columns not attached. Since hat attaches in multiple parts, then the situation in Fig. 7(b) may occur, or a similar situation with some parts translated. Thus the same tile set cannot be used for the hats in all gadgets (Fig. 8).

Thus each section must be assembled with only one south glue placed in the correct tile where the column attaches. Then m versions of that gadget are built (for each column attachment location) so that the section with glue g_i, where $1 \leq i \leq m$, exposed can attach whenever glue g_i is open on one of the columns. In order for the section piece to not attach shifted (Fig. 7(b)), the column must expose a corresponding glue for that horizontal position. This means for each horizontal position, we need m distinct deterministic tiles so that we can expose the correct g_i glue at the top of the column to attach the correct section without it being shifted. Thus, $\Omega(m^2)$ tile types are required.   □

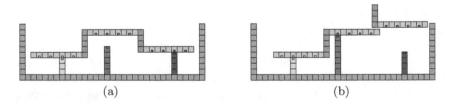

| (a) | (b) |

Fig. 7. (a) Strength-1 glues at $\tau = 2$ cannot be used, otherwise the hat may attach to the wrong gadget. (b) The hat can not be attached in separate pieces if the sections are the same for each gadget, since it may also attach shifted, and thus the walls prevent the rest of the hat from attaching.

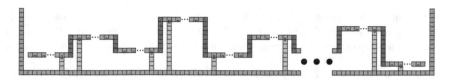

Fig. 8. Generalizing the shape for any τ by utilizing τ hats with τ glues per section.

Theorem 3. *For any $\tau_1, \tau_2 \in \{2, 3, \dots\}$ with $\tau_1 < \tau_2$, there exists a shape of size n with tile complexity gap $\Omega(n^{1/(2\tau_2+2)})$ between τ_1 and τ_2.*

Proof. This follows from a similar analysis as the proof of Theorem 2. Since there are τ_2 sections of the hat piece, then there are $\Theta(m^{2\tau_2})$ gadgets, each of width $\Theta(\tau_2 m)$. So the width of the shape is $\Theta(\tau_2 m^{2\tau_2+1})$, and the size of the shape is $n = \Theta(m^{1/(2\tau_2+1)} \log \tau_2 m)$. Following the same argument as given in the proof of Theorem 2, $\Omega(m^2)$ tile types are needed to assemble the gadget correctly for any $\tau_1 < \tau_2$. $\qquad\qquad\square$

5   Conclusion

There are a number of interesting directions to extend this work. First, while we have shown the UAV problem is coNP-complete, our reduction requires temperature to scale linearly in the assembly size. Since many systems of interest have small, and even constant temperature, we ask: does coNP-hardness hold for constant, or even logarithmic temperatures? When the model is extended to 3D, the answer is "Yes" for temperature $\tau = 2$ [4].

Our coNP-completeness result also pairs well with other recent results on verification problems in two-handed models of verification. For instance, that the *unique shape verification* or *USV problem* is coNP^{NP}-complete [22]. Similarly, in the more powerful *staged assembly model*, the UAV and USV problems are coNP^{NP}-hard and in PSPACE [22]. In this case, coNP^{NP}-hardness is known to hold even for $\tau = 2$ and constant stages and bins (additional complexity measures in the staged model), but characterizing the complexity as a function of the number of stages remains open.

References

1. Abel, Z., Benbernou, N., Damian, M., Demaine, E.D., Demaine, M.L., Flatland, R., Kominers, S.D., Schweller, R.T.: Shape replication through self-assembly and RNase enzymes. In: Proceedings of the Twenty-First Annual ACM-SIAM Symposium on Discrete Algorithms, pp. 1045–1064 (2010)
2. Adleman, L., Cheng, Q., Goel, A., Huang, M.-D.: Running time and program size for self-assembled squares. In: Proceedings of the 33rd Annual ACM Symposium on Theory of Computing (STOC), pp. 740–748 (2001)

3. Adleman, L.M., Cheng, Q., Goel, A., Huang, M.-D.A., Kempe, D., de Espanés, P.M., Rothemund, P.W.K.: Combinatorial optimization problems in self-assembly. In: Proceedings of the Thirty-Fourth Annual ACM Symposium on Theory of Computing, pp. 23–32 (2002)
4. Cannon, S., Demaine, E.D., Demaine, E.D., Eisenstat, S., Patitz, M.J., Schweller, R., Summers, S.M., Winslow, A.: Two hands are better than one (up to constant factors): self-assembly in the 2HAM vs. aTAM. In: Proceedings of 30th International Symposium on Theoretical Aspects of Computer Science (STACS). LIPIcs, vol. 20, pp. 172–184. Schloss Dagstuhl (2013)
5. Chen, H.-L., Doty, D.: Parallelism and time in hierarchical self-assembly. In: Proceedings of the 23rd Annual ACM-SIAM Symposium on Discrete Algorithms, SODA 2012, pp. 1163–1182. SIAM (2012)
6. Chen, H.-L., Doty, D., Seki, S.: Program size and temperature in self-assembly. Algorithmica **72**(3), 884–899 (2015)
7. Cheng, Q., Aggarwal, G., Goldwasser, M.H., Kao, M.-Y., Schweller, R.T., de Espanés, P.M.: Complexities for generalized models of self-assembly. SIAM J. Comput. **34**, 1493–1515 (2005)
8. Demaine, E.D., Patitz, M.J., Rogers, T.A., Schweller, R.T., Summers, S.M., Woods, D.: The two-handed tile assembly model is not intrinsically universal. Algorithmica **74**(2), 812–850 (2016)
9. Doty, D.: Producibility in hierarchical self-assembly. Nat. Comput. **15**(1), 41–49 (2016)
10. Doty, D., Lutz, J.H., Patitz, M.J., Schweller, R., Summers, S.M., Woods, D.: The tile assembly model is intrinsically universal. In: Proceedings of the 53rd IEEE Conference on Foundations of Computer Science (FOCS), pp. 302–310 (2012)
11. Doty, D., Patitz, M.J., Reishus, D., Schweller, R.T., Summers, S.M.: Strong fault-tolerance for self-assembly with fuzzy temperature. In: Proceedings of the 51st Annual IEEE Symposium on Foundations of Computer Science (FOCS 2010), pp. 417–426 (2010)
12. Doty, D., Patitz, M.J., Summers, S.M.: Limitations of self-assembly at temperature one. In: Deaton, R., Suyama, A. (eds.) DNA 2009. LNCS, vol. 5877, pp. 35–44. Springer, Heidelberg (2009). doi:10.1007/978-3-642-10604-0_4
13. Itai, A., Papadimitriou, C.H., Szwarcfiter, J.L.: Hamilton paths in grid graphs. SIAM J. Comput. **11**(4), 676–686 (1982)
14. Kao, M.-Y., Schweller, R.T.: Reducing tile complexity for self-assembly through temperature programming. In: Proceedings of the 17th Annual ACM-SIAM Symposium on Discrete Algorithms, SODA 2006, pp. 571–580 (2006)
15. Luhrs, C.: Polyomino-safe DNA self-assembly via block replacement. Nat. Comput. **9**(1), 97–109 (2010)
16. Maňuch, J., Stacho, L., Stoll, C.: Two lower bounds for self-assemblies at temperature 1. J. Comput. Biol. **16**(6), 841–852 (2010)
17. Meunier, P.-E.: The self-assembly of paths and squares at temperature 1. Technical report, arXiv (2013). http://arxiv.org/abs/1312.1299
18. Meunier, P.-E., Patitz, M.J., Summers, S.M., Theyssier, G., Winslow, A., Woods, D.: Intrinsic universality in tile self-assembly requires cooperation. In: Proceedings of the 25th Annual ACM-SIAM Symposium on Discrete Algorithms (SODA), pp. 752–771 (2014)
19. Patitz, M.J., Rogers, T.A., Schweller, R.T., Summers, S.M., Winslow, A.: Resiliency to multiple nucleation in temperature-1 self-assembly. In: Rondelez, Y., Woods, D. (eds.) DNA 2016. LNCS, vol. 9818, pp. 98–113. Springer, Cham (2016). doi:10.1007/978-3-319-43994-5_7

20. Rothemund, P.W.K., Winfree, E.: The program-size complexity of self-assembled squares (extended abstract). In: Proceedings of the 32nd ACM Symposium on Theory of Computing (STOC), pp. 459–468 (2000)
21. Schulman, R., Winfree, E.: Programmable control of nucleation for algorithmic self-assembly. SIAM J. Comput. **39**(4), 1581–1616 (2009)
22. Schweller, R., Winslow, A., Wylie, T.: Verification in staged tile self-assembly. In: Patitz, M.J., Stannett, M. (eds.) UCNC 2017. LNCS, vol. 10240, pp. 98–112. Springer, Cham (2017). doi:10.1007/978-3-319-58187-3_8
23. Seeman, N.C.: Nucleic-acid junctions and lattices. J. Theor. Biol. **99**, 237–247 (1982)
24. Seki, S., Ukuno, Y.: On the behavior of tile assembly system at high temperatures. Computability **2**(2), 107–124 (2013)
25. Soloveichik, D., Winfree, E.: Complexity of self-assembled shapes. SIAM J. Comput. **36**(6), 1544–1569 (2007)
26. Summers, S.M.: Reducing tile complexity for the self-assembly of scaled shapes through temperature programming. Algorithmica **63**(1), 117–136 (2012)
27. Winfree, E.: Algorithmic self-assembly of DNA. Ph.D. thesis, California Institute of Technology (1998)
28. Woods, D.: Intrinsic universality and the computational power of self-assembly. Philos. Trans. R. Soc. A **373**, 2015 (2046)

A DNA Neural Network Constructed from Molecular Variable Gain Amplifiers

Sherry Xi Chen and Georg Seelig[(✉)]

University of Washington, Seattle, WA, USA
gseelig@uw.edu

Abstract. Biological nucleic acids have important roles as diagnostic markers for disease. The detection of just one molecular marker, such as a DNA sequence carrying a single nucleotide variant (SNV), can sometimes be indicative of a disease state. However, a reliable diagnosis and treatment decision often requires interpreting a combination of markers via complex algorithms. Here, we describe a diagnostic technology based on DNA strand displacement that combines single nucleotide specificity with the ability to interpret the information encoded in panels of single-stranded nucleic acids through a molecular neural network computation. Our system is constructed around a single building block—a catalytic amplifier with a competitive inhibitor or "sink." In previous work, we demonstrated that such a system can be used to reliably detect SNVs in single stranded nucleic acids. Here, we show that these same building blocks can be reconfigured to create an amplification system with adjustable gain α. That is, the concentration of an output signal produced is exactly α times larger than the concentration of input added initially, and the value of α can be adjusted experimentally. Finally, we demonstrate that variable gain amplification and mismatch discrimination elements can be combined into a two-input neural network classifier. Together, our results suggest a novel approach for engineering molecular classifier circuits with predictable behaviors.

Keywords: DNA strand displacement · Linear classifier · Variable gain amplifier · Neural network · microRNA · Competitive inhibition

1 Introduction

Competitive inhibition is used throughout biology as a means for tuning the activity of enzymes and reshaping the response curves of signaling pathways or gene regulatory networks. For example, a competitive decoy RNA can convert an approximately linear relationship between the abundance of a small regulatory RNA and its mRNA target into an effectively sigmoidal, threshold linear one [1]. Similarly, inactivation of Notch by intracellular Delta ligand sets a threshold for the interaction of Notch with Delta in neighboring cells. The resulting nonlinearity ensures that cells reliably settle into one of two distinct cell fates, acting either as "senders" or "receivers" of the signal [2].

© Springer International Publishing AG 2017
R. Brijder and L. Qian (Eds.): DNA 23 2017, LNCS 10467, pp. 110–121, 2017.
DOI: 10.1007/978-3-319-66799-7_8

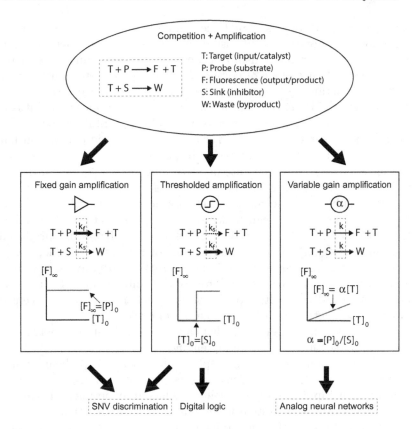

Fig. 1. Catalytic amplification can be combined with competitive inhibition to create a broad range of dynamic behaviors. Top: To qualitatively understand the different behaviors that can be achieved, we use a minimal one-step model of catalysis and inhibition. A catalyst (Target, T) can react with a substrate (Probe, P) to produce a product (Fluorescent product, F). Alternatively, T can also be bound irreversibly to a competitive inhibitor (the Sink, S), a reaction that produces an unreactive byproduct (waste, W). **Bottom left:** If the target has a strong kinetic preference for the amplification pathway, the sink has only a minimal impact on the overall reaction. In that limit, the target can convert all of the available probe to product. **Bottom middle:** If the competitive inhibition pathway is kinetically preferred, the sink acts as a threshold for the amplification reaction. If the initial concentration of the target exceeds that of the sink, all sink molecules will be used up and the remaining target molecules can trigger the amplification pathway. The gain is set by the amount of probe that is available since all probe molecules will be converted to product. Conversely, if the initial sink concentration is larger than the target concentration, all target molecules are irreversibly bound and no amplification reaction occurs. This competitive inhibition based on preferential binding of a target to a sink has been exploited to create digital logic gates and SNV discrimination systems. **Bottom right:** When the reaction rate constants for target binding to probe and sink are exactly the same, amplification occurs but the gain of the reaction is determined by the ratio of the initial concentrations of the probe to the sink. Importantly, the gain is independent of the initial amount of the target. Such variable gain amplification units form the basic building blocks for the molecular implementation of neural networks as shown here.

Common to these examples is that the competitor acts as a kinetically preferred threshold; only once it is exhausted can the productive reaction proceed, resulting in a non-linear relationship between the enzyme amount and product production rate. The degree of non-linearity is directly determined by levels of the different reactants, in particular the "threshold" concentration, and the kinetic rate constants. By concatenating multiple such non-linear units it is possible to create complex signaling networks with robust, digital behaviors.

Intriguingly, the same core reaction network—a catalytic reaction in parallel with a reaction that inhibits catalysis—responsible for producing non-linear input output curves can also generate perfectly linear ones [3] (see also Fig. 1). Specifically, when the catalyst has no kinetic preference between the inhibitor and productive substrate, the reaction network acts as a variable gain amplifier. That is, a given catalyst can undergo a fixed, but controllable number of catalytic cycles before being inactivated. The amount of product produced is a defined multiple (gain) of the initial amount of catalyst available; the gain is determined by the concentration ratio between the productive substrate and inhibitor (this parameter thus plays the role of the voltage in an electronic variable gain amplifier). The result, though not the mechanism, is reminiscent of a polymerase chain reaction with a fixed number of cycles.

The diversity of response functions that can be realized with competitive inhibition systems make them an intriguing engineering target (Fig. 1). For example, variable gain amplification or multiplication is a key ingredient of simple neural network models where different weights (synaptic strengths) are associated with different inputs to a given neuron. Thus an experimental realization of a fixed gain amplification system would enable the construction of molecular neural networks. However the fine-grained, quantitative control over reactant concentrations and rate constants required to achieve behaviors such as variable gain amplification may be difficult to realize in a biological system. Moreover, it may be challenging to create multiple, modular instances of the same motif, a requirement for the construction of multi-input neural networks.

DNA strand displacement cascades provide an alternative technology for the experimental realization of complex reaction pathways [4]. DNA strand displacement is a competitive hybridization reaction where an incoming DNA strand outcompetes an incumbent strand for binding to a complementary partner. The strand displacement mechanism was introduced to the DNA nanotechnology field by Yurke and co-workers as means to drive a DNA "tweezer," a DNA-based molecular motor, between an open and closed state [5]. Strand displacement is typically initiated through binding of short complementary toehold sequences and the reaction rate can be predictably tuned by controlling the length and sequence composition of the toehold [6,7]. Multiple such reactions can be concatenated to create multistep reaction cascades where the strand released in one strand displacement reaction acts as the input in a downstream reaction [8–10].

DNA strand displacement has already been used to create the key building blocks required for implementing variable gain amplifiers, including sequence-programmable catalytic amplification systems [8,10–12]. Moreover, competitive strand displacement was used by Li *et al.* [13] to convert differences in reaction rate constants between two DNA inputs into differences at the reaction end point, allowing them to build an SNV discrimination system. In recent work, we showed that combining a competitive sink with a catalytic reaction could result in quadratically better SNV discrimination than can be achieved with a non-catalytic probe or reporter [14]. Finally, Qian and Winfree used catalytic amplification together with competitive inhibition to create molecular logic gates with strongly non-linear input-output relationships that could be composed into multi-layered logic circuits [15] and neural networks [16].

Here, we build on these previous results to experimentally realize a DNA strand displacement-based variable gain amplifier. We then combine two such elements into a proof-of-principle neural network classifier that operates on two inputs with sequences of biological microRNA (miRNA). For our two-input classifier, we chose input sequences that differ by a single nucleotide to demonstrate that the expected output is produced even when very high specificity of reactions is required. The neural network architecture we describe here is distinct from the architecture used in prior work [16] in that neurons in the first layer do not act as digital units but can assign an arbitrary weight to an input. Moreover, the specific DNA architecture used also differs from a previously proposed alternative approach to implementing weighs in a neural network [17].

2 Results

2.1 DNA Implementation of a Variable Gain Amplifier

The catalytic probe P_{AMP} is implemented with an entropy driven amplification system as shown in Fig. 2A [12]. The inhibition system, or sink S_{DEG}, is implemented with a two stage cascade based on the same mechanism but using a truncated fuel as seen in Fig. 2B [14]. For experimental convenience we chose to incorporate a fluorophore quencher pair directly into the catalytic probe rather than using the translation scheme introduced in Ref. [12].

Both the amplification and inhibition reactions initiate with a reversible toehold exchange step; the activated probe or sink then react with their respective fuel species. The fuel for the catalytic probe displaces both the signal and releases the input. In contrast, the fuel for the sink cannot displace the input but rather irreversibly traps it. The catalytic and inhibition reactions are designed to have highly similar kinetics which is necessary to ensure that the gain is the same for all input concentrations [3]. Outer toeholds (Fig. 2, orange) on both probe and sink are identical by design. The inner toeholds (shown in pink for the amplification system, purple for the inhibition system) have different sequences to minimize crosstalk but similar binding energy to still ensure similar kinetics. A more detailed description and model of the sink can be found in Ref. [14].

2.2 Capturing Experimental Non-idealities

The model of Ref. [3] (see also Fig. 1) assumes that the amplifier is an ideal catalytic system. However, all strand displacement catalytic systems will deviate

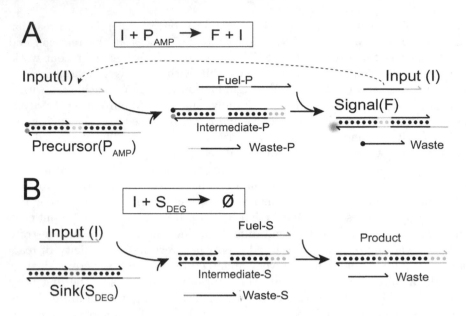

Fig. 2. DNA implementation of a linear classifier circuit. (A) We use a fluorescently labeled entropy driven catalytic system [12] as the amplifier. Input I reacts with the catalytic probe P_{AMP} to form waste and an intermediate complex, Intermediate-P. The latter then reacts with the fuel, Fuel-P to release the input I and also displace the quencher-labeled waste strand to form the double-stranded fluorescent signal species F. The input strand is then free to react with another amplifier complex P_{AMP} and repeat the catalytic cycle. **(B)** The sink is implemented with a two-stage reaction cascade using components almost identical to the amplification system to ensure similar kinetics. The sink S_{DEG} is identical to the amplifier at the domain level. However, the internal toehold sequence (purple) is distinct from that of the amplifier. The binding energies of the inner toeholds are similar to ensure linearity of amplification. The fuel species for the inhibition reaction, Fuel-S is designed not to release the input. Moreover, reaction with the sink does not result in a fluorescent signal. Sequences for all components can be found in Ref. [14]. (Color figure online)

from this assumption in several ways. First, DNA strand displacement systems exhibit "leak" whereby an output signal is produced even in the absence of an input. In our own experiments, this issue is compounded because we did not purify the DNA complexes. In practice, we thus expect that the final fluorescence signal has two components

$$F_{input} = F_{leak} + \alpha[I] \tag{1}$$

where F_{leak} is the background signal due to the leak and the second term represents the increase in signal due to the intended amplification reaction.

The gain α can be calculated from the initial concentrations of probe P_{AMP} and sink S_{DEG}:

$$\alpha = \frac{[P_{AMP}]_0}{[S_{DEG}]_0 + \frac{[P_{AMP}]_0}{C}}. \tag{2}$$

The constant C is the maximum turnover of the amplification system without any sink. This term reflects the experimental reality that strand displacement catalytic systems have an intrinsically limited turnover. The experimentally esti- mate this value, experiments can be performed with a large excess of probe P_{AMP} over the input I (and no sink). The effective turnover can then be calculated from the final fluorescent signal reached in that experiment.

Using this equation and an estimate for C we can predictably tune the gain α by, for example, changing the amount of sink S_{DEG} we add to the system for a fixed concentration of probe P_{AMP}. We also note that the amount of input should not exceed the amount of sink or probe to ensure proper operation.

2.3   Testing the Variable Gain Amplifier

To experimentally test our model, we designed two distinct variable gain sys- tems using the sequences of miRNAs let-7a and let-7c as inputs. These miRNA sequences differ in only a single nucleotide and we have previously used those same sequences to demonstrate specificity of a strand displacement-based SNV discrimination reaction [14]. Let-7 family miRNA have important roles in devel- opment and at least eight different family members are found in humans [18]. For our experiments, we used synthetic RNA oligonucleotides rather than biological miRNA.

For the let-7a system, the measured maximum turnover was approximately $C = 30$. We chose the initial concentrations $[P_{AMP}]_0 = 30\,\text{nM}$ and $[S_{DEG}]_0 = 9.3\,\text{nM}$ to achieve a gain $\alpha = 2.9$. We then performed experiments with five different concentrations of the let-7a RNA input: one experiment with input concentration of $1\,\text{nM}$, three with input concentration $2\,\text{nM}$ and one with input concentration $3\,\text{nM}$. The kinetics traces are shown on the left of Fig. 3A. As expected, the signal is higher with more input, and the three repeats with 2nM input show very similar kinetics, demonstrating good repeatability.

The net increment of signal due to the input is calculated by subtracting the signal due to the leak reaction from the total signal in the presence of the input, $F_{input} - F_{leak}$ (Eq. 1). This net steady state increment is plotted as a function of the input concentration on the right of Fig. 3A. We then fit the data to a straight line, in order to obtain a value for the experimentally observed gain α (the slope of the fitted line, Eq. 1). For our experiment, the slope was $\alpha = 2.86 \pm 0.06$ which agrees well with the target value of $\alpha = 2.9$.

We then performed a similar set of experiments with the let-7c system, using target gain $\alpha = 3.2$ ($C = 20$, $[P_{AMP}]_0 = 30\,\text{nM}$, $[S_{DEG}]_0 = 7.9\,\text{nM}$). The slope of the fitted line is $\alpha = 3.13 \pm 0.11$ in good agreement with the set point value.

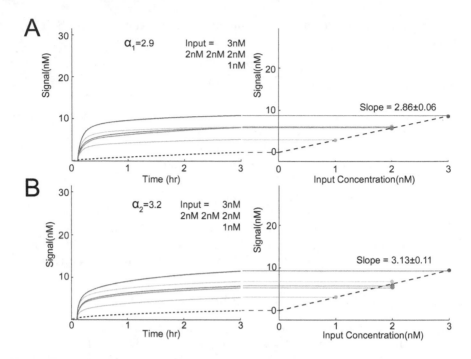

Fig. 3. Experimental result of an amplifier with varying input concentrations. (A) A DNA-based variable gain amplifier for an input with the let-7a RNA sequence. In the experiment, the gain is set to $\alpha = 2.9$. The input concentration varies from 1 nM to 3 nM. The graph on the left shows the reaction kinetics traces from a fluorescence time course experiment. The plot on the right shows the background-subtracted final signal as a function of the input concentration. The value for the slope 2.86 ± 0.06 of a line fit to the data agrees well with the target value for the gain. **(B)** A DNA-based variable gain ($\alpha = 3.2$) amplifier for the let-7c RNA sequence. Kinetics traces are shown on the left, background-subtracted end point values on the right. The slope of the fitted line is 3.13 ± 0.11 in agreement with the target value.

2.4   Varying the Gain

Next, we wanted to test whether we could predictably vary the gain. To this end, we performed experiments with varying amounts of sink S_{DEG} while fixing the input amount at 2 nM. For each experimental condition, we calculated the expected value of the gain α using Eq. 2.

Figure 4A (left) shows the kinetics traces for a set of experiments performed with the let-7a variable gain amplifier. On the right, background-corrected end-point fluorescence levels are shown as a function of the expected gain α (the target values are indicated next to the dots in the graph). If the predictions made with Eq. 2 are correct in call cases, we should be able to fit all data to a straight line with a slope that is equal to the amount of input. We found that for the let-7a system the best fit to the data was obtained with a slope of 2.03 ± 0.09 nM in good agreement with the prediction.

Performing a similar experiment with the let-7c system also resulted in good agreement between data and model (Fig. 4B). The best fit to the data was obtained with a slope of 1.99 ± 0.11 nM which agrees with the amount of input we added (2 nM).

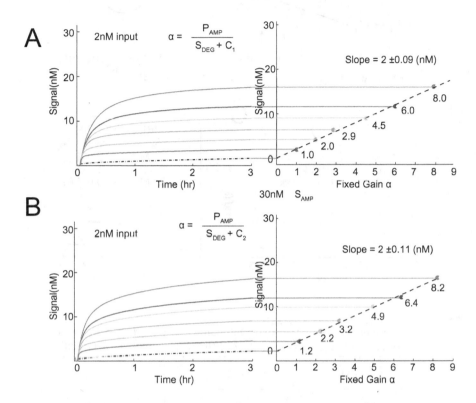

Fig. 4. Varying the gain α for a fixed amount of input. In the experiments, α is varied by changing the concentration of the sink (S_{DEG}). **(A)** The figure on the left shows the kinetics traces for a set of experiments performed with the let-7a variable gain amplifier and different sink concentrations. The final signal increment due to the input as a function of the gain α is shown on the right. The data can be fitted to a straight line with slope 2.03 ± 0.09 nM which agrees well with input concentration of 2 nM. **(B)** Variable gain amplifier for let-7c. Kinetics traces for various values of the gain are shown on the left, end points on the right. The slope of the linear fit of final signal vs. gain is 1.99 ± 0.11 nM which agrees with the real input concentration of 2 nM.

2.5   Classifier Design

Many problems in biology require classification of samples based on their gene expression profiles. For example, a decision tree classifier based on miRNA expression patterns can be used to identify cancer tissues of origin [19]. Although reliable classification often requires considering many molecular inputs, even

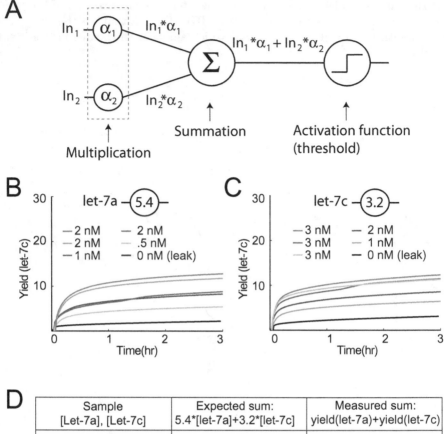

Fig. 5. A two-input neural network classifier composed of variable gain modules for inputs with sequences of let-7a and let-7c miRNAs. (A) A neural network representation of a linear two-input classifier. Here, we only implement the initial multiplication steps at the molecular level. (B) Fluorescence kinetics traces for the let-7a amplifier. Each amplifier module is functionalized with a unique fluorophore to allow independent analysis of both amplification reactions in the same test tube. (C) Fluorescence kinetics traces for Let-7c amplifier. (D) Comparison of the expected and experimental result. The experimental results are with in 5% of the expected result.

classifiers based on comparing the levels of only two genes can be of clinical relevance [20].

Here, we propose to build a simple two parameter classifier of the form

$$y = \alpha_1[RNA_1] + \alpha_2[RNA_2], \tag{3}$$

where $[RNA_1]$, $[RNA_2]$ are the concentration of the input RNAs, α_1 and α_2 are weights that can in principle be learned form labeled training samples and y is the score used to determine the class by comparing its value to a threshold value. A graphical representation of this system using a neural network representation is shown in Fig. 5A.

2.6   A Linear Classifier Circuit

To create a two-input linear classifier or neural network, we combined the two variable gain amplifier modules introduced above. The classifier we intended to construct realizes the function $y = 5.4$ [let-7a] $+ 3.2$ [let-7c]. We note that this particular choice of parameters is arbitrary and does not reflect a biologically relevant decision. However, this system still provides a good proof-of-principle demonstration of the computational system.

The let-7a and let-7c RNA sequences are first premixed in tubes for each sample and added to a mixture of let-7a (gain $\alpha_1 = 5.4$) and let-7c (gain $\alpha_2 = 3.2$) variable gain amplification systems. In this experiment, we use two different fluorophores to read out the reactions for let-7a and let-7c. As a result, we can observe the trace for each system in different channels (Fig. 5B, C) even though the reactions occur in parallel in the same test tube. The advantage of such a two-color setup is that it provides direct information about each sub-system which simplifies troubleshooting.

A comparison of the expected result and the actual experimental data for each combination of inputs is shown in Fig. 5D. The expected result is calculated by plugging the input concentrations into the equation $y = 5.4$ [let-7a] $+ 3.2$ [let-7c]. The experimental result is obtained by converting the fluorescence signal in each channel to a concentration using a calibration curve and then taking the sum of the signals of let-7a and let-7c. Remarkably, the experimental results are with in 5% of the expected result even though the sequences for let-7a and let-7c differ only in a single position. This good agreement is a direct consequence of the high specificity of the amplification mechanism [14].

3   Discussion

The work presented here demonstrates that variable gain amplifiers can be realized using a simple DNA strand displacement-based reaction network. Moreover, we showed that variable gain amplifiers can serve as multiplication elements that assign specific weights to an input in a molecular classifier.

We experimentally built a neural network that computed the weighted sum of two inputs. However, there a still several limitations to the experimental system presented here. First, rather than performing the summation at the molecular level, we used two distinct fluorescent reporters to read out the two amplification reactions individually and performed the summation *in silico*. This approach has the advantage that we can obtain additional information about the operation of the individual variable gain units but has the disadvantage that not all computation is performed in the test tube. Luckily, this limitation could easily be overcome by using entropy driven catalysts that produce identical single stranded outputs in both amplifier units. A single fluorescent reporter triggered by this output strand then provides a readout of the weighted sum of the inputs. Because of the modularity of the amplifier design which completely decouples input and output sequences, weights associated with any number of inputs could be summed up by operating multiple amplifiers with distinct inputs but identical outputs in parallel. To realize both negative and positive weights, two distinct sequences could be used: all variable gain amplifiers associated with a positive weight would release one type of output sequence while all amplifiers associated with a negative weight would release the other sequence.

A second limitation comes from the fact that we did not experimentally realize the activation function comparing the weighted sum of the inputs to some threshold value in order to solve the classification problem. Again, this limitation could be overcome through the use of "off-the-shelf" threshold units such as those introduced in Refs. [14,15] or by connecting the outputs to a consensus network [21].

Finally, we did not gel-purify the sink or entropy driven catalysts which considerably increases leak reactions. Simply using purified complexes would reduce the value of the leak constant and could also increase the intrinsic turnover, thus aligning the behavior of the variable gain amplifier units even more closely with that predicted for an ideal system.

Through this work, we experimentally demonstrated a novel primitive for molecular computation and a path towards building a class of linear classifiers or neural networks that has not yet been realized at the molecular level. Previous work on the construction of digital logic circuits using closely related molecular building blocks suggests that scaling up classifier size should be feasible [15]. In the longer term, such classification systems could result in novel technologies for analyzing biological samples, for example classifying patient samples based on the expression level of micro RNA or messenger RNA in blood or other biological liquids.

Acknowledgements. This work was supported through the NSF grant CCF-1317653 to GS.

References

1. Levine, E., Zhang, Z., Kuhlman, T., Hwa, T.: Quantitative characteristics of gene regulation by small RNA. PLoS Biol. **5**, e229 (2007)

2. Sprinzak, D., Lakhanpal, A., LeBon, L., Santat, L.A., Fontes, M.E., Anderson, G.A., Garcia-Ojalvo, J., Elowitz, M.B.: Cis-interactions between Notch and Delta generate mutually exclusive signalling states. Nature **465**, 86–90 (2010)
3. Zhang, D.Y., Seelig, G.: DNA-based fixed gain amplifiers and linear classifier circuits. In: Sakakibara, Y., Mi, Y. (eds.) DNA 2010. LNCS, vol. 6518, pp. 176–186. Springer, Heidelberg (2011). doi:10.1007/978-3-642-18305-8_16
4. Zhang, D.Y., Seelig, G.: Dynamic DNA nanotechnology using strand-displacement reactions. Nat. Chem. **3**, 103–113 (2011)
5. Yurke, B., Turberfield, A.J., Mills, A.P., Simmel, F.C., Neumann, J.L.: A DNA-fuelled molecular machine made of DNA. Nature **406**, 605–608 (2000)
6. Yurke, B., Mills, A.P.: Using DNA to power nanostructures. Genet. Program Evolvable Mach. **4**, 111–122 (2003)
7. Zhang, D.Y., Winfree, E.: Control of DNA strand displacement kinetics using toehold exchange. J. Am. Chem. Soc. **131**, 17303–17314 (2009)
8. Seelig, G., Yurke, B., Winfree, E.: Catalyzed relaxation of a metastable DNA fuel. J. Am. Chem. Soc. **128**, 12211–12220 (2006)
9. Seelig, G., Soloveichik, D., Zhang, D.Y., Winfree, E.: Enzyme-free nucleic acid logic circuits. Science **314**, 1585–1588 (2006)
10. Dirks, R.M., Pierce, N.A.: Triggered amplification by hybridization chain reaction. Proc. Natl. Acad. Sci. USA **101**, 15275–15278 (2004)
11. Turberfield, A.J., Mitchell, J., Yurke, B., Mills, A.P., Blakey, M., Simmel, F.C.: DNA fuel for free-running nanomachines. Phys. Rev. Lett. **90**, 118102 (2003)
12. Zhang, D.Y., Turberfield, A.J., Yurke, B., Winfree, E.: Engineering entropy-driven reactions and networks catalyzed by DNA. Science **318**, 1121–1125 (2007)
13. Li, Q., Luan, G., Guo, Q., Liang, J.: A new class of homogeneous nucleic acid probes based on specific displacement hybridization. Nucleic Acids Res. **30**, e5–e5 (2002)
14. Chen, S.X., Seelig, G.: An engineered kinetic amplification mechanism for single nucleotide variant discrimination by DNA hybridization probes. J. Am. Chem. Soc. **138**, 5076–5086 (2016)
15. Qian, L., Winfree, E.: Scaling up digital circuit computation with DNA strand displacement cascades. Science **332**, 1196–1201 (2011)
16. Qian, L., Winfree, E., Bruck, J.: Neural network computation with DNA strand displacement cascades. Nature **475**, 368–372 (2011)
17. Lakin, M.R., Stefanovic, D.: Supervised learning in adaptive DNA strand displacement networks. ACS Synth. Biol. **5**, 885–897 (2016)
18. Roush, S., Slack, F.J.: The let-7 family of microRNAs. Trends Cell Biol. **18**, 505–516 (2008)
19. Rosenfeld, N., Aharonov, R., Meiri, E., Rosenwald, S., Spector, Y., Zepeniuk, M., Benjamin, H., Shabes, N., Tabak, S., Levy, A., et al.: MicroRNAs accurately identify cancer tissue origin. Nat. Biotechnol. **26**, 462–469 (2008)
20. Price, N.D., Trent, J., El-Naggar, A.K., Cogdell, D., Taylor, E., Hunt, K.K., Pollock, R.E., Hood, L., Shmulevich, I., Zhang, W.: Highly accurate two-gene classifier for differentiating gastrointestinal stromal tumors and leiomyosarcomas. Proc. Natl. Acad. Sci. USA **104**, 3414–3419 (2007)
21. Chen, Y.J., Dalchau, N., Srinivas, N., Phillips, A., Cardelli, L., Soloveichik, D., Seelig, G.: Programmable chemical controllers made from DNA. Nat. Nanotechnol. **8**, 755–762 (2013)

A Stochastic Approach to Shortcut Bridging in Programmable Matter

Marta Andrés Arroyo[1], Sarah Cannon[2]([✉]), Joshua J. Daymude[3],
Dana Randall[2], and Andréa W. Richa[3]

[1] University of Granada, Granada, Spain
`martaandres@correo.ugr.es`
[2] College of Computing, Georgia Institute of Technology, Atlanta, USA
`sarah.cannon@gatech.edu, randall@cc.gatech.edu`
[3] Computer Science, CIDSE, Arizona State University, Tempe, USA
`{jdaymude,aricha}@asu.edu`

Abstract. In a *self-organizing particle system*, an abstraction of programmable matter, simple computational elements called *particles* with limited memory and communication self-organize to solve system-wide problems of movement, coordination, and configuration. In this paper, we consider stochastic, distributed, local, asynchronous algorithms for "shortcut bridging," in which particles self-assemble bridges over gaps that simultaneously balance minimizing the length and cost of the bridge. Army ants of the genus *Eticon* have been observed exhibiting a similar behavior in their foraging trails, dynamically adjusting their bridges to satisfy an efficiency tradeoff using local interactions [1]. Using techniques from Markov chain analysis, we rigorously analyze our algorithm, show it achieves a near-optimal balance between the competing factors of path length and bridge cost, and prove that it exhibits a dependence on the angle of the gap being "shortcut" similar to that of the ant bridges. We also present simulation results that qualitatively compare our algorithm with the army ant bridging behavior. The proposed algorithm demonstrates the robustness of the stochastic approach to algorithms for programmable matter, as it is a surprisingly simple generalization of a stochastic algorithm for compression [2].

1 Introduction

In developing a system of *programmable matter*, one endeavors to create a material or substance that can utilize user input or stimuli from its environment to change its physical properties in a programmable fashion. Many such systems have been proposed (e.g., DNA tiles, synthetic cells, and reconfigurable modular

S. Cannon—Supported in part by NSF DGE-1148903 and a grant from the Simons Foundation (#361047).
J.J. Daymude—Supported in part by NSF CCF-1637393.
D. Randall—Supported in part by NSF CCF-1637031 and CCF-1526900.
A.W. Richa—Supported in part by NSF CCF-1637393 and CCF-1422603.

© Springer International Publishing AG 2017
R. Brijder and L. Qian (Eds.): DNA 23 2017, LNCS 10467, pp. 122–138, 2017.
DOI: 10.1007/978-3-319-66799-7_9

robots) and each attempts to perform tasks subject to domain-specific capabilities and constraints. In our work on *self-organizing particle systems*, we abstract away from specific settings and envision programmable matter as a system of computationally limited devices (which we call *particles*) which can actively move and individually execute distributed, local, asynchronous algorithms to cooperatively achieve macro-scale tasks of movement and coordination.

The phenomenon of local interactions yielding emergent, collective behavior is often found in natural systems; for example, honey bees choose hive locations based on decentralized recruitment [3] and cockroach larvae perform self-organizing aggregation using pheromones with limited range [4]. In this paper, we present an algorithm inspired by the work of Reid et al. [1], who found that army ants continuously modify the shape and position of foraging bridges—constructed and maintained by their own bodies—across holes and uneven surfaces in the forest floor. Moreover, these bridges appear to stabilize in a structural formation which balances the "benefit of increased foraging trail efficiency" with the "cost of removing workers from the foraging pool to form the structure" [1].

We attempt to capture this inherent trade-off in the design of our algorithm for "shortcut bridging" in self-organizing particle systems (to be formally defined in Sect. 1.3). Our proposed algorithm for shortcut bridging is an extension of the stochastic, distributed algorithm for the *compression problem* introduced in [2], which shows that many fundamental elements of our stochastic approach can be generalized to applications beyond the specific context of compression. In particular, our stochastic approach may be of future interest in the molecular programming domain, where simpler variations of bridging have been studied. Groundbreaking works in this area, such as that of Mohammed et al. [5], focus on forming molecular structures that connect some fixed points; our work may offer insights on further optimizing the quality and/or cost of the resulting bridges.

Shortcut bridging is an attractive goal for programmable matter systems, as many application domains envision deploying programmable matter on surfaces with structural irregularities or dynamic topologies. For example, one commonly imagined application of smart sensor networks is to detect and span small cracks in infrastructure such as roads or bridges as they form; dynamic bridging behavior would enable the system to remain connected as the cracks form and to shift its position accordingly.

1.1 Related Work

When examining the recently proposed and realized systems of programmable matter, one can distinguish between *passive* and *active* systems. In passive systems, computational units cannot control their movement and have (at most) very limited computational abilities, relying instead on their physical structure and interactions with the environment to achieve locomotion (e.g., [6–8]). A large body of research in molecular self-assembly falls under this category, which has mainly focused on shape formation (e.g., [9–11]). Rather than focusing on constructing a specific fixed target shape, our work examines building dynamic bridges whose exact shape is not predetermined. Mohammed et al. studied the more relevant problem of connecting DNA origami landmarks with DNA nanotubes, using a carefully designed process of nanotube nucleation, growth, and diffusion to achieve and maintain the desired connections [5]. The most significant differences between their approach and ours is (*i*) the bridges we consider already connect their endpoints at the start, and focus on the more specific goal of optimizing their shape with respect to a parameterized objective function, and (*ii*) our system is active as opposed to passive.

Active systems, in contrast, are composed of computational units which can control their actions to solve a specific task. Examples include *swarm robotics*, various other models of modular robotics, and the *amoebot model*, which defines our computational framework (detailed in Sect. 1.2).

Swarm robotics systems usually involve a collection of autonomous robots that move freely in space with limited sensing and communication ranges. These systems can perform a variety of tasks including gathering [12], shape formation [13,14], and imitating the collective behavior of natural systems [15]; however, the individual robots have more powerful communication and processing capabilities than those we consider. *Modular self-reconfigurable robotic systems* focus on the motion planning and control of kinematic robots to achieve dynamic morphology [16], and *metamorphic robots* form a subclass of self-reconfiguring robots [17] that share some characteristics with our geometric amoebot model. Walter et al. have conducted some algorithmic research on these systems (e.g., [18,19]), but focus on problems disjoint from those we consider.

In the context of molecular programming, our model most closely relates to the *nubot* model by Woods et al. [20,21], which seeks to provide a framework for rigorous algorithmic research on self-assembly systems composed of active molecular components, emphasizing the interactions between molecular structure and active dynamics. This model shares many characteristics of our amoebot model (e.g., space is modeled as a triangular grid, nubot monomers have limited computational abilities, and there is no global orientation) but differs in that nubot monomers can replicate or die and can perform coordinated rigid body movements. These additional capabilities prohibit the direct translation of results under the nubot model to our amoebot model.

1.2   The Amoebot Model

We recall the main properties of the *amoebot model* [2,22], an abstract model for programmable matter that provides a framework for rigorous algorithmic research on nano-scale systems. We represent programmable matter as a collection of individual computational units known as *particles*. The structure of a particle system is represented as a subgraph of the infinite, undirected graph $G = (V, E)$, where V is the set of all possible locations a particle could occupy and E is the set of all possible atomic transitions between locations in V. For shortcut bridging (and many other problems), we assume the *geometric* amoebot model, in which $G = \Gamma$, the *triangular lattice* (Fig. 1a).

Each particle is either *contracted*, occupying a single location, or *expanded*, occupying or a pair of adjacent locations in Γ (Fig. 1b). Particles move via a series of *expansions* and *contractions*; in particular, a contracted particle may expand into an adjacent unoccupied location, and completes its movement by contracting to once again occupy a single location.

Two particles occupying adjacent locations in Γ are said to be *neighbors*. Each particle is *anonymous*, lacking a unique identifier, but can locally identify each of its neighbors via a collection of ports corresponding to edges incident to its location. We assume particles have a common *chirality*, meaning they share the same notion of *clockwise direction*, which allows them to label their ports in clockwise order. However, particles do not share a global orientation and thus may have different offsets for their port labels (Fig. 1c).

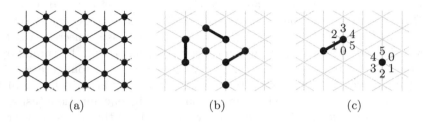

(a)   (b)   (c)

Fig. 1. (a) A section of the triangular lattice Γ; (b) expanded and contracted particles; (c) two non-neighboring particles with different offsets for their port labels.

Each particle has a constant-size, local memory that can read from for communication by it and its neighbors, so a particle's state (e.g., contracted or expanded) is visible to its neighbors. Due to the limitation of constant-size memory, a particle cannot know the total number of particles in the system or any estimate of it. We assume the standard asynchronous model from distributed computing [23], where progress is achieved through *atomic particle activations*. Once activated, a particle can perform an arbitrary, bounded amount of computation involving its local memory and the memories of its neighbors, and can perform at most one movement. A classical result under this model states that for any concurrent asynchronous execution of activations, there is a sequential ordering of activations producing the same result, provided conflicts that arise in the concurrent execution are resolved. In our scenario, conflicts arising from simultaneous memory writes or particle expansions into the same empty location are assumed to be resolved arbitrarily. Thus, while many particles may be activated at once in a realistic settings, it suffices to consider a sequence of activations in which only one particle is active at a time.

1.3   Problem Description

Just as the uneven surfaces of the forest floor affect the foraging behavior of army ants, the collective behavior of particle systems should change when Γ is non-uniform. Here, we focus on system behaviors when the vertices of Γ are either *gap* (unsupported) or *land* (supported) locations. A particle occupying some location in Γ can tell whether it is in the gap or on land. We also introduce *objects*, which are static particles that do not perform computation; these are used to constrain the particles to remain connected to certain fixed sites. In order to analyze the strength of the solutions our algorithm produces, we define the *weighted perimeter* $\overline{p}(\sigma, c)$ of a particle system configuration σ to be the summed edge weights of edges on the external boundary of σ, where edges on land have weight 1, edges in the gap have a fixed weight $c > 1$, and edges with one endpoint on land and one endpoint in the gap have weight $(1 + c)/2$.

In the *shortcut bridging problem*, we consider an instance $(L, O, \sigma_0, c, \alpha)$ where $L \subseteq V$ is the set of land locations, O is the set of (two) objects to be bridged between, σ_0 is the initial configuration of the particle system, $c > 1$ is a fixed weight for gap edges, and $\alpha > 1$ is the accuracy required of a solution. An instance is valid if (i) the objects of O and particles of σ_0 all occupy locations in L, (ii) σ_0 connects the objects, and (iii) σ_0 is connected, a notion formally defined in Sect. 2.1. An algorithm solves an instance if, beginning from σ_0, it reaches and remains (with high probability) in a set of configurations Σ^* such that any $\sigma \in \Sigma^*$ has perimeter weight $\overline{p}(\sigma, c)$ within an α-factor of its minimum value. The weighted perimeter balances in one function (using an appropriate weight for the land and gap perimeter edges) the trade-off observed in [1] between the competing objectives of establishing a short path between the fixed endpoints while not having too many particles in the gap. We focus on gap perimeter instead of the number of particles in the gap (which is perhaps a more natural analogy to [1]) because (1) the shortcut bridges produced with this metric more closely resemble the ant structures and (2) only particles on the perimeter of a configuration can move, and thus recognize the potential risk of being in the gap, justifying our focus on perimeter in the weight function.

In analogy to the apparatus used in [1] (see Fig. 3a), we are particularly interested in the special case where L forms a V-shape, O has two objects positioned at either base of L, and σ_0 lines the interior sides of L, as in Fig. 2a. However, our algorithm is not limited to this setting; for example, we show simulation results for an N-shaped land mass (Fig. 2b) in Sect. 5.

(a)   (b)

Fig. 2. Examples of L, O and σ_0 for instances of the shortcut bridging problem for which we present simulation results (Sect. 5). Light (brown) nodes are land locations, large (red) nodes are occupied by objects, and black nodes are occupied by particles. (Color figure online)

1.4   The Stochastic Approach to Self-organizing Particle Systems

In [2], we introduced a stochastic, distributed algorithm for compression in the amoebot model; here we extend that work to show that stochastic approach is in fact more generally applicable. The motivation underlying this Markov chain approach to programmable matter comes from statistical physics, where ensembles of particles reminiscent of the amoebot model are used to study physical systems and demonstrate that local micro-behavior can induce global macroscale changes to the system [24–26]. Like a spring relaxing, physical systems favor configurations that minimize energy. The energy function is determined by a *Hamiltonian* $H(\sigma)$. Each configuration σ has weight $w(\sigma) = e^{-B \cdot H(\sigma)}/Z$, where where $B = 1/T$ is inverse temperature and $Z = \sum_\tau e^{-B \cdot H(\tau)}$ is the normalizing constant known as the *partition function*.

For shortcut bridging, we introduce a Hamiltonian over particle system configurations so that the configurations of interest will have the lowest energy, and will design our algorithms to favor these low energy configurations. We assign each particle system configuration σ a Hamiltonian $H(\sigma) = \overline{p}(\sigma, c)$, its weighted perimeter. Setting $\lambda = e^B$, we get $w(\sigma, c) = \lambda^{-\overline{p}(\sigma, c)}/Z$. As λ gets larger (by increasing B, effectively lowering temperature), we increasingly favor configurations where $\overline{p}(\sigma, c)$ is small and the desired bridging behavior is exhibited. We prove (Theorem 1) that raising λ above $2 + \sqrt{2}$ suffices for the low energy configurations with small $\overline{p}(\sigma, c)$ to dominate the state space and overcome the entropy of the system. That is, for $\lambda > 2 + \sqrt{2}$, low energy configurations occur with sufficient frequency that we will find such configurations when we sample over the whole state space. The key tool used to establish this is a careful *Peierls argument*, used in statistical physics to study non-uniqueness of limiting Gibbs measures and to determine the presence of phase transitions and in computer science to establish slow mixing of Markov chains (see, e.g., [27], Chap. 15).

Compared to other algorithms for programmable matter and self-organizing particle systems, stochastic methods such as the compression algorithm of [2] and our shortcut bridging algorithm are nearly oblivious, more robust to failures, and require little to no communication between particles. Because each of these algorithms is derived from a stochastic process, powerful tools developed for Markov chain analysis can be employed to rigorously understand their behavior.

1.5   A Stochastic Algorithm for Shortcut Bridging

We present a Markov chain $\mathcal{M}$ for *shortcut bridging* in the geometric amoebot model which translates directly to a fully distributed, local, asynchronous algorithm $\mathcal{A}$. We prove that $\mathcal{M}$ (and by extension, $\mathcal{A}$) solves the shortcut bridging problem: for any constant $\alpha > 1$, the long run probability that $\mathcal{M}$ is in a configuration σ with $\overline{p}(\sigma, c)$ larger than α times its minimum possible value is exponentially small. We then specifically consider V-shaped land masses with an object on each branch of the V, and prove that the resulting bridge structures vary with the interior angle of the V-shaped gap being shortcut—a phenomenon also observed by Reid et al. [1] in the army ant bridges—and show in simulation that they are qualitatively similar to those of the ants (e.g., Fig. 3).

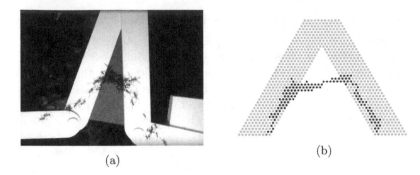

(a)

(b)

Fig. 3. (a) In this image from [1], army ants of the genus *Eticon* build a dynamic bridge which balances the benefit of a shortcut path with the cost of committing ants to the structure. (b) Our shortcut bridging algorithm also balances competing objectives and converges to similar configurations.

2 Background

2.1 Terminology for Particle Systems

For a particle P (resp., location ℓ), we use $N(P)$ (resp., $N(\ell)$) to denote the set of particles and objects[1] adjacent to P (resp., to ℓ). For adjacent locations ℓ and ℓ', we use $N(\ell \cup \ell')$ to denote the set $N(\ell) \cup N(\ell')$, not including particles or objects occupying either ℓ or ℓ'.

We define an *edge* of a particle configuration to be an edge of Γ where both endpoints are occupied by particles. When referring to a *path*, we mean a path in the subgraph of Γ induced by the locations occupied by particles. Two particles are *connected* if there exists a path between them, and a configuration is *connected* if all pairs of particles are. A *hole* in a configuration is a maximal finite component of adjacent unoccupied locations. We specifically consider connected configurations with no holes, and our algorithm, if starting at such a configuration, will maintain these properties.

Let σ be a connected configuration with no holes. The *perimeter* of σ, denoted $p(\sigma)$, is the length of the walk around the (single external) boundary of the particles. The *gap perimeter* of σ, denoted $g(\sigma)$, is the number of perimeter edges that are in the gap, where edges with one endpoint in the gap and one endpoint on land count as half an edge in the gap. Note that an edge may appear twice in the boundary walk, and thus may be counted twice in $p(\sigma)$ or $g(\sigma)$.

2.2 Markov Chains

Our distributed shortcut bridging algorithm is based on a Markov chain, so we briefly review the necessary terminology. A *Markov chain* is a memoryless

---

[1] The notion of a particle (resp., location) neighborhood $N(P)$ (resp., $N(\ell)$) has been extended from [2] to include objects.

stochastic process defined on a finite set of states Ω. The transition matrix P on $\Omega \times \Omega \to [0,1]$ is defined so that $P(x,y)$ is the probability of moving from state x to state y in one step, for any pair of states $x, y \in \Omega$. The t-step transition probability $P^t(x,y)$ is the probability of moving from x to y in exactly t steps.

A Markov chain is *ergodic* if it is *irreducible*, i.e., for all $x, y \in \Omega$, there is a t such that $P^t(x,y) > 0$, and *aperiodic*, i.e., for all $x, y \in \Omega$, g.c.d. $\{t : P^t(x,y) > 0\} = 1$. Any finite, ergodic Markov chain converges to a unique *stationary distribution* π given by, for all $x, y \in \Omega$, $\lim_{t \to \infty} P^t(x,y) = \pi(y)$. Any distribution π' satisfying $\pi'(x)P(x,y) = \pi'(y)P(y,x)$, for all $x, y \in \Omega$, (the *detailed balance condition*) and $\sum_{x \in \Omega} \pi'(x) = 1$ is the unique stationary distribution of the Markov chain (see, e.g., [28]).

Given a Markov chain and a desired stationary distribution π on Ω, the celebrated Metropolis-Hastings algorithm [29] defines appropriate transition probabilities for the chain so that π is its stationary distribution. Starting at $x \in \Omega$, pick a neighbor y in Ω uniformly with some fixed probability (that is the same for all x), and move to y with probability $\min\{1, \pi(y)/\pi(x)\}$; with the remaining probability stay at x and repeat. Using detailed balance, if the state space is connected then π must be the stationary distribution. While calculating $\pi(x)/\pi(y)$ seems to require global knowledge, this ratio can often be calculated using only local information when many terms cancel out. In our case, the Metropolis probabilities are simply $\min\{1, \lambda^{\overline{p}(x,c) - \overline{p}(y,c)}\}$; if x and y only differ by one particle P, as is the case with all moves of our algorithm, then $\overline{p}(x,c) - \overline{p}(y,c)$ can be calculated using only local information from the neighborhood of P.

3   A Stochastic Algorithm for Shortcut Bridging

Recall that for the shortcut bridging problem, we desire for our algorithm to achieve small weighted perimeter, where boundary edges in the gap cost more than those on land. The algorithm must balance the competing objectives of having a short path between the two objects while not forming too large of a bridge. We capture these two factors by preferring small perimeter and small gap perimeter, respectively. While these objectives may appear to be aligned rather than competing, decreasing the length of the overall perimeter increases the gap perimeter and vice versa in the problem instances we consider (e.g., Fig. 2).

Specifically, our Markov chain algorithm incorporates two bias parameters: λ and γ. The value of λ controls the preference for having a small perimeter, while γ controls the preference for having a small gap perimeter. In this paper, we only consider $\lambda > 1$ and $\gamma > 1$, which correspond to favoring a smaller perimeter and a smaller gap perimeter, respectively. Using a Metropolis filter, we ensure that our algorithm converges to a distribution over particle system configurations where the relative likelihood of the particle system being in configuration σ is $\lambda^{-p(\sigma)} \gamma^{-g(\sigma)}$, or equivalently, $\lambda^{-\overline{p}(\sigma,c)}$ for $c = 1 + \log_\lambda \gamma$. We note λ is the same parameter that controlled compression in [2], where particle configurations converged to a distribution proportional to $\lambda^{-p(\sigma)}$. That work showed that $\lambda > 1$ is not sufficient for compression to occur, so we restrict our attention to $\lambda > 2 + \sqrt{2}$, the regime where compression provably happens.

To ensure that during the execution of our algorithm the particles remain connected and hole-free, we introduce two properties every movement must satisfy. These properties help to guarantee the local connectivity structure in the neighborhood of a moving particle doesn't change; more details may be found in [2]. Importantly, these properties maintain system connectivity[2], prevent holes from forming, and ensure reversibility of the Markov chain. These last two conditions are necessary for applying established tools from Markov chain analysis. Let ℓ and ℓ' be adjacent locations in Γ, and let $\mathbb{S} = N(\ell) \cap N(\ell')$ be the particles adjacent to both; we note $|\mathbb{S}| = 0, 1$, or 2.

Property 1. $|\mathbb{S}| \in \{1, 2\}$ and every particle in $N(\ell \cup \ell')$ is connected to a particle in $\mathbb{S}$ by a path through $N(\ell \cup \ell')$.

Property 2. $|\mathbb{S}| = 0$; ℓ and ℓ' each have at least one neighbor; all particles in $N(\ell) \setminus \{\ell'\}$ are connected by paths within this set; and all particles in $N(\ell') \setminus \{\ell\}$ are connected by paths within this set.

Importantly, these properties are symmetric with respect to ℓ and ℓ' and can be locally checkable by an expanded particle occupying both ℓ and ℓ' (as in Lines 2–3 of the Markov chain process described below).

We can now introduce the Markov chain $\mathcal{M}$ for an instance $(L, O, \sigma_0, c, \alpha)$ of shortcut bridging. For input parameter $\lambda > 2 + \sqrt{2}$, set $\gamma = \lambda^{c-1}$, and beginning at initial configuration σ_0, which we assume has no holes,[3] repeat:

1. Select a particle P uniformly at random from among all n particles; let ℓ denote its location. Choose a neighboring location ℓ' and $q \in (0, 1)$ uniformly. Let σ be the configuration with P at ℓ and σ' the configuration with P at ℓ'.
2. If ℓ' is unoccupied, then P expands to occupy both ℓ and ℓ'. Otherwise, return to step 1.
3. If (i) $|N(\ell)| \neq 5$, (ii) ℓ and ℓ' satisfy Property 1 or Property 2, and (iii) $q < \lambda^{p(\sigma)-p(\sigma')}\gamma^{g(\sigma)-g(\sigma')}$, then P contracts to ℓ'. Otherwise, P contracts back to ℓ.

Although $p(\sigma) - p(\sigma')$ and $g(\sigma) - g(\sigma')$ are values defined at system-level scale, we show these differences can be calculated locally.

Lemma 1. *The values of $p(\sigma) - p(\sigma')$ and $g(\sigma) - g(\sigma')$ in Step 3(iii) of $\mathcal{M}$ can be calculated using information involving only ℓ, ℓ', and $N(\ell \cup \ell')$.*

Proof. These values only need to be calculated if $3(i)$ and $3(ii)$ are both true. By a result of [2], $p(\sigma) - p(\sigma') = |N(\ell')| - |N(\ell)|$, which is computable with only local information.

Note $g(\sigma)$ is also the number of particles that are on the perimeter and in the gap, counted with appropriate multiplicity if a particle is on the perimeter

---

[2] Since particles treat objects as static particles, the particle system may actually disconnect into several components which remain connected through objects.

[3] If σ_0 has holes, our algorithm will eliminate them and they will not reform [2]; for simplicity, we focus only on the behavior of the system after this occurs.

more that once. Given a particle Q, let $G(Q)$ be 1 if Q is in a gap location and 0 if land; define $G(\ell)$ for a location ℓ similarly. Let $\delta(Q, \sigma)$ be the number of times Q appears on the perimeter of σ. Then $g(\sigma) = \sum_{Q \in p(\sigma)} G(Q)\delta(Q, \sigma)$. Define $\Delta(Q) = \delta(Q, \sigma) - \delta(Q, \sigma')$. For particles not in $\{P\} \cup N(\ell \cup \ell')$, $\Delta(Q) = 0$ as the neighborhood of Q will be identical in σ and σ'. Because steps $3(i)$ and $3(ii)$ are true, inspection shows this implies $\Delta(P) = 0$. Then:

$$g(\sigma) - g(\sigma') = \sum_{Q \in N(\ell \cup \ell')} G(Q)\Delta(Q) + \delta(P, \sigma)(G(\ell) - G(\ell')).$$

The second term above is calculable with only local information; for $Q \in N(\ell \cup \ell')$, to find $\Delta(Q)$ only Q's neighbors in this set need to be considered. If Q is adjacent to ℓ and not ℓ', $\Delta(Q) = -1$ if it has two neighbors in $N(\ell)$, $\Delta(Q) = 1$ if it has no neighbors in $N(\ell)$, and $\Delta(Q) = 0$ otherwise. If Q is adjacent to ℓ' but not ℓ, the opposite is true. If Q is adjacent to ℓ and ℓ', then $\Delta(Q) = 0$ if Q has zero or two neighbors in $N(\ell \cup \ell')$; $\Delta(Q) = 1$ if Q has a common neighbor with ℓ' but not ℓ; and $\Delta(Q) = -1$ if Q has a common neighbor with ℓ but not ℓ'. In all cases $\Delta(Q)$, and thus $g(\sigma) - g(\sigma')$, can be found with only local information. □

The state space Ω of $\mathcal{M}$ is the set of all configurations reachable from σ_0 via valid transitions of $\mathcal{M}$. We conjecture this includes all connected, hole-free configurations of n particles connected to both objects, but proving all such configurations are reachable from σ_0 is not necessary for our results. (The proof of the corresponding result in [2] does not generalize due to the presence of objects).

3.1   From $\mathcal{M}$ to a Distributed, Local Algorithm

While $\mathcal{M}$ is a Markov chain with centralized control of the particle system, one can transform $\mathcal{M}$ into a distributed, local, asynchronous algorithm $\mathcal{A}$ that each particle runs individually. The full details of this construction are given in [2], and we give a high level description here. When a particle is activated, it randomly chooses one of its six neighboring locations, checks if moving there is valid, and locally determines how the move will affect the global weight function $\lambda^{-p(\sigma)}\gamma^{-g(\sigma)}$. If the weight will increase, the particle performs the move; otherwise the particle only moves with some probability less than 1.

Specifically, in Step 1 of $\mathcal{M}$, a particle is chosen uniformly at random to be activated; to mimic this random activation sequence in a local way, we assume each particle has its own Poisson clock with mean 1 and becomes active after some random delay drawn from e^{-t}. During its activation, a contracted particle P occupying location ℓ chooses a neighboring location ℓ' and a real value $q \in (0, 1)$ uniformly at random[4], expanding into ℓ' if it is unoccupied, just as in $\mathcal{M}$. However, unlike in $\mathcal{M}$, the expansion and contraction movements of P

---

[4] Note only a constant number of bits are needed to produce q, as λ and γ are constants and a particle move changes perimeter and gap perimeter by at most a constant.

are necessarily split into two activations, since in the amoebot model a central assumption is that a particle can perform at most one movement per activation (see Sect. 1.2). Since P's two activations are not necessarily consecutive, P must be able to resolve conflicts with any other particles that may expand into its neighborhood before it becomes activated again and contracts. We accomplish this by introducing a system of Boolean flags maintained by all expanded particles. If P is the only expanded particle in its neighborhood, it stores a boolean flag $f = TRUE$ in its memory; otherwise, it sets $f = FALSE$. When P is activated again (now occupying both ℓ and ℓ'), it checks its flag f. If it is $FALSE$, P contracts back to ℓ, since some other particle in its neighborhood activated and expanded earlier. Otherwise, if f is $TRUE$, P checks the conditions in Step 3 of $\mathcal{M}$ and contracts either to ℓ or ℓ' accordingly. This ensures that at most one particle in a local neighborhood is moving at a time, mimicking the sequential nature of particle moves during the execution of Markov chain $\mathcal{M}$.

While this shows our Markov chain $\mathcal{M}$ can be translated into a fully local distributed algorithm with the same behavior, such an implementation is not always possible in general. Any Markov chain for particle systems that inherently relies on non-local moves of particles or has transition probabilities relying on non-local information cannot be executed by a local, distributed algorithm. Additionally, most distributed algorithms for amoebot systems are not stochastic; see, e.g., the mostly deterministic algorithms in [22, 30].

3.2   Properties of Markov Chain $\mathcal{M}$

We now show some useful properties of $\mathcal{M}$. Our first two claims follow from work in [2] and basic properties of Markov chains and our particle systems.

Lemma 2. *If σ_0 is connected and has no holes, then at every iteration of $\mathcal{M}$, the current configuration is connected and has no holes.*

Lemma 3. *$\mathcal{M}$ is ergodic.*

As $\mathcal{M}$ is finite and ergodic, it converges to a unique stationary distribution, and we can find that distribution using detailed balance.

Lemma 4. *The stationary distribution of $\mathcal{M}$ is given by*

$$\pi(\sigma) = \lambda^{-p(\sigma)}\gamma^{-g(\sigma)}/Z,$$

where $Z = \sum_{\sigma \in \Omega} \lambda^{-p(\sigma)}\gamma^{-g(\sigma)}$.

Proof. Properties 1 and 2 ensure that particle P moving from location ℓ to location ℓ' is valid if and only if P moving from ℓ' to ℓ is. This implies for any configurations σ and τ, $P(\sigma, \tau) > 0$ if and only if $P(\sigma, \tau) > 0$. Using this, the lemma can easily be verified via detailed balance.   □

As referenced above, this stationary distribution can be expressed in an alternate way using weighted perimeter.

Lemma 5. *For* $c = 1 + \log_\lambda \gamma$, *the stationary distribution of* $\mathcal{M}$ *is given by*

$$\pi(\sigma) = \lambda^{-\overline{p}(\sigma,c)}/Z,$$

where $Z = \sum_{\sigma \in \Omega} \lambda^{-\overline{p}(\sigma,c)}$.

Proof. This follows immediately from the definition of $\overline{p}(\sigma, c)$. $\qquad\square$

Theorem 1. *Consider an execution of Markov chain* $\mathcal{M}$ *on state space* Ω, *with* $\lambda > 2 + \sqrt{2}$, $\gamma > 1$, *and stationary distribution* π, *where starting configuration* σ_0 *has* n *particles. For any constant* $\alpha > \frac{\log(\lambda)}{\log(\lambda) - \log(2+\sqrt{2})} > 1$, *the probability that a configuration* σ *drawn at random from* π *has* $\overline{p}(\sigma, 1 + \log_\lambda \gamma) > \alpha \cdot \overline{p}_{min}$ *is exponentially small in* n, *where* $\overline{p}_{min}$ *is the minimum weighted perimeter of a configuration in* Ω.

Proof. This mimics the proof of α-compression in [2], though additional insights and care were necessary to accommodate the difficulties introduced by considering weighted perimeter instead of perimeter.

Given any configuration σ, let

$$w(\sigma) := \pi(\sigma) \cdot Z = \lambda^{-p(\sigma)}\gamma^{-g(\sigma)} = \lambda^{-\overline{p}(\sigma, 1 + \log_\lambda \gamma)}.$$

For a set of configurations $S \subseteq \Omega$, we let $w(S) = \sum_{\sigma \in S} w(\sigma)$. Let $\sigma_{min} \in \Omega$ be a configuration of n particles with minimal weighted perimeter $\overline{p}_{min}$, and let S_α be the set of configurations with weighted perimeter at least $\alpha \cdot \overline{p}_{min}$. We show:

$$\pi(S_\alpha) = \frac{w(S_\alpha)}{Z} < \frac{w(S_\alpha)}{w(\sigma_{min})} \le \zeta^{\sqrt{n}},$$

where $\zeta < 1$. The first equality follows from Lemma 5; the next inequality follows from the definitions of Z, w, and σ_{min}. We focus on the last inequality.

We stratify S_α into sets of configurations with the same weighted perimeter; there are at most $\mathcal{O}(n^2)$ such sets, as the total perimeter and gap perimeter can each take on at most $\mathcal{O}(n)$ values. Label these sets $A_1, A_2, \ldots, A_m$ in order of increasing weighted perimeter, where m is the total number of distinct weighted perimeters possible for configurations in S_α. Let $\overline{p}_i$ be the weighted perimeter of all configurations in set A_i; since $A_i \subseteq S_\alpha$, we have $\overline{p}_i \ge \alpha \cdot \overline{p}_{min}$.

We note $w(\sigma) = \lambda^{-\overline{p}_i}$ for every $\sigma \in A_i$, so to bound $w(A_i)$ it only remains to bound $|A_i|$. Any configuration with weighted perimeter $\overline{p}_i$ has perimeter $p \le \overline{p}_i$, and a result from [2] which exploits a connection between particle configurations and self-avoiding walks in the hexagon lattice shows that the number of connected hole-free particle configurations with perimeter p is at most $f(p)(2+\sqrt{2})^p$, for some subexponential function f. Letting p_{min} denote the minimum possible (unweighted) perimeter of a configuration of n particles, we conclude that

$$w(A_i) \le \lambda^{-\overline{p}_i} \cdot \sum_{p=p_{min}}^{\overline{p}_i} f(p)\left(2 + \sqrt{2}\right)^p \le \lambda^{-\overline{p}_i} f'(\overline{p}_i)\left(2 + \sqrt{2}\right)^{\overline{p}_i},$$

where $f'(\overline{p}_i) = \sum_{p=p_{min}}^{\overline{p}_i} f(p)$ is necessarily also a subexponential function because it is a sum of at most a linear number of subexponential terms. So,

$$w(S_\alpha) = \sum_{i=1}^{m} w(A_i) \le \sum_{i=1}^{m} f'(\overline{p}_i) \left(\frac{2+\sqrt{2}}{\lambda}\right)^{\overline{p}_i} \le f''(n) \left(\frac{2+\sqrt{2}}{\lambda}\right)^{\alpha \cdot \overline{p}_{min}},$$

where $f''(n) = \sum_{i=1}^{m} f'(\overline{p}_i)$ is a subexponential function because $\overline{p}_i = \mathcal{O}(n)$, $m = \mathcal{O}(n^2)$, and f' is subexponential. The last inequality follows because $\lambda > 2 + \sqrt{2}$ and $\overline{p}_i \ge \alpha \overline{p}_{min}$ by assumption. Finally, because $w(\sigma_{min}) = \lambda^{-\overline{p}_{min}}$,

$$\frac{w(S_\alpha)}{w(\sigma_{min})} \le f''(n) \left(\frac{2+\sqrt{2}}{\lambda}\right)^{\alpha \cdot \overline{p}_{min}} \lambda^{\overline{p}_{min}} = f''(n)\zeta^{\overline{p}_{min}},$$

where $\zeta = \lambda \left(\frac{2+\sqrt{2}}{\lambda}\right)^{\alpha} < 1$ whenever $\alpha > \frac{\log(\lambda)}{\log(\lambda)-\log(2+\sqrt{2})}$. We have $\overline{p}_{min} \ge \sqrt{n}$ because any n particles must have perimeter at least $\sqrt{n}$. This suffices to show there is a constant $\zeta < 1$ and a subexponential function $f''(n)$ such that

$$\pi(S_\alpha) < f''(n)\zeta^{\sqrt{n}},$$

which proves the theorem.   □

As we see in the following corollary, to solve an instance $(L, O, \sigma_0, c, \alpha)$ of the shortcut-bridging problem, one just needs to run algorithm $\mathcal{M}$ with carefully chosen parameters λ and γ.

Corollary 1. *The distributed algorithm associated with Markov chain $\mathcal{M}$ can solve any instance $(L, O, \sigma_0, c, \alpha)$ of the shortcut-bridging problem.*

Proof. It suffices to run the distributed algorithm associated with $\mathcal{M}$ starting from configuration σ_0 with parameters $\lambda > (2 + \sqrt{2})^{\frac{\alpha}{\alpha-1}}$ and $\gamma = \lambda^{c-1}$. Then it holds that $\alpha > \frac{\log(\lambda)}{\log(\lambda)-\log(2+\sqrt{2})} > 1$, so by Theorem 1 the system reaches and remains with all but exponentially small probability in a set of configurations with weighted perimeter $\overline{p}(\sigma, c) \le \alpha \cdot \overline{p}_{min}$, where $\overline{p}_{min}$ is the minimum weighted perimeter of a configuration in Ω.   □

4   Dependence of Bridge Structure on Gap Angle

Specifically, we consider V-shaped land masses (e.g., Fig. 2a) of various angles. We prove that our shortcut bridging algorithm exhibits a dependence on the internal angle θ of the gap that is similar to that of the army ant bridging process observed by Reid et al. [1]. When the internal angle θ is sufficiently small, with high probability the bridge constructed by the particles stays close to the bottom of the gap (away from the apex of angle θ). Furthermore, when θ is large and λ and γ satisfy certain conditions (made explicit in Theorem 3), with high probability the bridge stays close to the top of the gap. Both of these results are proven using a Peierls argument and careful analysis of the geometry of the gap. Due to space constraints, we merely state our main results and omit the proofs, while noting that they are far from trivial.

Theorem 2. *Let $\lambda > 2 + \sqrt{2}$ and $\gamma > 1$. Then there exists θ_1 such that for all $\theta < \theta_1$, the probability at stationarity of $\mathcal{M}$ that the bridge structure is strictly above the midpoint of the gap is exponentially small in n, the number of particles. In particular, $\theta_1 = 2\tan^{-1}\left(\log_{\lambda\gamma}\left(\lambda/\left(2+\sqrt{2}\right)\right)/\sqrt{3}\right)$.*

Theorem 3. *For each $\lambda > 2 + \sqrt{2}$ and $\gamma > (2+\sqrt{2})^4\lambda^4$, there is a constant $\theta_2 > 60°$ such that for all $\theta \in (60°, \theta_2)$, the probability at stationarity of $\mathcal{M}$ that the bridge structure goes through or below the midpoint of the gap is exponentially small in n. In particular, $\theta_2 = 2\tan^{-1}\left[\frac{1}{2\sqrt{3}}\frac{\log(\gamma\lambda^{-4})}{\log(2+\sqrt{2})} - \frac{1}{\sqrt{3}}\right]$.*

5   Simulations

In this section, we show simulation results of our algorithm running on a variety of instances. Figure 4 shows snapshots over time for a bridge shortcutting a V-shaped gap with internal angle $\theta = 60°$ and biases $\lambda = 4, \gamma = 2$. Qualitatively, this bridge matches the shape and position of the army ant bridges in [1]. Figure 5 shows the resulting bridge structure when the land mass is N-shaped. Lastly, Fig. 6 shows the results of an experiment which held λ, γ, and the number of iterations of $\mathcal{M}$ constant, varying only the internal angle of the V-shaped land mass. The particle system exhibited behavior consistent with the theoretical

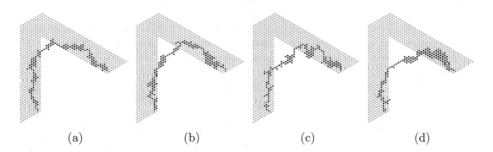

(a)   (b)   (c)   (d)

Fig. 4. A particle system using biases $\lambda = 4$ and $\gamma = 2$ to shortcut a V-shaped land mass with $\theta = 60°$ after (a) 2 million, (b) 4 million, (c) 6 million, and (d) 8 million iterations of Markov chain $\mathcal{M}$, beginning in configuration σ_0 shown in Fig. 2a.

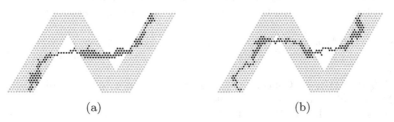

(a)   (b)

Fig. 5. A particle system using $\lambda = 4$ and $\gamma = 2$ to shortcut an N-shaped land mass after (a) 10 million and (b) 20 million steps of $\mathcal{M}$, beginning in σ_0 of Fig. 2b.

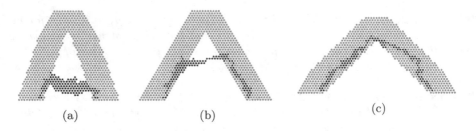

(a)   (b)   (c)

Fig. 6. A particle system using biases $\lambda = 4$ and $\gamma = 2$ to shortcut a land mass with angle (a) 30°, (b) 60°, and (c) 90° after 20 million iterations of $\mathcal{M}$. For a given angle, the land mass and initial configuration were constructed as described in Sect. 4.

results in Sect. 4 and the army ant bridges, shortcutting closer to the bottom of the gap when θ is small and staying almost entirely on land when θ is large.

These simulations demonstrate the successful application of our stochastic approach to shortcut bridging. Moreover, experimenting with variants suggests this approach may be useful for other related applications in the future.

References

1. Reid, C.R., Lutz, M.J., Powell, S., Kao, A.B., Couzin, I.D., Garnier, S.: Army ants dynamically adjust living bridges in response to a cost-benefit trade-off. Proc. Natl. Acad. Sci. **112**(49), 15113–15118 (2015)
2. Cannon, S., Daymude, J.J., Randall, D., Richa, A.W.: A Markov chain algorithm for compression in self-organizing particle systems. In: Proceedings of 2016 ACM Symposium on Principles of Distributed Computing (PODC 2016), pp. 279–288 (2016)
3. Camazine, S., Visscher, K.P., Finley, J., Vetter, S.R.: House-hunting by honey bee swarms: collective decisions and individual behaviors. Insectes Soc. **46**(4), 348–360 (1999)
4. Jeanson, R., Rivault, C., Deneubourg, J.L., Blanco, S., Fournier, R., Jost, C., Theraulaz, G.: Self-organized aggregation in cockroaches. Anim. Behav. **69**(1), 169–180 (2005)
5. Mohammed, A.M., Šulc, P., Zenk, J., Schulman, R.: Self-assembling DNA nanotubes to connect molecular landmarks. Nat. Nanotechnol. **12**, 312–316 (2017)
6. Woods, D.: Intrinsic universality and the computational power of self-assembly. In: Proceedings of Machines, Computations and Universality (MCU 2013), pp. 16–22 (2013)
7. Angluin, D., Aspnes, J., Diamadi, Z., Fischer, M.J., Peralta, R.: Computation in networks of passively mobile finite-state sensors. Distrib. Comput. **18**(4), 235–253 (2006)
8. Reid, C.R., Latty, T.: Collective behaviour and swarm intelligence in slime moulds. FEMS Microbiol. Rev. **40**(6), 798–806 (2016)
9. Douglas, S.M., Dietz, H., Liedl, T., Högberg, B., Graf, F., Shih, W.M.: Self-assembly of DNA into nanoscale three-dimensional shapes. Nature **459**, 414–418 (2009)

10. Cheung, K.C., Demaine, E.D., Bachrach, J.R., Griffith, S.: Programmable assembly with universally foldable strings (moteins). IEEE Trans. Rob. **27**(4), 718–729 (2011)
11. Wei, B., Dai, M., Yin, P.: Complex shapes self-assembled from single-stranded DNA tiles. Nature **485**, 623–626 (2012)
12. Cieliebak, M., Flocchini, P., Prencipe, G., Santoro, N.: Distributed computing by mobile robots: gathering. SIAM J. Comput. **41**(4), 829–879 (2012)
13. Flocchini, P., Prencipe, G., Santoro, N., Widmayer, P.: Arbitrary pattern formation by asynchronous, anonymous, oblivious robots. Theoret. Comput. Sci. **407**(1), 412–447 (2008)
14. Rubenstein, M., Cornejo, A., Nagpal, R.: Programmable self-assembly in a thousand-robot swarm. Science **345**(6198), 795–799 (2014)
15. Chazelle, B.: Natural algorithms. In: Proceedings of 2009 ACM-SIAM Symposium on Discrete Algorithms (SODA 2009), pp. 422–431 (2009)
16. Yim, M., Shen, W.-M., Salemi, B., Rus, D., Moll, M., Lipson, H., Klavins, E., Chirikjian, G.S.: Modular self-reconfigurable robot systems. IEEE Robot. Autom. Mag. **14**(1), 43–52 (2007)
17. Chirikjian, G.: Kinematics of a metamorphic robotic system. In: Proceedings of 1994 International Conference on Robotics and Automation (ICRA 1994), vol. 1, pp. 449–455 (1994)
18. Walter, J.E., Welch, J.L., Amato, N.M.: Distributed reconfiguration of metamorphic robot chains. In: Proceedings of 2000 ACM Symposium on Principles of Distributed Computing (PODC 2000), pp. 171–180 (2000)
19. Walter, J.E., Brooks, M.E., Little, D.F., Amato, N.M.: Enveloping multi-pocket obstacles with hexagonal metamorphic robots. In: Proceedings of 2004 IEEE International Conference on Robotics and Automation (ICRA 2004), pp. 2204–2209 (2004)
20. Woods, D., Chen, H.-L, Goodfriend, S., Dabby, N., Winfree, E., Yin, P.: Active self-assembly of algorithmic shapes and patterns in polylogarithmic time. In: Proceedings of 4th Innovations in Theoretical Computer Science Conference (ITCS 2013), pp. 353–354 (2013)
21. Chen, M., Xin, D., Woods, D.: Parallel computation using active self-assembly. Nat. Comput. **14**(2), 225–250 (2015)
22. Derakhshandeh, Z., Gmyr, R., Strothmann, T., Bazzi, R., Richa, A.W., Scheideler, C.: Leader election and shape formation with self-organizing programmable matter. In: Phillips, A., Yin, P. (eds.) DNA 2015. LNCS, vol. 9211, pp. 117–132. Springer, Cham (2015). doi:10.1007/978-3-319-21999-8_8
23. Lynch, N.: Distributed Algorithms. Morgan Kauffman, San Francisco (1996)
24. Baxter, R.J., Enting, I.G., Tsang, S.K.: Hard-square lattice gas. J. Stat. Phys. **22**, 465–489 (1980)
25. Restrepo, R., Shin, J., Tetali, P., Vigoda, E., Yang, L.: Improving mixing conditions on the grid for counting and sampling independent sets. Probab. Theory Relat. Fields **156**, 75–99 (2013)
26. Blanca, A., Galvin, D., Randall, D., Tetali, P.: Phase coexistence for the hard-core model on $\mathbb{Z}^2$. In: 17th International Workshop on Randomization and Computation (RANDOM 2013), pp. 379–394 (2013)
27. Levin, D.A., Peres, Y., Wilmer, E.L.: Markov Chains and Mixing Times. American Mathematical Society, Providence (2009)

28. Bruguière, C., Tiberghien, A., Clément, P.: Introduction. In: Bruguière, C., Tiberghien, A., Clément, P. (eds.) Topics and Trends in Current Science Education. CSER, vol. 1, pp. 3–18. Springer, Dordrecht (2014). doi:10.1007/978-94-007-7281-6_1
29. Hastings, W.K.: Monte carlo sampling methods using Markov chains and their applications. Biometrika **57**(1), 97–109 (1970)
30. Derakhshandeh, Z., Gmyr, R., Richa, A.W., Scheideler, C., Strothmann, T.: Universal coating for programmable matter. Theoret. Comput. Sci. **671**, 56–68 (2017)

A Minimal Requirement for Self-assembly of Lines in Polylogarithmic Time

Yen-Ru Chin, Jui-Ting Tsai, and Ho-Lin Chen[✉]

National Taiwan University, Taipei, Taiwan
holinchen@ntu.edu.tw

Abstract. Self-assembly is the process in which small and simple components assemble into large and complex structures without explicit external control. The nubot model generalizes previous self-assembly models (e.g. aTAM) to include active components which can actively move and undergo state changes. One main difference between the nubot model and previous self-assembly models is its ability to perform exponential growth.

In the paper, we study the problem of finding a minimal set of features in the nubot model which allows exponential growth to happen. We only focus on nubot systems which assemble a long line of nubots with a small number of supplementary layers. We prove that exponential growth is not possible with the limit of one supplementary layer and one state-change per nubot. On the other hand, if two supplementary layers are allowed, or the disappearance rule can be performed without a state change, then we can construct nubot systems which grow exponentially.

1 Introduction

Self-assembly is the process in which small and simple components assemble into large and complex structures without explicit external control. The concept of self-assembly arises from nature. For instance, mineral crystallization and cell division in the embryo developing process can both be viewed as self-assembly processes. As the size of a system approaches the molecular scale, precise direct external control becomes prohibitively costly, if not impossible. As a result, molecular self-assembly has become an important tool for molecular computation, nano-scale fabrication and nano-machines. DNA has received much attention as a substrate for molecular self-assembly due to its combinatorial nature and simple, predictable geometric structure. DNA self-assembly has been used for many different applications including performing computation [3, 22, 30], constructing molecular patterns [8, 12, 20, 21, 24, 37], and building nano-scale machines [4, 9, 14, 23, 25, 35].

The first two authors make equal contribution to the paper and are listed in alphabetical order.

H.-Y. Chen—Research supported in part by MOST grant number 104-2221-E-002-045-MY3.

R. Brijder and L. Qian (Eds.): DNA 23 2017, LNCS 10467, pp. 139–154, 2017.
DOI: 10.1007/978-3-319-66799-7_10

In many different applications, DNA strands first assemble into small structures called "tiles" which assembly according to simple rules [13]. The first self-assembly model, abstract Tile Assembly Model (aTAM), is proposed by Rothemund and Winfree [19,30]. Many tile systems have been designed under aTAM, including systems that build counters [2,7] and squares [11,15,19], perform Turing-universal computation [30], and produce arbitrary computable shapes [16,26]. Many tile systems has also been successfully implemented in lab experiments [3,6,21,31].

Compared to the self-assembly systems built in nano-technology, the nano-components in nature are much more active: they sense and process environmental cues; upon interaction, their internal structures may change; they can both passively diffuse and actively move. DNA has been used to build many different active components including autonomous walkers [14,17,18,23,25,27,34], logic and catalytic circuits [22,29,33,36], and triggered assembly of linear [10,28] and dendritic structures [33]. It is a natural idea to use these active DNA components in artificial self-assembly systems.

The nubot model [32] generalizes previous self-assembly models to include active components which can actively move and undergo state changes. Rules are applied asynchronously and in parallel in the model. One main difference between the nubot model and previous self-assembly models is its ability to perform exponential growth. It has been shown that there exists a set of nubot systems that grow a line of length n in time $O(\log n)$ using $O(\log n)$ states [32]. The nubot model assumes that each monomer (nubot) can undergo an arbitrary number of state changes. Also, the nubot model assumes that the structures are rigid and do not deform even when two parts of the structure are moving relative to each other using a movement rule. Although these assumptions can be quite natural to some applications (e.g. robotics), implementing this model at the molecular level is not an easy task.

In the paper, we study the problem of finding a minimal set of features in the nubot model which allows exponential growth to happen. Previously, it has been shown that nubot systems can still perform exponential growth even if the active movement rules are disabled [5]. However, the construction relies very heavily on the rigidity of structures. We consider the nubot model with the following two restrictions. First, we consider nubot systems in which every monomer can only make a certain number of state changes. Second, we only focus on nubot systems which assemble a long line of nubots. All movements must be in the direction parallel to the line. Only a fixed number of supplementary layer is allowed to help the growth of the line. When we only look at this type of restricted nubot system, even if the structure is not rigid enough, as long as the structure still locally looks like a line, all of our system construction will still work. When the state change is unlimited, it has been shown that one supplementary layer is enough to allow exponential growth [32].

Our result: With the limit of one supplementary layer and one state-change per nubot, we prove that exponential growth is not possible. The expected length at time t is upper bounded by $O(t)$. On the other hand, if two supplementary layers are allowed, or the disappearance rule can be performed without a state change (a more formal definition will be in Sect. 2), then we can construct systems which grow exponentially and they are verified by a nubot simulator [1].

2   Model

2.1   The Nubot Model

In this subsection, we present a simplified description of the nubot model proposed in [32]. The basic units of the system are called nubot monomers, which are placed on a two-dimensional triangular grid, and thus the position can be represented as a 2-dimensional vector. To illustrate the relative position of adjacent nubot monomers, we define a set of *directions* $\mathcal{D} = \{\text{NE}, \text{E}, \text{SE}, \text{SW}, \text{W}, \text{NW}\}$. More formally, let S be a finite set of monomer states, a *nubot monomer*, or simply a *nubot* m is defined to be a pair $(s, \boldsymbol{p})$, where $s \in S$ and $\boldsymbol{p}$ is a grid point.

These nubot monomers are connected via three types of bonds. One is a null bond, another is a rigid bond, and the other is a flexible bond. More formally, a *bond* b is a triple (m_1, m_2, t) where m_1 and m_2 are two adjacent nubot monomers, and t is a ternary variable representing a bond type. Specifically, $t = 0$ represents a null bond, $t = 1$ represents a rigid bond, and $t = 2$ represents a flexible bond. In this paper, for all nubot system constructions, we only use rigid bonds.

A *configuration* is a set of nubots, their locations, and the set of all bonds between every pair of adjacent nubots. A *(connected) component* is a maximal set of adjacent monomers where every pair of monomers in the set is connected by a path consisting of monomers bound by either flexible or rigid bonds.

Definition 1 (Rule). *A rule (reaction) describes potential ways to change the configuration, and is of the form* $(s_1, s_2, b, \boldsymbol{u}) \rightarrow (s_1', s_2', b', \boldsymbol{u}')$, *where* $s_1, s_2, s_1', s_2' \in S$, $\boldsymbol{u}, \boldsymbol{u}' \in \mathcal{D}$ *and* b, b' *are ternary variables of bonds.*

The rule presentation above is slightly modified from what is used in the nubot simulator [1]. When a rule is represented as the above form, it means that whenever two nubots with states s_1 and s_2 are connected with each other by a bond b, and the direction pointing from s_1 to s_2 is $\boldsymbol{u}$, then the rule can be applied to the two nubots. After the rule being applied, the nubot with state s_1 and s_2 changes their states to s_1' and s_2' respectively, the bond connecting them changes to b', and the direction pointing from s_1' to s_2' becomes $\boldsymbol{u}'$.

There are four types of rules: appearance, disappearance, movement and state change. The formal definition of these rules are the following:

Definition 2 (Appearance). *The appearance of a nubot models the attachment of a nubot onto the main structure, and must be accompanied by a bond. The appearance rule is denoted as* $(s_1, empty, 0, \boldsymbol{u}) \rightarrow (s_1', s_2', 1, \boldsymbol{u})$ *or* $(empty, s_2, 0, \boldsymbol{u}) \rightarrow (s_1', s_2', 1, \boldsymbol{u})$.

Definition 3 (Movement). *The movement rule is denoted as* $(s_1, s_2, 1, \boldsymbol{u}) \rightarrow$ $(s_1', s_2', 1, \boldsymbol{u}')$ *where* $\boldsymbol{u} \neq \boldsymbol{u}'$. *The nubots with state* s_1 *and* s_2 *react with each other, change their states, and change their relative position from* $\boldsymbol{u}$ *to* $\boldsymbol{u}'$. *Furthermore, a minimal set of nubots which must be moved together to avoid breaking bonds (called the* movable set*) will also be moved. A more detailed description on how to find the movable set can be found in [32].*

While changing the relative position between the two nubots, we set the first one as a pivot and move the second one and its movable set.

In the original nubot model, there is another type of movement rule called agitation. Agitation can be applied to any nubot moving towards any direction. In all of our system constructions, all bonds are rigid, thus agitation never affects the system. In the impossibility proof, agitations can be simulated by movement rules and do not add any power to the system. Therefore, for the rest of our paper, we assume that agitation never happens, but our results still hold even if we allow agitation.

Definition 4 (Disappearance). *The disappearance models the detachment of a nubot from the main structure. When the disappearance rule happens on a nubot, it breaks all bonds with all the neighboring nubots and transitions to the empty state. The disappearance rule is denoted as* $(s_1, s_2, b, \boldsymbol{u}) \rightarrow (s_1', empty, 0, \boldsymbol{u})$ *or* $(s_1, s_2, b, \boldsymbol{u}) \rightarrow (empty, empty, 0, \boldsymbol{u})$.

Definition 5 (State change). *A state change indicates the interactions of a single nubot with its neighbor and is performed when the condition described in a certain state change rule is satisfied. A state change rule is denoted as* $(s_1, s_2, b, \boldsymbol{u}) \rightarrow (s_1', s_2', b, \boldsymbol{u})$ *where* $s_1 \neq s_1'$ *or* $s_2 \neq s_2'$.

Appearance, disappearance or movement rules may include state change at the same time. In this way, the neighboring nubots can perceive the information and trigger other functions.

Definition 6 (Nubot system). *A nubot system* $\mathcal{N} = (c_0, \mathcal{R})$ *is a pair where* c_0 *is an initial configuration and* $\mathcal{R}$ *is a rule set.*

A nubot system evolves as a asynchronous, continuous-time Markov chain. Technically, if there are k rules which are able to be applied to the configuration c_i, then the probability of any rule being applied to c_i is $1/k$, and the time needed for that chosen rule to be applied to c_i is an exponential random variable with expected value equal to $1/k$.

2.2 Our Setting

In this subsection, we describe our restrictions to the original nubot system.

Layered Nubot Model. In a layered nubot model, the goal is to construct a line on the x-axis ($y = 0$). All nubots are only allowed to move along the x-direction. The *main layer* denotes the line $y = 0$ on the triangular grid. The line $y = i$ is called a *supplementary layer* for every $i \neq 0$. In an *n-layered nubot model*, all nubot systems can only use the main layer plus $n - 1$ supplementary layers. We also make an extra assumption that nubots on the main layer never disappear and thus the growth is restricted to the main layer in our following constructions.

In our work, we focus on 2-layered and 3-layered nubot models. For the 3-layered nubot model, our system construction only uses the layers $y = 1$ and $y = -1$. Although the line doubling construction in [5] is achieved by breaking the nubot assembly into two pieces, since it is not possible to build such structures in the 2 or 3-layered nubot models, the agitation won't affect the system once the configuration is disconnected. It is reasonable for us to make an assumption that the nubot system is fully connected.

N-change and N'-change nubot model. Basically, both the N-change and N'-change nubot model mean that every monomer can only undergo at most N state changes. The only difference is on the disappearance rules.

The definition in the original nubot model does not directly tell us how to implement multiple nubot-bond deletion in molecules. In the original nubot model, when two nubots A and B react with each other, causing B to disappear, B just goes away without any state change. However, in order to explain how to cut the bonds, one explanation may be changing to a special state, which tries to cut all bonds with adjacent nubots in practice. This does not affect the original nubot model, since there is no limit on the number of state changes each nubot can make. In this paper, we make a restriction that the only way for a nubot to disappear is changing its state into a special state called a "waste state".

Definition 7 (Waste state). *A waste state W is a special state satisfying the following properties: first in all disappearance rules, the nubot that disappears must be in the waste state; second, nubots in waste states do not participate in any reaction except disappearance; third, a nubot in the waste state can react with all neighbors in all directions and disappear.*

An *N-change* nubot model is a nubot model in which every nubot can undergo at most N state changes, excluding the step of changing into the waste state. This is the number of state changes calculated using the original nubot model. An *N'-change* nubot model is a nubot model in which every nubot can undergo at most N state changes, including the step of changing into the waste state. This is more restrictive, but seems to be a reasonable model unless the actual implementation has a special way of deleting nubots.

In the rest of this paper, we use upper case letters to represent the nubots on the main layer and lower case letters for the nubots on the supplementary layer. The state change remaining for each nubot is represented as a subscript of the name of state. Take $A_0 - B_1$ for example, 0 and 1 represent there is no state change left for the state A and 1 state change left for the state B on the main layer.

Definition 8 (Reaction sequence). *A reaction sequence, denoted as* $(c_t)_{t=1}^{\infty}$, *is a series of configurations through time where c_0 is the initial configuration of the system, and c_i is the configuration right after the i-th rule is applied to c_{i-1} for $i > 0$.*

A reaction sequence is *valid* in a nubot system Γ if all rules (reactions) are in the nubot system. A reaction sequence is *proper* in a nubot system Γ if it is valid and for all valid reaction sequences starting from c_0, all configurations consist of exactly one connected component.

In N or N'-change models, we assume that the initial (seed) configuration of any nubot system consists of 2 nubots A, B on the main layer with 1 state change left. Furthermore, any nubot is not allowed to appear on the main layer to the left of A or to the right of B throughout the whole reaction sequence. Therefore, the only way to increase the length of the line is to insert nubots in the middle. We want to construct nubot systems in which the expected number of total insertions is exponential in time.

3   Result

Considering a Markov process of one insertion site growing into 2^k insertion sites, we may classify the growing process into k stages, and note the expected time of i-th insertion-site-growing step of any stage j as T_{ji}, as shown in Fig. 1.

Let X_i be the total time consumed when the i-th insertion site is generated in stage k:

$$X_1 = T_{11} + T_{21} + \cdots + T_{k1}$$
$$X_2 = T_{11} + T_{21} + \cdots + T_{k2} \tag{1}$$
$$X_{2^k} = T_{12} + T_{24} + \cdots + T_{k(2^k)}$$

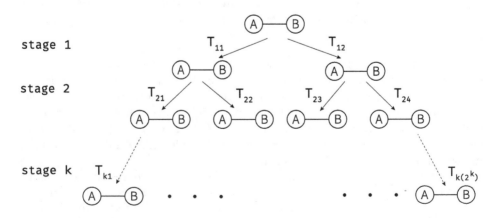

Fig. 1. Tree view

We define a *regular system* with the following property: there are $O(1)$ steps for one insertion site to grow into two insertion sites, regardless of how many reactions happen initially. Namely, $T_{ji} = O(1)$ $\forall j, i$.

The next theorem shows that the expected time of $\max X_i$ is $O(k)$

Lemma 1. *For any regular system,* $\mathbb{E}\left[\max_i X_i\right] = O(k)$

Proof. The proof that a line of length 2^k can be constructed in $O(k)$ time is similar to the proofs in [32].

Let t be the max expected time for any configuration to generate a new insertion site. For regular systems,

$$\mathbb{E}[T_{ji}] \leq t, \ \forall j, i \tag{2}$$

By Markov's inequality,

$$\Pr[T_{ji} \geq 2t] \leq \frac{1}{2} \tag{3}$$

By the definition of regular systems, if the configuration doesn't generate an insertion site at time $2t$, we can wait for another $2t$.

$$\Pr[T_{ji} \geq 4t \mid T_{ji} \geq 2t] \leq \frac{1}{2} \tag{4}$$

Therefore,

$$\Pr[T_{ji} \geq 2tm] \leq \frac{1}{2^m} \tag{5}$$

Construct

$$\begin{aligned} X_1' &= (Z_{11} + 2t) + (Z_{21} + 2t) + \cdots + (Z_{k1} + 2t) \\ &= Z_{11} + Z_{21} + \cdots + Z_{k1} + 2kt \end{aligned} \tag{6}$$

where Z_{ji} are geometric random variables with $\mathbb{E}[Z_{ji}]$ equal to $4t$.

The C.D.F. of Z_{ji}

$$F_{Z_{ji}}(u) = 1 - 2^{-\frac{u}{2t}} \tag{7}$$

so that we can bound X_i using X_i'.

Using Chernoff bound,

$$\begin{aligned} &\Pr\left[X_i' - 2kt > (1 + \delta) \times 2kt\right] \\ &= \Pr[Z_{11} + Z_{21} + \cdots + Z_{k1} > (1 + \delta) \times 2kt] \\ &\leq \left(\frac{1 + \delta}{e^\delta}\right)^k \end{aligned} \tag{8}$$

For $M \geq 0$ and any j, i,

$$\begin{aligned} &F_{Z_{ji}}(M) \leq F_{T_{ji}}(M) \\ &\Rightarrow \Pr[T_{ji} > M] \leq \Pr[Z_{ji} > M] \\ &\Rightarrow \Pr[X_i > M] \leq \Pr[X_i' > M] \end{aligned} \tag{9}$$

Consider max X_i and from Eqs. 8 and 9,

$$\Pr\left[\max X_i > (2 + \delta) \times 2kt\right]$$
$$\leq \Pr\left[\max X_i' > (2 + \delta) \times 2kt\right]$$
$$\leq 2^k \times \left(\frac{1 + \delta}{e^\delta}\right)^k \tag{10}$$
$$\leq e^{-0.2k}, \text{ for } \delta \geq 2$$

Therefore,

$$\mathbb{E}[\max X_i] = O(k) \tag{11}$$

□

3.1 Upper Bound

Here, we show that exponential growth is possible if any of the restrictions are slightly relaxed. More specifically, we construct nubot systems for both 2-layered 1-change nubot model and 3-layered 1'-change nubot model which can grow exponentially. Both systems constructed in this subsection use $O(1)$ total states and will grow exponentially to an unbounded final length.

Theorem 1. *In 3-layered, 1'change nubot system, there exists a set of nubot rules that allow nubots to grow exponentially to an infinite length of line using $O(1)$ states. Furthermore, time needed for the system to grow to length k is $O(\log k)$. The complete set of rules are omitted due to space constraints.*

Proof. Prove by construction:

The design of 3-layered 1'-change nubot which can grow exponentially includes two sub-procedures, which we call `lengthen_lr` and `lengthen_rl`, which are different from their directions of growing on the supplementary layer, as shown in Fig. 2. The main function of them are to lengthen the main layer at a constant rate. Although they are not the crucial steps of the ability to grow exponentially, they provide a method to make the distance between insertion sites far enough and guarantee that the growing processes of different insertion sites do not interfere each other.

The procedure `lengthen_lr` starts with two connected nubots $A_0 - B_1$ and is to insert nubots, $C_1 - D_0$ inside them. In order to create space, a, b, c attaches to A_0 one by one. After that, B_1 reacts with c, changes its state into B_0 and breaks bond with A_0, forming a "bridge", as described in step 2. Two movement rules apply, one between A_0 and a, the other between B_0 and c so that A_0 and B_0 are able to swing just like two "legs". When they are separated, C_1 and D_1 may attach, as described in step 3.

Note that on the supplementary layer, b attaches to a, and c attaches to b. When breaking the bridge, out algorithm is such that we intend that there are no nubots staying on the supplementary layer without any connection to the main layer. Therefore, we choose c to be changed into a waste state. To do this, we must let D know the connection of C instead of letting C know the connection of D. More precisely, in the procedure we design, D_1 detects the presence of C_1, connecting C_1, and then changes its state to D_0. D_0 then turns c into a waste state, breaking the original bridge, as shown in step 4 and 5. Note that we still draw nubot c in the step 5 on the top of Fig. 2 because it is possible that another nubot c attaches to b again after c is turned into a waste state. As a result, the procedure `lengthen_lr` leads to the configuration where C_1 − D_0 is inserted between A_0 and B_0.

After that, `lengthen_rl`, which is symmetric to `lengthen_lr`, may be applied to C_1 − D_0 and will insert A_0 − B_1. The configuration returns to step 1 of `lengthen_lr` and enables the main layer to keep growing, as shown in the lower row of Fig. 2.

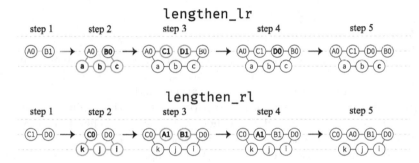

Fig. 2. Lengthen sub-procedure

The two sub-procedures mentioned above will be used to construct the final design of 3-layered nubot model. Figure 3 shows the brief process. We present the growing process in the following passage:

With the initial configuration of two nubots, A_0 and B_1, in order to create space for insertion, a, b, c attaches to A_0 one by one with the help of the upper layer to connect with B_1. After they are connected, B_1 reacts with c, changes its state to B_0 and breaks the bond with A_0, forming a "bridge", as described in step 2.

Two movement rules apply, one between A_0 and a, the other between B_0 and c. When they are separated, C_1 and D_1 may attach, as described in step 3. After C_1 appears, e, f, g attaches to C_1 one by one on the lower layer so as to reach D_1. On reaching, D_1 reacts with g, changes its state to D_0, indicating that there is another bridge (on the lower layer) connecting two parts of the monomers, as described in step 4.

At this moment, D_0 is very important because the forming of D_0 shows that there are now 2 "bridges" both connecting A_0 and B_0. Because of the existence of 2 bridges, D_0 is able to destroy the upper layer by reacting with c and turning c into waste state, as shown in step 5.

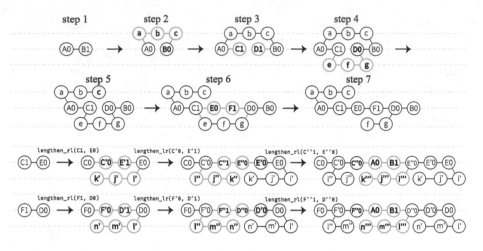

Fig. 3. 3-layered 1'-change nubot system

After C_1 and D_0 are separated, E_1 and F_1 may attach. E_1 detects the presence of F_1, connecting F_1, changing itself into E_0, and turns e into a waste state, which breaks the lower bridge, as shown in step 6 and 7.

Note that there are 2 potential insertion sites on the main layer at this moment: one is $C_1 - E_0$ and the other is $F_1 - D_0$. Therefore, if we are able to insert $A_0 - B_1$ into both of the insertion sites and make them operate independently just like they are starting from step 1 in Fig. 3, they will grow exponentially.

The last 2 rows illustrate how to insert $A_0 - B_1$ into the two insertion sites. We use the sub-procedures, `lengthen_lr` and `lengthen_rl`, which is described in Fig. 2 to complete these subtasks. Take the case of $C_1 - E_0$ for example, `lengthen_rl`, `lengthen_lr`, and then `lengthen_rl` are executed sequentially, which insert $C'_0 - E'_1$, $C''_1 - E''_0$, and finally $A_0 - B_1$, respectively, inside the original insertion site. The similar approach is used in the case of $F_1 - D_0$. As a result, the distance between each $A_0 - B_1$ is at least 3, which is far enough in our design and guarantees that the growing process of one insertion site won't get stuck by others, and each of them is able to operate independently, which leads to exponential growth.

Because the system constructed is a regular system, the theorem follows from lemma 1.   □

Theorem 2. *In 2-layered, 1-change nubot model, there exists a set of nubot rules that allow nubots to grow exponentially to an infinite length of line using $O(1)$ states. Furthermore, time needed for the system to grow to length k is $O(\log k)$.*

Proof. Prove by construction:

As shown in Fig. 4 and Table 1, we assume an initial configuration of two nubots, A_0 and B_1. $A_0 - B_1$, as a potential insertion site, will start to open up by forming a linkage through the supplementary layer (step 2). Once nubot B

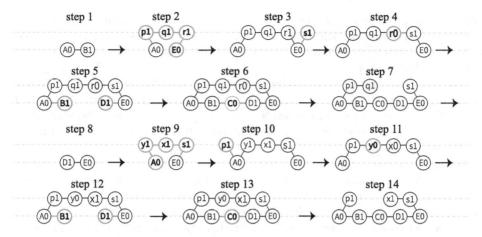

Fig. 4. 2-layered 1-change nubot system.

changes its state to E_0, movement rules apply to nubot A and E will generate 2 empty spots on the main layer.

However, we try to generate an extra empty space on the main layer so that another insertion site is able to form. We add an appear rule to the northeast of nubot E (step 3). Contrary to the 1'-change system, we allow 2 state changes in the auxiliary layer including the change into the waste state, so nubot s can link to nubot r, and nubot r itself can change its state and break the bond with nubot E (step 4), successfully generating an extra empty spot on the main layer.

Finally, nubots B, C attach to nubot A, nubot D attaches to nubot E, and makes a rigid bond with nubot C by changing the state of nubot C (step 5 and 6). Nubot C then clears the nubots in the auxiliary layer by changing the state of nubot r into waste state (step 7).

On the main layer, there are now two insertion sites, $A_0 - B_1$ and $D_1 - E_0$, separated by C_0. Step 8 to 14 describe the insertion of $D_1 - E_0$, basically it's the symmetric mechanism to step 1 to 7 explained above, initiating from the opposite direction.

Because the system constructed is a regular system, the theorem follows from lemma 1.   □

3.2   Lower Bound

In this subsection, we prove that nubot systems in 2-layered 1'-change nubot model cannot grow exponentially, as stated in the following theorem.

Theorem 3. *In any 2-layered 1'-change nubot system, the expected number of nubots appearing on the main layer before time T is at most $3kT$, where k is the number of appearance rules.*

Proof. The proof essentially enumerates all reachable configurations and shows that no configuration with 3 empty spots on the main layer will ever be generated. All nubot systems mentioned in the subsection are in the 2-layered 1'-change nubot model.

In order to enumerate all configurations, we identify some special configurations and classify these configurations into different sets. Eventually, we will show that there is no way to generate 3 empty spots on the main layer starting from any of these special configurations listed. All sets of special configurations are listed in Fig. 5.

The set S_1 contains all configurations in which two nubots A_0 and B_0 on the main layer are connected by three nubots p, q, r on the supplementary layer. Removing any of the three nubots p, q, r or any bond between them disconnects the structure into two pieces. All nubots on the main layer left of A_0 and right of B_0 are nubots that cannot make any further state changes. Notice that there might be other nubots on the supplementary layer and between A_0 and B_0. We call this structure $A_0 - p - q - r - B_0$ a bridge. The bonds between these nubots can be either flexible or rigid. We divide configurations into different sets depending on their bridge structures. All bridge structures are listed in Fig. 5.

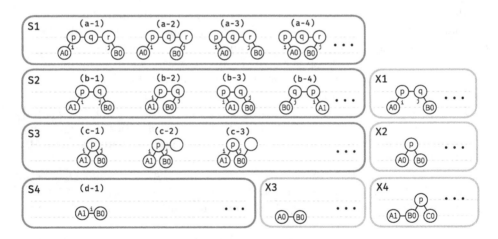

Fig. 5. A configuration is in a particular set if it has the bridge structure described in the figure. All bonds can be either rigid or flexible. The numbers written on each nubot denote the number of state changes remaining.

Given a configuration set S and a positive number k, we also define a set 3-feasible(S, k) to be all valid sequence of configurations starting from configurations in S and produces any configuration with three empty spots on the main layer in k configuration changes.

Definition 9. *3-feasible$(S, k) = \{(c_t)_{t=1}^k \mid c_0 \in S$ and c_k is the first configuration with 3 empty spots on the main layer$\}$, where S is a set of configurations.*

Table 1. Rules of exponential growth in the 2-layered 2'-change nubot model. The notations (State1, State2, Bond, Direction, State1', State2', Bond', Direction') in this table correspond to the definitions in Sect. 2.

No.	State1	State2	Bond	Direction	State1'	State2'	Bond'	Direction'
1	A_0	Empty	0	NW	A_0	p_1	1	NW
2	p_1	Empty	0	E	p_1	q_1	1	E
3	q_1	Empty	0	E	q_1	r_1	1	E
4	E_0	Empty	0	NE	E_0	s_1	1	NE
5	A_0	Empty	0	E	A_0	B_1	1	E
6	E_0	Empty	0	W	E_0	D_1	1	W
7	B_1	Empty	0	E	B_1	C_1	1	E
8	s_1	Empty	0	W	s_1	x_1	1	W
9	x_1	Empty	0	W	x_1	y_1	1	W
Connect								
10	C_1	D_1	0	E	C_0	D_1	1	E
11	B_1	r_1	0	NE	E_0	r_1	1	NE
12	A_0	E_0	1	E	A_0	E_0	0	E
13	r_1	s_1	0	E	r_0	s_1	1	E
14	E_0	r_0	1	NW	E_0	r_0	0	NW
15	D_1	y_1	0	NW	A_0	y_1	1	NW
16	p_1	y_1	0	E	p_1	y_0	1	E
17	A_0	y_0	1	NE	A_0	y_0	0	NE
Movement								
18	E_0	r_1	1	NE	E_0	r_1	1	NW
19	A_0	y_1	1	NW	A_0	y_1	1	NE
20	A_0	p_1	1	NE	A_0	p_1	1	NW
1	A_0	p_1	1	NW	A_0	p_1	1	NE
22	E_0	s_1	1	NE	E_0	s_1	1	NW
23	E_0	s_1	1	NW	E_0	s_1	1	NE
Disappear								
24	C_0	r_0	0	NE	C_0	Waste	0	NE
25	C_0	y_0	0	NW	C_0	Waste	0	NW
26	p_1	p_1	0	E	Waste	p_1	0	E
27	s_1	s_1	0	W	Waste	s_1	0	W
28	A_0	r_1	0	NW	A_0	Waste	0	NW
29	A_0	r_0	0	NW	A_0	Waste	0	NW
30	E_0	y_1	0	NE	E_0	Waste	0	NE
31	E_0	y_0	0	NE	E_0	Waste	0	NE
32	E_0	r_0	1	NE	Waste	r_0	0	NE
33	A_0	y_0	1	NW	Waste	y_0	0	NW

Definition 10. *A reaction sequence in 3-feasible(S, k) is a valid reaction sequence of a nubot system Γ if the reaction from c_{i-1} to c_i is a reaction in Γ. A reaction sequence in 3-feasible(S, k) is a proper reaction sequence of a nubot system Γ if it is valid and all valid reaction sequences of Γ starting from state c_0 do not cut the structure into two disconnected components.*

Notice that starting from any special configuration in some set S_i defined in Fig. 5, any configuration sequence 3-feasible(S_i, k) must eventually break the bridge structure. This can be done in two ways: either making a nubot on the bridge disappear or having two nubots on the bridge reacting with each other and cut the bond between them. The former one requires a reaction R_1 that changes the nubot into a waste state. The latter one requires one of the reacting nubots to change state (reaction R_2) before the reaction, otherwise the cutting-bond reaction may directly apply to the initial configuration c_0 and cut the structure into two disconnected components. Given a configuration sequence 3-feasible(S_i, k), we call the first reaction which breaks the bridge structure and the corresponding state change reaction (R_1 or R_2) *breaking reactions*. Notice that the two breaking reactions can always happen consecutively. Furthermore, if the reactions R_1 and R_2 can directly apply to the initial configuration c_0, disappearance rule/cutting bond may directly follow and the structure may still become disconnected. Suppose that the first configuration that R_1 or R_2 may apply is c_i, then we define the *critical reaction* of the sequence 3-feasible(S, k) to be the reaction from c_{i-1} to c_i.

We show that for any 2-layered 1'-change nubot system, there will never be three empty spots on the main layer simultaneously. Therefore, if there are k appearance rules in the system, the expected rate at which appearance rules happen on the main layer is at most $3k$. The proof essentially enumerates all configurations reachable from the initial configuration and thus is omitted due to space constraints.   □

References

1. Nubot-simulator (2014). https://github.com/domardfern/Nubot-Simulator
2. Adleman, L., Cheng, Q., Goel, A., Huang, M.-D.: Running time and program size for self-assembled squares. In: Proceedings of the 33rd Annual ACM Symposium on Theory of Computing, pp. 740–748 (2001)
3. Barish, R.D., Rothemund, P.W.K., Winfree, E.: Two computational primitives for algorithmic self-assembly: copying and counting. Nano Lett. **5**(12), 2586–2592 (2005)
4. Bishop, J., Klavins, E.: An improved autonomous DNA nanomotor. Nano Lett. **7**(9), 2574–2577 (2007)
5. Chen, H.-L., Doty, D., Holden, D., Thachuk, C., Woods, D., Yang, C.-T.: Fast algorithmic self-assembly of simple shapes using random agitation. In: Murata, S., Kobayashi, S. (eds.) DNA 2014. LNCS, vol. 8727, pp. 20–36. Springer, Cham (2014). doi:10.1007/978-3-319-11295-4_2
6. Chen, H.-L., Schulman, R., Goel, A., Winfree, E.: Error correction for DNA self-assembly: preventing facet nucleation. Nano Lett. **7**, 2913–2919 (2007)

7. Cheng, Q., Goel, A., Moisset, P.: Optimal self-assembly of counters at temperature two. In: Proceedings of the 1st Conference on Foundations of Nanoscience: Self-Assembled Architectures and Devices, pp. 62–75 (2004)
8. Dietz, H., Douglas, S., Shih, W.: Folding DNA into twisted and curved nanoscale shapes. Science **325**, 725–730 (2009)
9. Ding, B., Seeman, N.: Operation of a DNA robot arm inserted into a 2D DNA crystalline substrate. Science **384**, 1583–1585 (2006)
10. Dirks, R.M., Pierce, N.A.: Triggered amplification by hybridization chain reaction. Proc. Natl. Acad. Sci. **101**(43), 15275–15278 (2004)
11. Doty, D.: Randomized self-assembly for exact shapes. In: Proceedings of the 50th Annual IEEE Symposium on Foundations of Computer Science, pp. 85–94 (2009)
12. Douglas, S., Dietz, H., Liedl, T., Hogberg, B., Graf, F., Shih, W.: Self-assembly of DNA into nanoscale three-dimensional shapes. Nature **459**, 414–418 (2009)
13. Fu, T.-J., Seeman, N.C.: DNA double crossover structures. Biochemistry **32**, 3211–3220 (1993)
14. Green, S., Bath, J., Turberfield, A.: Coordinated chemomechanical cycles: a mechanism for autonomous molecular motion. Phys. Rev. Lett. **101**, 238101 (2008)
15. Kao, M.-Y., Schweller, R.: Reducing tile complexity for self-assembly through temperature programming. In: Proceedings of the 17th Annual ACM-SIAM Symposium on Discrete Algorithms, pp. 571–580 (2006)
16. Lagoudakis, M., LaBean, T.: 2D DNA self-assembly for satisfiability. In: Proceedings of the 5th DIMACS Workshop on DNA Based Computers in DIMACS Series in Discrete Mathematics and Theoretical Computer Science, vol. 54, pp. 141–154 (1999)
17. Pei, R., Taylor, S., Stojanovic, M.: Coupling computing, movement, and drug release (2007)
18. Reif, J.H., Sahu, S.: Autonomous programmable DNA nanorobotic devices using DNAzymes. In: Proceedings of the Thirteenth International Meeting on DNA Based Computers. Memphis, TN, June 2007
19. Rothemund, P., Winfree, E.: The program-size complexity of self-assembled squares (extended abstract). In: Proceedings of the 32nd Annual ACM Symposium on Theory of Computing, pp. 459–468 (2000)
20. Rothemund, P.W.K.: Folding DNA to create nanoscale shapes and patterns. Nature **440**, 297–302 (2006)
21. Rothemund, P.W.K., Papadakis, N., Winfree, E.: Algorithmic self-assembly of DNA sierpinski triangles. PLoS Biol. **2**, 424–436 (2004)
22. Seelig, G., Soloveichik, D., Zhang, D., Winfree, E.: Enzyme-free nucleic acid logic circuits. Science **314**, 1585–1588 (2006)
23. Sherman, W.B., Seeman, N.C.: A precisely controlled DNA bipedal walking device. Nano Lett. **4**, 1203–1207 (2004)
24. Shih, W.M., Quispe, J.D., Joyce, G.F.A.: A 1.7-kilobase single-stranded DNA that folds into a nanoscale octahedron. Nature **427**, 618–621 (2004)
25. Shin, J.-S., Pierce, N.A.: A synthetic DNA walker for molecular transport. J. Am. Chem. Soc. **126**, 10834–10835 (2004)
26. Soloveichik, D., Winfree, E.: Complexity of self-assembled shapes. SIAM J. Comput. **36**, 1544–1569 (2007)
27. Tian, Y., He, Y., Chen, Y., Yin, P., Mao, C.: A DNAzyme that walks processively and autonomously along a one-dimensional track. Angew. Chem. **44**, 4355–4358 (2005)

28. Venkataraman, S., Dirks, R.M., Rothemund, P.W.K., Winfree, E., Pierce, N.A.: An autonomous polymerization motor powered by DNA hybridization. Nat. Nanotechnol. **2**, 490–494 (2007)
29. Win, M.N., Smolke, C.D.: Higher-order cellular information processing with synthetic rna devices. Science **322**(5900), 456 (2008)
30. Winfree, E.: Algorithmic Self-Assembly of DNA. Ph.D. thesis, California Institute of Technology, Pasadena (1998)
31. Winfree, E., Liu, F., Wenzler, L., Seeman, N.: Design and self-assembly of two-dimensional DNA crystals. Nature **394**, 539–544 (1998)
32. Woods, D., Chen, H.-L., Goodfriend, S., Dabby, N., Winfree, E., Yin, P.: Active self-assembly of algorithmic shapes and patterns in polylogarithmic time. In: Proceedings of the 4th Conference on Innovations in Theoretical Computer Science, ITCS 2013, pp. 353–354 (2013)
33. Yin, P., Choi, H.M.T., Calvert, C.R., Pierce, N.A.: Programming biomolecular self-assembly pathways. Nature **451**, 318–322 (2008)
34. Yin, P., Turberfield, A.J., Sahu, S., Reif, J.H.: Designs for autonomous unidirectional walking DNA devices. In: Proceedings of the 10th International Meeting on DNA Based Computers. Milan, Italy, June 2004
35. Yurke, B., Turberfield, A., Mills Jr., A., Simmel, F., Neumann, J.: A DNA-fuelled molecular machine made of DNA. Nature **406**, 605–608 (2000)
36. Zhang, D.Y., Turberfield, A.J., Yurke, B., Winfree, E.: Engineering entropy-driven reactions and networks catalyzed by DNA. Science **318**, 1121–1125 (2007)
37. Zhang, Y., Seeman, N.: Construction of a DNA-truncated octahedron. J. Am. Chem. Soc. **116**(5), 1661 (1994)

Robust Detection in Leak-Prone Population Protocols

Dan Alistarh[1,5], Bartłomiej Dudek[2], Adrian Kosowski[3], David Soloveichik[4], and Przemysław Uznański[1(✉)]

[1] ETH Zürich, Zürich, Switzerland
{dan.alistarh,przemyslaw.uznanski}@inf.ethz.ch
[2] University of Wrocław, Wrocław, Poland
bartlomiej.dudek@cs.uni.wroc.pl
[3] Inria Paris and IRIF, Université Paris Diderot, Paris, France
adrian.kosowski@inria.fr
[4] University of Texas, Austin, TX, USA
david.soloveichik@utexas.edu
[5] IST Austria, Klosterneuburg, Austria

Abstract. In contrast to electronic computation, chemical computation is noisy and susceptible to a variety of sources of error, which has prevented the construction of robust complex systems. To be effective, chemical algorithms must be designed with an appropriate error model in mind. Here we consider the model of chemical reaction networks that preserve molecular count (population protocols), and ask whether computation can be made robust to a natural model of unintended "leak" reactions. Our definition of leak is motivated by both the particular spurious behavior seen when implementing chemical reaction networks with DNA strand displacement cascades, as well as the unavoidable side reactions in any implementation due to the basic laws of chemistry. We develop a new "Robust Detection" algorithm for the problem of fast (logarithmic time) single molecule detection, and prove that it is robust to this general model of leaks. Besides potential applications in single molecule detection, the error-correction ideas developed here might enable a new class of robust-by-design chemical algorithms. Our analysis is based on a non-standard hybrid argument, combining ideas from discrete analysis of population protocols with classic Markov chain techniques.

1 Introduction

A major challenge in designing autonomous molecular systems is to achieve a sufficient degree of error tolerance despite the error-prone nature of the chemical substrate. While considerable effort has focused on making the chemistry itself

D. Alistarh—Supported by an SNF Ambizione Fellowship.

A. Kosowski—Supported by Inria project GANG, ANR project DESCARTES, and NCN grant 2015/17/B/ST6/01897.

D. Soloveichik—Supported by NSF grants CCF-1618895 and CCF-1652824.

R. Brijder and L. Qian (Eds.): DNA 23 2017, LNCS 10467, pp. 155–171, 2017.
DOI: 10.1007/978-3-319-66799-7_11

more robust, here we look at the possibility of developing chemical algorithms that are inherently resilient to the types of error encountered. Before designing robust chemical algorithms, we must decide on a good error model that is relevant to the systems we care about. In this paper we focus on a very simple and general error model that is motivated both by basic laws of chemistry as well as by implementation artifacts in strand displacement constructions for chemical reaction networks. We begin by listing the types of errors we aim to capture.

Leaks due to Law of Catalysis. A fundamental law of chemical kinetics is that for every catalyzed reaction, there is an uncatalyzed one that occurs at a (often much) slower rate. By a *catalytic* reaction, we mean a reaction that involves some species X but does not change its amount; this species is called a *catalyst* of that reaction. For example, the reaction $X + Y \to X + Z$ is catalytic, and species X is the catalyst since its count remains unchanged by the firing of this reaction. By the law of catalysis, reaction $X + Y \to X + Z$ must be accompanied by a (slower) *leak* reaction $Y \to Z$. (A more general formulation of the law of catalysis is that if any sequence of reactions does not change the net count of X, then there is a pathway that has the same effect on all the other species, but can occur in the absence of X (possibly much slower). Thus for example, if $X + Y \to W$ and $W \to X + Z$ are two reactions, then there must also be a leak reaction $Y \to Z$. Formally defining catalytic cycles and catalysts is non-trivial and is beyond the scope of this paper [1].)

Leaks due to Law of Reversibility. Another fundamental law of chemical kinetics is that any reaction occurs also in the reverse direction at some (possibly much slower) rate. In other words, reaction $X + Y \to Z + W$ must be accompanied by $Z + W \to X + Y$. (The degree of reaction reversibility is related to the free-energy use, such that irreversible reactions would require "infinite" free energy.)

Leaks due to Spurious Activation in Strand-Displacement Cascades. Arbitrary chemical reaction networks can in principle be implemented with DNA strand displacement cascades [2,3]. Implementations based on strand displacement also suffer from the problem of leaks [4]. The implementation of a reaction like $X + Y \to Z + W$ consists of "fuel" complexes present in excess, that hold Z and W sequestered. A cascaded reaction of the fuel complex with X and Y results in the release of active Z and W. Leaks in this case consist of Z and W becoming spuriously activated, even in the absence of active X or Y.

Importantly, for a catalytic reaction such as $X + Y \to X + Z$, it is possible to design a strand displacement implementation that does not leak the catalyst X. This implementation would release the same exact molecule of X as was consumed to initiate the process, as in the catalytic system described in [5]. Since fuels do not hold X sequestered, X cannot be produced in the leak process (although Z can).

Modeling Reactions and Leaks. Note that in all cases above, we can guarantee that a species does not leak if it is exclusively a catalyst in every reaction it occurs in. This allows us some handle on the leak. In particular, we will ensure that the species we are trying to detect (called D below) will be a catalyst in every reaction that involves it. Otherwise, there might be a leak pathway to generating D itself—which is fundamentally irrecoverable.

We express the implementations below in the *population protocol* formalism [6]. That is, we consider a system with n molecules (aka nodes), which interact uniformly at random in a series of discrete steps. A population protocol is given as a set of reactions (aka transition rules) of the form

$$A + B \to C + D.$$

Note that unlike general reaction networks, population protocols conserve total molecular count since molecules never combine or split. For this reason, compared to general chemical reaction networks, this model is easier to analyze.

Given the set of reactions defining a protocol, we partition the species into *catalytic* states, which never change count as a consequence of any reaction, and *non-catalytic*, otherwise. Crucially, we model *leaks* as spurious reactions which can consume and create arbitrary non-catalytic species. More formally, a leak is a reaction of the type

$$S \to S',$$

where S and S' denote arbitrary non-catalytic species. In the following, we do not make any assumptions on the way in which these leak transitions are chosen (i.e., they could in theory be chosen *adversarially*), but we assume an upper bound on the *rate* at which leaks may occur in the system, controlled by a parameter β.

Leak-Robust Detection. A computationally simple task which already illustrates the difficulty of information processing in such an error-prone system is *single molecule detection*. Consider a solution of n molecules, in which a single molecule D may or may not be present. Intuitively, the goal is to generate large-scale (in the order of n) change in the system, depending on whether or not D is present or absent. Our time complexity measure is *parallel time*, defined as the number of pairwise interactions, divided by n. This measure of time naturally captures the parallelism of the system, where each molecule can participate in a constant number of interactions per unit time. Subject to leaks, our goal is to design the chemical interaction rules (formalized as a population protocol) to satisfy the following behavior. If D is present then it is detected fast, in logarithmic parallel time, and that the output is probabilistically "stable" in the sense that sampled at a random future time the system is in the "detected configuration" with high probability. By contrast, if D is absent, then the system sampled at a random future time should be in the "undetected configuration" with high probability. This basic task has several variations, for instance signal amplification or approximate counting of D.

We first develop some intuition about this problem, by considering some strawman approaches.

A first trivial attempt would be to have neutral molecules become "detectors" (state T) as soon as they encounter D, that is,

$$D + N \to D + T.$$

This approach suffers from two fatal flaws. First, it is *slow*, in that detection takes *linear* parallel time. Second, it has no way from recovering from leaks of the type $N \rightarrow T$.

A second attempt could try to implement an epidemic-style detection of D, that is:

$$D + N \rightarrow D + T$$

$$T + N \rightarrow T + T.$$

This approach is *fast*, i.e. converges in *logarithmic* parallel time in case D is present. However, if D is not present, the algorithm converges to a *false positive* state: a leak of the type $N \rightarrow T$ brings the system to an all-T state, despite the absence of D. One could try to add a "neutralization" pathway by having T turn back to N after a constant number of interactions, but a careful analysis shows that this approach also fails to recover from leaks of the type $N \rightarrow T$.

Thus, it is not clear whether leak-resistant detection is possible in population protocols (or more generally chemical reaction networks). There has been considerable work in the algorithmic community on diffusion based models, e.g. [7]. However, such results do not seem to apply to this setting, since leak models have not been considered previously, and none of the known techniques are robust to leaks. In particular, it appears that techniques for *deterministic* computation in population protocols do not carry over in the presence of leaks. More generally, this seems to create an unfortunate gap between the algorithmic community, which designs and analyzes population protocols in leak-free models, and more practically-minded research, which needs to address such implementation issues.

Contribution. In this paper, we take a step towards bridging this gap. We provide a general algorithmic model of leaks, and apply it to the detection problem. Specifically, our immediate goal is to elucidate the question of whether efficient, leak-robust detection is possible.

We prove that the answer is yes. We present a new algorithm, called *Robust-Detect*, which guarantees the following. Assume that the rate at which leaks occur is upper bounded by $\beta/n \ll 1/n$, and that we return the output mapping (detect/non-detect) of a randomly chosen molecule after $O(\log n)$ parallel time. Then the probability of a *false negative* is at most $1/e + o(1)$, and the probability of a *false positive* is at most β. (Note that as the total molecular count n increases, the chance that a particular interaction involves D decreases linearly with n. Thus the leak rate must also decrease linearly with n, or else the leaks will dominate. Alternatively, we can view some fixed leak rate as establishing an upper bound on the molecular count n, see below.)

Algorithm Description. We now sketch the intuition behind the algorithm and its properties, leaving the formal treatment to Sects. 4 and 5. Fix a parameter $s \geq 1$, to be defined later. We define a set of "detecting" species $X_1, \ldots, X_s$, arranged in consecutive levels. Whenever a molecule meets D, it moves to the highest "alert" level, X_1. Since leaks might produce this species as well, we decay it gracefully across s levels. More precisely, whenever a molecule at level X_i meets another molecule at level X_j, both molecules move to state $X_{\min(i,j)+1}$.

A molecule which would move beyond level X_s following a reaction becomes neutral, i.e. moves to species N. Nodes in state X_i with $i < s$ turn N into X_{i+1}, whereas molecules in state X_s also become neutral when interacting with N.

Analysis. Intuitively, the algorithm's dynamics for the case where a single molecule is in state D are as follows. The counts of molecules in state X_i tend to increase *exponentially* with the alert level i, up to levels $\approx \log n$, when the count becomes a constant fraction of n. However, once level $\log n$ is reached, these counts decrease *doubly exponentially*. Thus, it suffices to set $s = \log n$ to obtain that a fraction of at least $(1 - 1/e)$ molecules are in one of the alert states X_i in case D is present. It is not hard to prove that leaks cannot meaningfully affect the convergence behavior in this case.

The other interesting case is when D is not present, but leaks may occur, leading to possible false positives. Intuitively, we can model this case as one where states X_1 at the highest alert level simply are created at a lower rate $\beta/n \ll 1/n$. A careful analysis of this setting yields that the probability of a *false positive* (D detected, but not present) in this case is at most β, corresponding to the leak rate parameter.

Our analysis technique works by characterizing the stationary behavior of the Markov chain corresponding to the algorithm, and the convergence properties (mixing time) of this chain. For technical reasons, the analysis uses a non-standard hybrid argument, combining ideas from discrete analysis of population protocols with classic Markov chain techniques. The argument proves that the algorithm always stabilizes to the correct output in logarithmic parallel time.

The analysis further highlights a few interesting properties of the algorithm. First, if the detectable species D is present in a higher count $k > 1$, then the algorithm effectively skips the first $\log k$ levels, and thus requires $\log(n/k) + O(\log \log n)$ states. Second, it is not necessary to know the exact value of $\log n$, as the counts of species past this threshold decrease doubly exponentially.

Alternative Formulations. An alternative view of this protocol is as solving the following related *amplification* problem: we are given a signal of strength (rate) ϕ, and the algorithm's behavior should reflect whether this strength is below or above some threshold. The detection problem requires us to differentiate thresholds set at β/n and $1/n$, for constant $\beta \ll 1$, but our analysis applies to more general rates.

Above, we have assumed that the leak rate decreases linearly with n, to separate from the case where a single instance of D is present. However, it is also reasonable to consider that the leak rate is *fixed*, say, upper bounded by a constant λ. In this case, the analysis works as long as the number of molecules n satisfies $\lambda \ll 1/n$.

Self-stabilization. Our algorithm is self-stabilizing in the sense that if the count of D changes due to some external reason, the output quickly adapts (within logarithmic parallel time). This is particularly interesting if the algorithm is used in the context of a "control module" for a cell detecting D and the amount of

D changes over time. Note that strawman solutions considered above cannot be "untriggered" once D has been detected, and thus cannot adapt to a changing input.

2   Related Work

There is much work on attempting to decrease error in the underlying chemical substrate. A famous example includes kinetic proofreading [8]. In the context of DNA strand displacement systems in particular, leak reduction has been a prevailing topic [9]. Despite the importance of handling leaks, there are few examples of non-trivial *algorithms*, where leaks are handled through computation embedded in chemistry. One algorithm that appears to be able to handle errors is *approximate majority* [10], originally analyzed in a model where a fraction of the nodes are Byzantine, in that they can change their reported state in an adversarial way. Potentially due to its robustness properties, the approximate majority algorithm appears to be widely used in biological regulatory networks [11], and it was also one of the first chemical reaction network algorithms implemented with strand displacement cascades [4].

Our algorithm can be viewed as a timed, self-stabilizing version of rumor spreading. For analysis of simple rumor-spreading, see [12]. Other work include fault-tolerant rumor spreading [13], push-pull models [7] and self-stabilizing broadcasting [14]. A rumor-spreading formulation of the molecule detection problem is also considered in recent work [15], which relies on a different source amplification mechanism based on oscillator dynamics. This protocol [15] is self-stabilizing in a weaker (probabilistic) sense compared to the algorithms from this paper and does not provide leak robustness guarantees.

3   Preliminaries

3.1   Population Protocols with Leaks

Population Protocols. We start from a standard population protocol model, where n molecules (nodes) interact uniformly at random in a series of discrete steps. In our formulation, in each step, a coin is flipped to decide whether the current interaction is a regular reaction or a leak reaction. In the former case, two molecules are picked uniformly at random, and interact according to the rules of the protocol. In the latter case, a leak reaction occurs (see below).

A population protocol is given as a set of reactions (transition rules) of the form

$$A + B \rightarrow C + D,$$

(where some of A, B, C, D might be the same). We (arbitrarily) match the first reactant (A) with the first product (C), and the second reactant (B) with the second product (D), and think of A as changing state to C, and B as changing state to D. If the two molecules picked to interact do not have a corresponding interaction rule, then they don't change state and we call this a null interaction. Population protocols are a special case of the stochastic chemical reaction networks kinetic model (e.g., [16](A.4)).

Catalytic and Non-Catalytic Species. Given a set of reactions, we define the set of *catalytic* species as the set of states which never change as a consequence of any reaction. That is, for every reaction, the species is present in the same count both in the input and the output of the reaction. For example, in the reactions

$$A + C \to B + C$$
$$A + B \to A + D$$

we call C *catalytic*. Note that A acts as a catalyst in the second reaction, but its count is changed by the first reaction, thus it is not overall catalytic. All species whose count is modified by some reaction are called non-catalytic. Note that it is possible that a species is never created, but disappears as a consequence of an interaction. For example, in the reaction

$$L + L \to A + B,$$

L is such as species. We define such species as *non-catalytic*, since their creation is possible by the law of reversibility, and thus they can leak.

An Algorithmic Model of Leaks. A leak is a reaction of the type

$$S \to S'$$

where S and S' are arbitrary non-catalytic species produced by the algorithm. Note that the input and output species of a leak may be the same (although in that case the reaction is trivial). In the following, we make no assumptions on the way in which the input and output of a leak reaction are chosen—we assume that they are chosen adversarially. Instead, we assume an absolute bound on the probability of a leak.

We assume that each reaction is either a *leak reaction* or a *normal reaction*, which follows the algorithm. We formalize this as follows.

Definition 1. *Given an algorithm, defined by a set of reactions, the set of catalysts is the set of species whose count does not change as a consequence of any reaction. A leak is a spurious reaction, which changes an arbitrary non-catalytic species to an arbitrary non-catalytic species. The leak rate β/n is the probability that any given interaction is a leak reaction.*

3.2 The Detection Problem

In the following, we consider the following *detection* task: we are given a distinct species D, whose presence or absence must be detected by the algorithm, in the presence of leaks. More precisely, if the species D is present, then the algorithm should stabilize to a state in which molecules map to output value "detect". Otherwise, if D is not present, then the algorithm should stabilize to a state in which molecules map to output value "non-detect". To observe the algorithm's output, we sample a molecule at random, and return its output mapping. (Alternatively, to boost accuracy, we can take a number of samples, and return the majority output mapping.) We require that species D are catalytic.

4   The Robust-Detect Algorithm

Description. As given in the problem statement, we assume that there exists a distinguished species D, which is to be detected, and which never changes state. Our algorithm implements a chain of detection species $X_1, \ldots, X_s$, for some parameter s, each of which maps to output "detect", but with decreasing "confidence". Further, we have a neutral species N, which maps to output "non-detect". We assume that the parameter $s = \lceil \log n \rceil$, and that initially all molecules are in state N. We specify the transitions below, and provide the intuition behind them.

Algorithm 1

$$D + X_i \rightarrow D + X_1, \quad \forall i \in \{2, \ldots, s\}$$
$$D + N \rightarrow D + X_1$$
$$X_s + X_s \rightarrow N + N$$
$$X_s + N \rightarrow N + N$$
$$X_i + X_j \rightarrow X_{\min(i,j)+1} + X_{\min(i,j)+1}, \quad \forall i, j \in \{1, 2, \ldots, s-1\}$$
$$X_i + N \rightarrow X_{i+1} + X_{i+1}, \quad \forall i \in \{1, 2, \ldots, s-1\}$$

The intuition behind the algorithm is as follows. The "detecting" species $X_1, \ldots, X_s$ are arranged in consecutive levels. Whenever a molecule meets D, it moves to the highest "alert" level, X_1. Since leaks might produce this species as well, we decay it gracefully across s levels. After going through these levels, a molecule moves to neutral state N, in case it is not brought back either by meeting D, or some molecule at a lower alert level. For this, whenever two of these species X_i and X_j meet, they both move to level $\min(i,j)+1$. This reaction has the double purpose of both decaying the alert level of the molecule at the lower level, and of bringing back the molecule with the higher alert level. Further, whenever a molecule at level X_i meets a neutral molecule N, it advances its level by 1. At the same time, neutral molecules are turned into detector molecules whenever meeting some molecule at an alert level smaller than s.

Intuitive Dynamics. Roughly, the chain of alert levels have the property that, for the first $\sim \log n$ levels, the count roughly *doubles* with level index. At the same time, past this point, counts exhibit a steep (doubly exponential) drop, so that a small constant fraction of molecules are always neutral. The presence of D acts like a trigger, which maintains the chain in "active" state. The analysis in the next section makes this intuition precise. These dynamics are illustrated in Fig. 1.

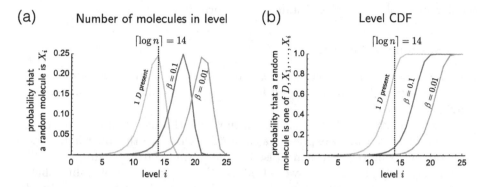

Fig. 1. Steady state probabilities of the Robust-Detect algorithm for $n = 10^4$ molecules. Three conditions are evaluated: (blue) 1 molecule of D is present and no leak (leak parameter $\beta = 0$); (orange, red) no D is present but with worst-case leak for false-positives (leak reactions $X_i \to X_1$ and $N \to X_1$) (orange: leak parameter $\beta = 0.01$, red: leak parameter $\beta = 0.1$). (a) The probabilities of each level i. (b) The cumulative probabilities of levels $\leq i$, capturing the probability that a random molecule is in a "detect" state. Note that it is enough to set the number of levels $s = 14 = \lceil \log n \rceil$ to have both false positive and false negative error probabilities small, although for smaller leak rates ($\beta = 0.01$) increasing s beyond $\log n$ can help better distinguish true and false positives. Numerical probabilities are computed using Eqs. (1) and (2).

5   Analysis

Overview. We divide the analysis of the detection algorithm into two parts. First, we derive stationary probabilities of the underlying Markov chain of transitions of particles, by solving recursively the equations following from the underlying dynamics. Later, we derive optimal bounds on the mixing time of this Markov chain—that is we show that probability distribution of states at every time $t \geq cn \log n$ (for some constant c) is almost the same as the stationary distribution.

Simplified Algorithm. For the purpose of analysis, let us consider a following rephrasing of the detection algorithm: molecule states are $D, X_1, X_2, X_3, \ldots$, and interactions are as follows:

Algorithm 2

$$D + X_i \to D + X_1,$$
$$X_i + X_j \to X_{\min(i,j)+1} + X_{\min(i,j)+1}.$$

This algorithm uses infinite number of states, thus it is useful only for purposes of theoretical analysis. However, it captures the behavior of the original algorithm in the following way: if in Algorithm 2 all states $X_{s+1}, X_{s+2}, \ldots$ are collapsed to N, the transitions are equivalent to Algorithm 1. However, formulation of Algorithm 2 is oblivious to parameter s, thus captures simultaneously the dynamics of all possible instances of Algorithm 1.

5.1 Stationary Analysis

Let us consider an initial state when $k \geq 0$ instances of state D are present, with special attention given to $k = 0$ and $k = 1$. Those molecules do not change their state.

We can imagine tracking a particular molecule through its state transitions, such that its state can be expressed as a Markov chain. In the following, we will focus on analyzing the stationary distribution of this Markov chain.

For any $i \in \{1, 2, \ldots, s\}$, we let $p_i^\star$ be the stationary probability that a molecule chosen uniformly at random is in state X_i. We let $p_0^\star = \frac{k}{n}$ be the (stationary) probability that the molecule is in the state D. Let $p_{\leq i}^\star = p_0^\star + \ldots + p_i^\star$ be the probability that a molecule is in any of the states $D, X_1, \ldots, X_i$.

Let us now analyze these stationary probabilities.

Stable State with No Leaks. We first analyze the simplified case where no leaks occur. A molecule u is in one of states $D, X_1, \ldots, X_i$ at time t, in two cases:

- It was in state $D, X_1, \ldots, X_i$ at time $t - 1$, and did not get selected for a reaction, which occurs with probability $1 - 2/n$.
- It got selected for a reaction with element u', and either u or u' was in one of states $D, X_1, \ldots, X_{i-1}$.

Hence, by stationarity, we get that

$$p_{\leq i}^\star = p_{\leq i}^\star \left(1 - \frac{2}{n}\right) + \frac{2}{n} \cdot (1 - (1 - p_{\leq(i-1)}^\star)^2).$$

From this we get that

$$1 - p_{\leq i}^\star = (1 - p_{\leq(i-1)}^\star)^2,$$

which solves to

$$p_{\leq i}^\star = 1 - \left(1 - \frac{k}{n}\right)^{2^i}. \tag{1}$$

This gives us following estimates: if $k \geq 1$, then $p_i^\star \approx 2^{i-1}\frac{k}{n}$ for $i \leq \log(n/k)$. Additionally, for $i = \log(n/k) + 1 + j$, $p_i^\star \approx e^{-2^j}$. Thus, for $i \geq \log(n/k) + \Theta(\log \log n)$ in all practicalities $p_i^\star \approx 0$.

To analyze the probability of detection when $k = 1$, we sum probabilities for all i from 0 to $s = \lceil \log n \rceil$

$$\Pr[\text{detect}] = \sum_{i=0}^{s} p_i^{\star} = p_{\leq s}^{\star} \geq 1 - \left(1 - \frac{1}{n}\right)^n \geq 1 - \frac{1}{e}.$$

Probability of False Positives with Leaks. A useful side effect of the previous analysis is that we also get probability bounds for detection in the case where D is *not* present, i.e. false positives. We model this case as follows. Assume that there exists an upper bound λ on the probability that a certain reaction is a leak. Examining the structure of the algorithm, we note that the worst-case adversarial application of leaks would be if this probability is entirely concentrated into leaks which produce species X_1.

To preserve molecular count, we assume the following simplified leak model, which is equivalent to the general one, but easier to deal with in the confines of our algorithm.

Each reaction is a *leak* with probability $\lambda = \beta/n$, where $\beta \ll 1$ is a small constant. If a reaction is a leak, it selects a molecule at random, and transforms it into an arbitrary state. In this case, we will assume adversarially that all leaked molecules are transformed into state X_1. Notice that the assumption that $\beta \ll 1$ is required to separate this setting from the case where D is present in the system, where the probability of producing state X_1 is $2/n$.

We continue with calculations of $p_0^{\star}, p_1^{\star}, \ldots$ for the above formulation. Note that the recurrence relation for $p_{\leq i}^{\star}, i \geq 1$ is changed as follows:

- If at that round there was no leak, the transition probabilities are as previously. This happens with probability $1 - \frac{\beta}{n}$.
- If there was a leak, then the molecule either is selected as a leaked molecule (this happens with probability $\frac{1}{n} \cdot \frac{\beta}{n}$) or it was not selected as a leaked molecule, but it was already in the proper state (probability $\frac{\beta}{n} \cdot \frac{n-1}{n} p_{\leq i}^{\star}$).

The recursive formulation gives

$$p_{\leq i}^{\star} = \left(p_{\leq i}^{\star}\left(1 - \frac{2}{n}\right) + \frac{2}{n}\left(1 - (1 - p_{\leq(i-1)}^{\star})^2\right)\right)\left(1 - \frac{\beta}{n}\right) + \left(\frac{1}{n} + \frac{n-1}{n}p_{\leq i}^{\star}\right)\frac{\beta}{n}.$$

Which is equivalent to

$$1 - p_{\leq i}^{\star} = \frac{\left(1 - \frac{\beta}{n}\right)}{\left(1 - \frac{\beta}{2n}\right)}(1 - p_{\leq(i-1)}^{\star})^2$$

leading to (using estimate $(1 - \frac{\beta}{n})/(1 - \frac{\beta}{2n}) \approx (1 - \frac{\beta}{2n})$)

$$p_{\leq i}^{\star} \approx 1 - \left(1 - \frac{\beta}{2n}\right)^{1+2+\ldots+2^{i-1}} = 1 - \left(1 - \frac{\beta}{2n}\right)^{2^i - 1}. \tag{2}$$

This gives us following estimates: $p_i^\star \approx 2^{i-2}\frac{\beta}{n}$ for $i \le \log(2n/\beta)$. Additionally, for $i = \log(2n/\beta) + 1 + j$, $p_i^\star \approx e^{-2^j}$. Thus, for $i \ge \log(2n/\beta) + \Theta(\log\log n)$ in all practicalities $p_i^\star \approx 0$.

This immediately implies that

$$\Pr[\text{detect}] = \sum_{i=0}^{s} p_i^\star = p_{\le s}^\star \le 1 - \left(1 - \frac{\beta}{2n}\right)^{2n} = 1 - \frac{1}{e^\beta} \approx \beta,$$

which means that the probability that a randomly chosen molecule is in detect state when chosen uniformly at random is at most β.

Probability of False Negatives with Leaks. Under the same leak model, it is easy to notice that the "best" adversarial strategy for our algorithm in case D is present is to concentrate all leaks to create the neutral species N (or X_∞ in case of Algorithm 2). It is easy to see that this just decreases the total probability of detect states by the leak probability $\lambda = \beta/n$. More formally, we compute once again stationary probabilities. The recurrent relation is

$$p_{\le i}^\star = \left(p_{\le i}^\star \left(1 - \frac{2}{n}\right) + \frac{2}{n} \cdot (1 - (1 - p_{\le(i-1)}^\star)^2)\right)\left(1 - \frac{\beta}{n}\right) + \frac{\beta}{n} \cdot \frac{n-1}{n} p_{\le i}^\star.$$

Using estimate $(1 - \frac{\beta}{n})/(1 - \frac{\beta}{2n}) \approx (1 - \frac{\beta}{2n})$ we reach

$$p_{\le i}^\star = \left(1 - \frac{\beta}{2n}\right)(1 - (1 - p_{\le(i-1)}^\star)^2).$$

Thus we have for the first $\log(n/k)$ levels the dampening factor of $(1 - \beta/(2n))$ per level (compared to the leakless case). It can be easily shown by induction that

$$\left(1 - \frac{\beta}{2n}\right)^i \left(1 - \frac{k}{n}\right)^{2^i} \le p_{\le i}^\star \le \left(1 - \frac{k}{n}\right)^{2^i}.$$

The estimates for $p_i^\star$ follow from the leakless case, after taking into the account the composed dampening factor:

$$\Pr[\text{detect}] = \sum_{i=0}^{s} p_i^\star = p_{\le s}^\star \ge \left(1 - \frac{1}{e}\right) \cdot \left(1 - \frac{\beta}{2n}\right)^{\log n} = 1 - \frac{1}{e} - \mathcal{O}\left(\frac{\log n}{n}\beta\right).$$

Finally, we summarize the results in this section as follows:

Theorem 1. *Assuming leak rate β/n for $\beta \ll 1$, Robust-Detect guarantees the following.*

- *The probability of a false positive is at most β.*
- *The probability of a false negative is at most $1/e + \mathcal{O}(\beta \cdot (\log n)/n)$.*

Notice that these probabilities can be boosted by standard sampling techniques.

5.2   Convergence Analysis

We now proceed with an analysis of the convergence speed of the previously described protocols. To avoid separate analysis for each of the aforementioned cases (no leaks, false positives, false negatives) and to be independent from all possible initializations of the algorithm, we first start with showing that, under no leaks and with no D present, all states $X_1, \ldots, X_c$ are quickly killed.

In this section, it is more convenient to use t to refer to the total number of interactions, rather than parallel time. To convert to parallel time, one needs to divide by n, the number of molecules.

Lemma 1. *Assume arbitrary (adversarial) initial state in $t = 0$ and evolution with no leaks ($\beta = 0$) and no D is present. For any $c(n) \geq 1$, there is $t = \mathcal{O}(n \cdot (c(n) + \log n))$ such that with probability $1 - 1/n^{\Theta(1)}$ (with high probability) there is no molecule in any of the states $X_1, X_2, \ldots, X_{c(n)}$ after t interactions.*

Proof. We assign a potential to each molecule, based on the state it is currently in: $\Phi(X_i) = 3^{-i}$. We also define a global potential Φ_t as sum of all molecular potentials after t interactions. Observe, that when two molecules interact, following rule $X_i + X_j \to X_{\min(i,j)+1} + X_{\min(i,j)+1}$, then there is:

$$\Phi(X_{\min(i,j)+1}) + \Phi(X_{\min(i,j)+1}) \leq 2/3 \cdot (\Phi(X_i) + \Phi(X_j)),$$

which can be interpreted that each interacting molecule loses at least $1/3$ of its potential. Since each molecule participates in an interaction with probability $\frac{2}{n}$ in each round, the following bound holds:

$$\mathbf{E}[\Phi_{t+1} - \Phi_t | \Phi_t] \geq \cdot \sum_v \Pr(v \text{ interacts in round } t) \cdot \frac{1}{3} \Phi_t(v) = \frac{2}{3n} \Phi_t,$$

$$\mathbf{E}[\Phi_{t+1} | \Phi_t] \leq \left(1 - \frac{2}{3n}\right) \Phi_t.$$

Substituting $\Phi_0 \leq n$ and fixing $t \geq \frac{3}{2} n \ln(n \cdot 3^{c(n)} \cdot n^{\Theta(1)}) = \mathcal{O}(n(c(n) + \log n + \Theta(\log n)))$ we have

$$\mathbf{E}[\Phi_t] \leq \left(1 - \frac{2}{3n}\right)^t \cdot n \leq e^{-\ln(n \cdot 3^{c(n)} \cdot n^{\Theta(1)})} \cdot n = 3^{-c(n)} \cdot \frac{1}{n^{\Theta(1)}}.$$

By Markov's inequality, this means that there is no molecule in any of the states $X_1, X_2, \ldots, X_c$ with probability at least $1 - n^{-\Theta(1)}$, that is with high probability. $\square$

We mention one additional useful property of Algorithm 2, that its actions on population are decomposable with respect to levels. That is, define $\mathrm{LEVEL}_t(u) = i$ if molecule u at time t is in state X_i, and $\mathrm{LEVEL}_t(u) = 0$ if it is in state D.

Observation 1. *Let* $\{u_1, u_2, \ldots, u_n\}, \{v_1, v_2, \ldots, v_n\}, \{w_1, w_2, \ldots, w_n\}$ *be 3 disjoint populations each on n molecules, following evolution defined by Algorithm 2. Moreover, let their evolutions be coupled: at each time t, in each population the corresponding molecules interact (i.e., the interaction is $u_i + u_j$, $v_i + v_j$, $w_i + w_j$ in the three populations for some i, j).*

If $\forall_i \text{LEVEL}_0(u_i) = \min(\text{LEVEL}_0(v_i), \text{LEVEL}_0(w_i))$, *then at any time $t > 0$:* $\forall_i \text{LEVEL}_t(u_i) = \min(\text{LEVEL}_t(v_i), \text{LEVEL}_t(w_i))$.

This observation can be naturally generalized to more than 3 populations. As shown below, the observation implies that to analyze detection under noisy start, we can decouple starting noise from detected particle and analyze evolution under those two separately. Denote by $p_i(t)$ and $p_{\leq i}(t)$ the probability for a randomly picked molecule after t interactions to be in the state X_i or $D, X_1, \ldots, X_i$ respectively.

Theorem 2. *Fix arbitrary leak model (i.e. no leaks, false-positives, false-negatives) and arbitrary concentration of D. For any $c \geq 1$, and $t = \Omega(n \cdot (c + \log n))$, there is*

$$\left| p_{\leq c}^\star - p_{\leq c}(t) \right| \leq 1/n^{\Theta(1)},$$

where $p^\star$ is the stationary probability distribution of the identical process.

Proof. First, for simplicity we collapse all states $X_{c+1}, X_{c+2}, \ldots$ into N, since it has no effect on $p_{\leq c}$ distributions. Consider a population of size n, under no leaks, no D, evolution. By Lemma 1, in $\tau = \mathcal{O}(n \cdot (c + \log n))$ steps it reaches all-N state, regardless of initial configuration, with high probability. Thus evolution of any population $\{u_i\}$, under no leaks, with D present, is a coupling (as in Observation 1) of following evolutions:

- initial configuration of population $\{u_i\}$, with each D replaced with N;
- for every timestep t_i such that D interacted with X_i or N creating X_1, we couple a population with corresponding molecule set to X_1 and every other molecule set to N, shifted in time so its evolution starts at time t_i.

Observe, that evolution of population of both types will reach all-N state in τ steps, with high probability. Thus, conditioned on this high probability, the configuration at any $t \geq \tau$ is the result of coupling of all-N (result of evolution of first type population) with possibly several configurations of the second type, where at each timestep $t' \in [t - \tau, t]$ such population was created independently with some probability only depending on n and k. However, the coupling we just described is invariant from the choice of t, as long as $t \geq \tau$. Thus, for any $t_1, t_2 \geq \tau$, there is $|p_{\leq c}(t_1) - p_{\leq c}(t_2)| \leq 1/n^{\Theta(1)}$. Since $p_{\leq c}^\star = \lim_{t \to \infty} \frac{1}{t} \sum_{i=1}^{t} p_{\leq c}(i)$, the claimed bound follows.

To take into account errors, we say that whenever there is a leak changing state of molecule v to some state S at time t, we change state of v at that time in all existing populations to N, and create new population where v has state S, and all other molecules are in N state. The same reasoning as in the error-less case follows, since switching molecules to N state it only speeds up convergence of populations to all-N state and since populations created due to leaks are created at each step with the same probability depending only on n and error model.   □

6 Simulation Results

We simulated the Robust-Detect algorithm (Algorithm 1) using a modified version of the CRNSimulatorSSA Mathematica package [17]. Figure 2 shows the shape of typical trajectories when there is one molecule in state D ($k = 1$), compared with no molecules in state D ($k = 0$) but with the worst-case leak for false-positives. Note that D is quickly detected if present, and if absent the system exhibits random perturbations that are quickly extinguished and are clearly distinguishable from the true positive case.

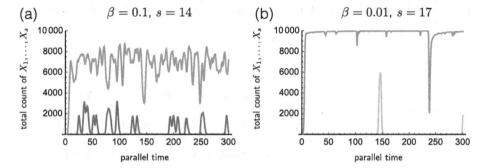

Fig. 2. Typical time-evolution of the Robust-Detect algorithm (Algorithm 1). Three colors correspond to the three conditions described in Fig. 1: (blue) 1 molecule of D is present and no leak (leak parameter $\beta = 0$); (orange, red) no D is present but with worst-case false-positive leak $X_i \to X_1$ and $N \to X_1$ (orange: leak parameter $\beta = 0.01$, red: leak parameter $\beta = 0.1$). All X_i states map to output value "detect", and thus we plot the sum of all their counts. (a) $s = 14$ layers, $\beta = 0.1$ (red). (b) $s = 17$ layers, $\beta = 0.01$ (orange). See Fig. 1 for the corresponding steady state probabilities. Note that with smaller leak ($\beta = 0.01$), it is possible to better distinguish true positives and false positives by increasing the number of layers (from 14 to 17). In all cases there are $n = 10^4$ molecules, and the initial configuration is all molecules in neutral state N. Parallel time (number of interactions divided by n) corresponds to the natural model of time where each molecule can interact with a constant number of other molecules per unit time. (Color figure online)

7 Conclusions

We have considered the problem of modeling and withstanding leaks in chemical reaction networks, expressed as population protocols. We have presented an arguably simple algorithm which is probabilistically correct under assumptions on the leak rate, and converges quickly to the correct answer.

Beyond the specific example of robust detection, we hope that our results motivate more systematic modeling of leaks, and further work on algorithmic techniques to withstand them. As such errors appear to spring from the basic laws of chemistry, their explicit treatment appears to be necessary. The authors found it surprising that many of the algorithmic techniques developed in the context of deterministically correct population protocols might not carry over to implementations, due to their inherent non-robustness to leaks.

In future work, we plan to perform an exhaustive examination of which of the current algorithmic techniques could be rendered leak-robust, and whether known algorithms can be modified to withstand leaks via new techniques. Another interesting avenue for future work is *lower bounds* on the set of computability or complexity of fundamental predicates in the leak model. Finally, we would like to examine whether our robust detection algorithm can be implemented in strand displacement systems.

Acknowledgments. We thank Lucas Boczkowski and Luca Cardelli for helpful comments on the manuscript.

References

1. Gopalkrishnan, M.: Catalysis in reaction networks. Bull. Math. Biol. **73**(12), 2962–2982 (2011)
2. Soloveichik, D., Seelig, G., Winfree, E.: DNA as a universal substrate for chemical kinetics. Proc. Natl. Acad. Sci. **107**(12), 5393–5398 (2010)
3. Cardelli, L.: Two-domain DNA strand displacement. Math. Struct. Comput. Sci. **23**(02), 247–271 (2013)
4. Chen, Y.-J., Dalchau, N., Srinivas, N., Phillips, A., Cardelli, L., Soloveichik, D., Seelig, G.: Programmable chemical controllers made from DNA. Nat. Nanotechnol. **8**(10), 755–762 (2013)
5. Zhang, D.Y., Turberfield, A.J., Yurke, B., Winfree, E.: Engineering entropy-driven reactions and networks catalyzed by DNA. Science **318**(5853), 1121–1125 (2007)
6. Angluin, D., Aspnes, J., Diamadi, Z., Fischer, M., Peralta, R.: Computation in networks of passively mobile finite-state sensors. Distrib. Comput. **18**, 235–253 (2006). Preliminary version appeared in PODC 2004
7. Karp, R.M., Schindelhauer, C., Shenker, S., Vöcking, B.: Randomized rumor spreading. In: 41st Annual Symposium on Foundations of Computer Science, FOCS 2000, pp. 565–574, IEEE Computer Society (2000)
8. Hopfield, J.J.: Kinetic proofreading: a new mechanism for reducing errors in biosynthetic processes requiring high specificity. Proc. Natl. Acad. Sci. **71**(10), 4135–4139 (1974)
9. Thachuk, C., Winfree, E., Soloveichik, D.: Leakless DNA strand displacement systems. In: Phillips, A., Yin, P. (eds.) DNA 2015. LNCS, vol. 9211, pp. 133–153. Springer, Cham (2015)
10. Angluin, D., Aspnes, J., Eisenstat, D.: A simple population protocol for fast robust approximate majority. Distrib. Comput. **21**(2), 87–102 (2008)
11. Cardelli, L.: Morphisms of reaction networks that couple structure to function. BMC Syst. Biol. **8**(1), 84 (2014)
12. Pittel, B.: On spreading a rumor. SIAM J. Appl. Math. **47**(1), 213–223 (1987)

13. Doerr, B., Doerr, C., Moran, S., Moran, S.: Simple and optimal randomized fault-tolerant rumor spreading. Distrib. Comput. **29**(2), 89–104 (2016)
14. Boczkowski, L., Korman, A., Natale, E.: Minimizing message size in stochastic communication patterns: fast self-stabilizing protocols with 3 bits. In: Proceedings of the Twenty-Eighth Annual ACM-SIAM Symposium on Discrete Algorithms, SODA 2017, pp. 2540–2559
15. Dudek, B., Kosowski, A.: Spreading a confirmed rumor: A case for oscillatory dynamics, CoRR, vol. abs/1705.09798 (2017)
16. Soloveichik, D.: Robust stochastic chemical reaction networks and bounded tau-leaping. J. Comput. Biol. **16**(3), 501–522 (2009)
17. http://users.ece.utexas.edu/~soloveichik/crnsimulator.html

Inferring Parameters for an Elementary Step Model of DNA Structure Kinetics with Locally Context-Dependent Arrhenius Rates

Sedigheh Zolaktaf[1], Frits Dannenberg[2], Xander Rudelis[2], Anne Condon[1],
Joseph M. Schaeffer[3], Mark Schmidt[1], Chris Thachuk[2], and Erik Winfree[2(✉)]

[1] University of British Columbia, Vancouver, BC, Canada
[2] California Institute of Technology, Pasadena, CA, USA
winfree@caltech.edu
[3] Autodesk Research, San Francisco, CA, USA

Abstract. Models of nucleic acid thermal stability are calibrated to a wide range of experimental observations, and typically predict equilibrium probabilities of nucleic acid secondary structures with reasonable accuracy. By comparison, a similar calibration and evaluation of nucleic acid kinetic models to a broad range of measurements has not been attempted so far. We introduce an Arrhenius model of interacting nucleic acid kinetics that relates the activation energy of a state transition with the immediate local environment of the affected base pair. Our model can be used in stochastic simulations to estimate kinetic properties and is consistent with existing thermodynamic models. We infer parameters for our model using an ensemble Markov chain Monte Carlo (MCMC) approach on a training dataset with 320 kinetic measurements of hairpin closing and opening, helix association and dissociation, bubble closing and toehold-mediated strand exchange. Our new model surpasses the performance of the previously established Metropolis model both on the training set and on a testing set of size 56 composed of toehold-mediated 3-way strand displacement with mismatches and hairpin opening and closing rates: reaction rates are predicted to within a factor of three for 93.4% and 78.5% of reactions for the training and testing sets, respectively.

1 Introduction

Although nucleic acids are commonly synthesized and applied in various settings, it remains difficult to predict the kinetics of their interaction and conformational change. Accurate models of nucleic acid kinetics are desirable for biological and biotechnological applications, such as understanding the various roles of RNA within the cell and the design of sensitive molecular probes. Within the field of molecular programming, hairpin motifs and toehold-mediated strand displacement are commonly used to implement autonomous devices such as DNA walkers and logic gates. Models of nucleic acid thermal stability have been extensively calibrated to experimental data [4,16] and enable secondary

© Springer International Publishing AG 2017
R. Brijder and L. Qian (Eds.): DNA 23 2017, LNCS 10467, pp. 172–187, 2017.
DOI: 10.1007/978-3-319-66799-7_12

structure software such as RNAsoft, ViennaRNA, RNAstructure, NUPACK, and mfold [3,12,26,27,29] to efficiently predict the equilibrium probabilities of nucleic acid secondary structures. In comparison, a similar extensive calibration and evaluation of nucleic acid kinetic models has not been attempted so far, despite the development of kinetic models and simulation software such as Multistrand and Kinefold [7,9,21,22,25]. Of particular interest is a study by Srinivas et al., which demonstrates that the Metropolis model of Multistrand is incompatible with observations of toehold-mediated strand displacement [23].

We develop a nucleic acid kinetic model based on Arrhenius dynamics that surpasses the performance of the Metropolis model. States in our continuous-time Markov chain (CTMC) model correspond to non-pseudoknotted secondary structures and each transition in the model corresponds to either the opening or closing of a base pair. A key difference with the Metropolis model is the use of activation energy, which depends on the immediate local environment surrounding the affected base pair. To calibrate and evaluate the Arrhenius and the Metropolis models, we compile a dataset of 376 experimentally determined reaction rate constants that we source from existing publications and cover a wide range of reactions, including hairpin closing and opening, bubble closing, helix association and dissociation, toehold-mediated 3-way strand displacement, and toehold-mediated 4-way strand exchange (see Fig. 1). To efficiently infer parameters and to obtain posterior parameter distributions, we use an ensemble Markov chain Monte Carlo (MCMC) approach. Similar to the Metropolis model, our model is consistent with existing thermodynamic models and Gillespie's stochastic simulation algorithm can be used to estimate kinetic rate constants for a variety of reactions. However, obtaining precise predictions using explicit stochastic simulation is computationally expensive, making MCMC parameter inference difficult. Instead we employ a reduced state space approach, enabling reaction rate constants to be computed efficiently and exactly using a sparse

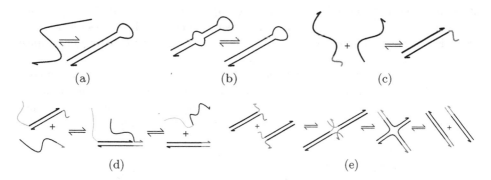

(a)   (b)   (c)

(d)   (e)

Fig. 1. Five types of reactions that we simulate and for which reaction rate constants have been measured. (a) Hairpin closing and opening. (b) Bubble closing. (c) Helix association and dissociation. (d) Toehold-mediated 3-way strand displacement. (e) Toehold-mediated 4-way strand exchange.

matrix representation. Our state space is based on 'zipper models' that were investigated previously to model DNA hybridization [11].

Our results are encouraging and suggest that the new Arrhenius model is applicable to a wide range of DNA dynamic interactions and can be efficiently trained with our framework. The rest of this paper is organized as follows. Section 2 describes preliminaries and the Metropolis kinetic model, Sect. 3 introduces our Arrhenius kinetic model, Sect. 4 introduces our kinetic dataset, Sect. 5 introduces our inference framework, Sect. 6 describes our results comparing the inferred parameters to the database of experimental measurements, Sect. 7 discusses the limitations of our approach and directions for future research, and in Sect. 8 we describe details of the methods we used.

2   Preliminaries

In this section, we briefly discuss the type of reactions we are interested in modeling, and we discuss the Metropolis kinetic model (Sect. 2.1).

When DNA strands interact, base pairs form and break stochastically under the influence of thermal noise, resulting in a highly stochastic back-and-forth dynamic process. When two strands share a mutual base pair, we regard the strands as connected and we define a complex to be a set of connected strands. A single complex can have many different secondary structures. Similar to Kinfold [9] and Multistrand [20, 21], we model the kinetics of interacting DNA strands as a CTMC, where the state space S is a set of non-pseudoknotted secondary structures. Transitions between states correspond to the forming or breaking of a single base pair, which may be called an elementary step. For example, in Fig. 2, state i can transition to states h and j. The rate at which a transition triggers is determined by a kinetic model, that is, the Metropolis or the Arrhenius model, and we distinguish between unimolecular and bimolecular transitions. Because all transitions in our model are reversible, we group transitions into pairs of forward and reverse reactions; a transition in the model is called bimolecular if a complex grows or shrinks by one strand, and is called unimolecular otherwise. As a result, successful helix association and helix dissociation both require at least one bimolecular transition to trigger, despite the latter reaction being strictly first order.

Experimentally observable reactions involve pathways of multiple elementary step transitions, are also inherently reversible, and thus can be classified similarly. We are interested in modeling both unimolecular and bimolecular reactions. In a unimolecular reaction, a complex of strands is altered through the

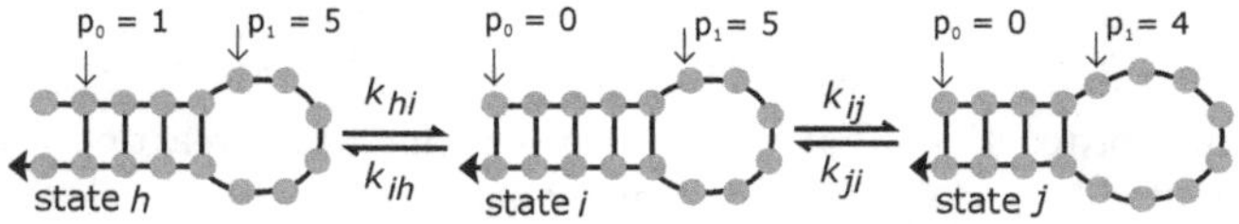

Fig. 2. State i can transition to states h and j. See Sect. 5.1 for definitions of the pointers p_0 and p_1.

formation or disruption of base pairs, but all strands in the complex remain connected. An example of a unimolecular reaction is hairpin closing (Fig. 1a), where a DNA strand hybridizes itself and forms a hairpin loop. Another example of a unimolecular reaction is bubble closing (Fig. 1b). Helix association (Fig. 1c) is a bimolecular reaction. Toehold-mediated 3-way strand displacement (Fig. 1d) is another example of a bimolecular reaction, where one of the strands in a duplex is replaced by the invader strand. The duplex consists of an incumbent strand and a complementary strand. In addition to the hybridized domain, the incumbent strand also contains an unhybridized region called a toehold. The invading strand binds to the toehold region of the substrate and then displaces the incumbent strand via three-way branch migration. Another bimolecular example is toehold-mediated 4-way strand exchange (Fig. 1e), where two duplexes simultaneously exchange strands via four-way branch migration.

2.1   The Metropolis Kinetic Model

The Metropolis model is one of the kinetic rate models implemented in Multistrand [20, 21]. The Multistrand model considers a finite set of strands in a fixed volume ('the box') and defines the energy of a state as the sum of the standard free energy for each complex and a volume-dependent entropy term. To ensure that simulations converge to the Boltzmann distribution over the states at equilibrium, the transition rates between any two adjacent states i and j must satisfy detailed balance:

$$k_{ij}/k_{ji} = \exp\left\{-\left(\Delta G^0_{\text{box}}(j) - \Delta G^0_{\text{box}}(i)\right)/RT\right\} \tag{1}$$

where k_{ij} is the transition rate from state i to state j, $\Delta G^0_{\text{box}}(i)$ is the free energy of state i, R is the gas constant, and T is the temperature. For a state i containing $\mathcal{N}$ strands and $\mathcal{M}$ complexes, the free energy is

$$\Delta G^0_{\text{box}}(i) = \sum_{c=1}^{\mathcal{M}} \Delta G^0_{\text{complex}}(c) + (\mathcal{N} - \mathcal{M})\Delta G^0_{\text{volume}} \tag{2}$$

where $\Delta G^0_{\text{complex}}(c)$ is the difference in Gibbs free energy of complex c relative to the reference state and standard buffer conditions ($[\text{Na}^+] = 1$ M), and $\Delta G^0_{\text{volume}} = -RT\ln u$ is the loss of entropy resulting from fixing the position of a strand of concentration u relative to the standard concentration (1 M). Unimolecular transition rates are given by

$$k_{ij} = \begin{cases} k_{\text{uni}} & \text{if } \Delta G^0_{\text{box}}(j) < \Delta G^0_{\text{box}}(i) \\ k_{\text{uni}} \exp\left(\frac{\Delta G^0_{\text{box}}(i) - \Delta G^0_{\text{box}}(j)}{RT}\right) & \text{otherwise} \end{cases} \tag{3}$$

where $k_{\text{uni}} > 0$ is the unimolecular rate constant (units: s^{-1}). For bimolecular transitions $i \rightarrow j$ where two previously unconnected strands form a mutual base pair, the rate is given as

$$k_{ij} = k_{\text{bi}}u \tag{4}$$

and the rate of dissociation for the bimolecular transition $j \to i$ is given by

$$k_{ji} = k_{\mathrm{bi}} e^{-\frac{\Delta G^0_{\mathrm{box}}(i) - \Delta G^0_{\mathrm{box}}(j) + \Delta G^0_{\mathrm{volume}}}{RT}} \times \mathrm{M} \tag{5}$$

where $k_{\mathrm{bi}} > 0$ is the bimolecular rate constant (units: $\mathrm{M^{-1}s^{-1}}$). We treat $\theta = \{\ln k_{\mathrm{uni}}, \ln k_{\mathrm{bi}}\}$ as 2 free parameters in the model that we calibrate to experimental measurements. We emphasize that the rate of dissociation, Eq. 5, is independent of concentration u and $\Delta G^0_{\mathrm{volume}}$, which follows from the definition of the free energy in a state (Eq. 2).

3   The Arrhenius Kinetic Model

In our Arrhenius kinetic model, the activation energy of each transition depends on the immediate context of the closing or opening base pair. Our classification incorporates some, but not all, factors that may affect the activation energy of a transition. For example, the activation energy might depend on the strand sequence, but modeling this dependence would increase the number of free parameters, and we anticipate to have insufficient experimental evidence to accurately distinguish all relevant factors. However, we emphasize that transition rates in the model still depend on the nucleotide sequence via the nearest neighbor model of base pair stability that determines the free energy of a complex (see Eqs. 3 and 5).

Consider a reaction where a base pair is formed or broken, and denote by $l, r \in \mathcal{C}$ one half of the local context on either side of the base pair. Our model differentiates between seven different half contexts

$$\mathcal{C} = \{\mathrm{stack, loop, end, stack+loop, stack+end, loop+end, stack+stack}\} \tag{6}$$

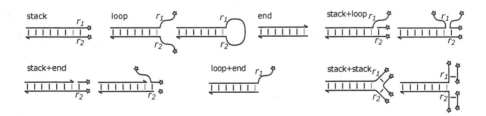

Fig. 3. The right side of the red base pair forms one half of the local context. The classification of the half context depends on the pairing status of the two bases r_1 and r_2 (if they exist) immediately to the right side of the base pair: stack means r_1 and r_2 form a base pair with each other, loop means that neither r_1 nor r_2 forms a base pair, end means that neither r_1 nor r_2 exists, stack+loop means that both r_1 and r_2 exist and one of the bases forms a base pair with another base while the other does not, stack+end means that only one of r_1 or r_2 exists and forms a base pair, loop+end means that only one of r_1 or r_2 exists and it does not form a base pair, and stack+stack means that both r_1 and r_2 exist and they both form base pairs with other bases. Stars indicate the possible continuation of the strands, which may be connected to other starred strands, provided the resulting complex is non-pseudoknotted.

so that the set of local contexts is given by $\mathcal{C} \times \mathcal{C}$. The different half contexts are shown in Fig. 3. The Arrhenius model is equal to the Metropolis model (Eqs. 3, 4 and 5), except that we now re-define $k_{\mathrm{uni}} : \mathcal{C} \times \mathcal{C} \to \mathbb{R}_{>0}$ and $k_{\mathrm{bi}} : \mathcal{C} \times \mathcal{C} \to \mathbb{R}_{>0}$ by setting

$$k_{\mathrm{uni}}(l, r) = k_l k_r \qquad k_l = A_l \exp\left(-E_l/RT\right) \qquad k_r = A_r \exp\left(-E_r/RT\right) \qquad (7)$$

$$k_{\mathrm{bi}}(l, r) = \alpha k_{\mathrm{uni}}(l, r) \tag{8}$$

where A_l, A_r are Arrhenius rate constants, E_l, E_r are activation energies, and α is a bimolecular scaling constant. We treat $\theta = \{\ln A_l, E_l \mid \forall l \in \mathcal{C}\} \cup \{\alpha\}$ as 15 free parameters that we fit to data.

4   Dataset

We compiled a dataset of experimentally determined reaction rate constants, extracting 376 reaction rate constants published in the literature. Each data point in our dataset is annotated with a reaction temperature and the concentration of Na^+ and Mg^{2+} cations in the buffer. The dataset is partitioned into a training set of size 320, which we call $\mathcal{D}_{\mathrm{train}}$, and a testing set with size 56, which we call $\mathcal{D}_{\mathrm{test}}$. The training set covers a wide range of observations, in terms of both reaction types and half contexts. The testing set includes both unimolecular and bimolecular reactions. An overview of our dataset is given in Table 1.

Table 1. Dataset of experimentally measured reaction rate constants. The † sign indicates that the experiment was performed without Na^+ in the buffer, in which case our model computes the free energy as if 50 mM $[Na^+]$ is present (in addition to Mg^{2+}).

$\mathcal{D}_{\mathrm{train}}$	$[Na^+]$ /M	$[Mg^{2+}]$ /mM	T / °C	Source
Hairpin closing and opening	0.1		10–49	Fig. 4 of Bonnet et al. [6]
	0.1–0.5		10–49	Fig. 6 of Bonnet et al. [6]
	0.25		18–49	Fig. 3.28 of Bonnet [5]
	0.137		20	Fig. 3 of Kim et al. [14]
Bubble closing	0.1		25–45	Fig. 4 of Altan-Bonnet et al. [2]
Association and dissociation	1.0		4–68	Fig. 6 of Morrison and Stols [17]
	0.05†	4	30–55	Fig. 6a of Reynaldo et al. [19]
Toehold-mediated 3-way strand displacement	0.05†	4	30–55	Fig. 6b of Reynaldo et al. [19]
	0.05†	12.5	25	Fig. 3b of Zhang and Winfree [28]
Toehold-mediated 4-way strand exchange	0.05†	12.5	25	Table 5.2 of Dabby [8]
$\mathcal{D}_{\mathrm{test}}$				
Hairpin closing and opening	0.137		10–60	Fig. 5a, b of Kim et al. [14]
Toehold-mediated 3-way strand displacement with mismatches	0.05†	10	23	Fig. 2d of Machinek et al. [15]

5   Modeling Framework

We augmented the Multistrand software [20,21] to implement the new Arrhenius model using the full state space of all non-pseudoknotted secondary structures. Given values for the 15 free parameters, a sufficient number of stochastic simulations could be run to estimate the models prediction for an experimental reaction of interest. Unfortunately, obtaining low error bars on this estimate is prohibitively slow, and thus is not feasible within the inner loop of parameter inference procedures. To address this limitation, we developed a computational framework in which we obtain fast, exact predictions for a feasible approximation of the full Multistrand state space. Specifically, we use a reduced state space that is a strict subset of the full state space, enabling sparse matrix computations of mean first passage times, from which reaction rate constants are predicted. With this computation in the inner loop, we used two methods for training the model. The first is a maximum a priori (MAP) approach that optimizes a single set of parameters, and the second is based on MCMC that produces an ensemble of parameter sets. In the latter case, a posterior parameter probability density is computed.

5.1   State Space

In this section, we describe our reduced state space. In the future, our aim is to train the model using a larger set of non-pseudonotted secondary structures. In either case, the number of states in the model directly affects the computational cost of inference through the set of linear equations (Eq. 10 in Sect. 5.2) that is solved for each reaction at each iteration of the parameter search. In this study, the largest state space in the training data is toehold-mediated 4-way strand exchange and contains 14,438 states.

In our reduced state space, base pairs are permitted to form if and only if they occur in either the initial or final state of our simulation. For example, during the simulation of duplex hybridization, only base pairs that are consistent with the perfect alignment of the two strands are permitted to form. We further prune the state space by only allowing base pairs to form or break at the edge of a hybridized domain.

A separate state space $\mathcal{S}_r$ is constructed for each reaction r that we wish to model (Fig. 1). Each state corresponds to a set of indices $\langle p_0, p_1, ... \rangle \in \mathcal{S}_r$, where the indices indicate the begin and end points of the hybridized domains. The maximum number of continuously hybridized domains is precisely defined for each reaction r. For example, the state space for hairpin closing and opening (Fig. 1a) and hybridization (Fig. 1c) only contain one hybridized domain. In such cases, the state description requires only two indices, and the length of the hybridized domain is given by $p_1 - p_0$. In Fig. 2, we show the pointers for the states h, i, and j in the state space for hairpin closing and opening. In each transition, one of the pointers is incremented or decremented. Specifically, state i can transition to state h by incrementing p_0 and it can transition to state j by decrementing p_1. We restrict $0 \leq p_0 \leq p_1 \leq m$, where m is the length of the stem in the closed state. If $p_0 = p_1$, then the domain is absent in the given state. A full description of the state space is given in the online appendix.

5.2   Estimating Mean First Passage Times with Exact Solvers

Given a parametrized kinetic model, we describe how to compute the mean first passage time of a CTMC with state space $\mathcal{S}$ using a sparse matrix representation. Let the mean first passage time t be the average time it takes to reach one of a set of final states $\mathcal{S}_{\text{final}}$ from an initial state i_0. For a first order reaction r, the reaction rate constant is found as $\hat{k}_r = \frac{1}{t}$. For a second order reaction, the reaction rate constant is computed as $\hat{k}_r = \frac{1}{t}\frac{1}{u}$ where u is the initial concentration of the reactants in the simulation [20]. A bimolecular reaction may be effectively first order or second order under the given conditions, depending on the time scale of the unimolecular portion of the reaction pathway relative to the overall reaction time. In our reaction kinetics dataset, all bimolecular reactions are second order in the forward direction.

Let the random variable T_i^{final} represent the time required to reach any state in $\mathcal{S}_{\text{final}}$ starting in state $i \in \mathcal{S}$, where $T_i^{\text{final}} = 0$ for $i \in \mathcal{S}_{\text{final}}$. The time required to reach $\mathcal{S}_{\text{final}}$ starting in i is equal to the initial holding time in state i, which we call h_i, plus the time required to hit $\mathcal{S}_{\text{final}}$ starting in the next visited state. h_i is distributed exponentially with exit rate $k_i = \sum_{j \in S} k_{ij}$. The probability to move to state j is directly proportional to the transition rate, so that $P(i \to j) = \frac{k_{ij}}{k_i}$. Therefore, the mean first passage time is found as [24]

$$\mathbb{E}[T_i^{\text{final}}] = \frac{1}{k_i} + \sum_{j \in S} \frac{k_{ij}}{k_i}\mathbb{E}[T_j^{\text{final}}]. \tag{9}$$

Multiplying Eq. 9 by the exit rate k_i and applying $k_i = \sum_{j \in S} k_{ij}$ then yields

$$\sum_{j \in S} k_{ij}(\mathbb{E}[T_j^{\text{final}}] - \mathbb{E}[T_i^{\text{final}}]) = -1. \tag{10}$$

Equation 10 permits a sparse matrix representation $\mathbf{K}t = -1$ for a rate matrix $\mathbf{K}$ and solution vector t, where $\mathbf{K}_{ij} = k_{ij}$ for $i \neq j$, $\mathbf{K}_{ii} = -\sum_{j \in S} k_{ij}$, and $t_i = \mathbb{E}[T_i^{\text{final}}]$. To compute first passage times for a distribution over initial states $\mathcal{S}_{\text{init}}$ rather than an individual state, the weighted average of the first passage time is computed.

5.3   Estimating the Unnormalized Posterior Distribution of the Parameters

Let θ be the set of parameters in a kinetic model. For a given experimentally observable reaction r, the predicted reaction rate constant $\hat{k}_r$ will deviate from the experimental measurement k_r. We define the error of the prediction to be the $\log_{10}$ difference, $\epsilon_r = \log_{10} k_r - \log_{10} \hat{k}_r$. To produce a measure of likelihood for our parameter valuation, we assume ϵ_r is normally distributed with an unbiased mean and variance σ^2, so that $\epsilon_r \sim N(0, \sigma^2)$. We treat σ as a nuisance parameter. For reaction r the likelihood function is given as

$$P(r|\theta, \sigma) = \frac{1}{\sqrt{2\pi\sigma^2}} \exp\left\{ -\left(\log_{10} k_r - \log_{10} \hat{k}_r\right)^2 \Big/ 2\sigma^2 \right\} \tag{11}$$

and the likelihood function over the set of training data is given as

$$P(\mathcal{D}_{\text{train}}|\theta, \sigma) = \prod_{r \in \mathcal{D}_{\text{train}}} P(r|\theta, \sigma)$$

$$= \exp\left\{ -\frac{\sum_{r \in \mathcal{D}_{\text{train}}} \left(\log_{10} k_r - \log_{10} \hat{k}_r\right)^2}{2\sigma^2} - \frac{n}{2} \log 2\pi\sigma^2 \right\} \tag{12}$$

where n is the number of observations in $\mathcal{D}_{\text{train}}$. To define the probability of the parameters given the data we need to assume prior distributions for θ and σ. During preliminary fitting, a number of parameter values were found to be divergent, which we explain as follows. For a fixed temperature T and a fixed local context (l, r), there are many assignments of A_l, E_l and A_r, E_r that result in nearly equal transition rates $k_{\text{uni}}(l, r) = A_l A_r \exp\{-(E_l + E_r)/RT\}$ (we expand Eq. 7) that result in similar model predictions $\hat{k}_r$. This allows dissimilar valuations for E and A to have nearly equal (log)likelihood scores (Eq. 12). The problem becomes even more apparent when we consider the intrinsic measurement error on k_r (for example, a standard deviation of 22% was reported by Machinek et al. [15]), the limited range of temperatures (see Table 1) inherent to our observations, and the relative frequency of the different half contexts appearing in each simulation (see the online appendix). In practice, $k_{\text{uni}}(l, r)$ is well constrained for many different $l, r \in \mathcal{C}$. As is common in data-fitting applications, we assume a regularization prior that improves the stability of the estimation. We assume that all parameters in θ are independent and identically Gaussian distributed with mean 0 and variance $\frac{1}{\lambda}$. In our inference, we use $\lambda = 0.02$, and the predictive quality of the model does not change for minor adjustments to λ. For the nuisance parameter σ, we use a non-informative Jeffreys prior [13]. Under these assumptions, the posterior distribution is proportional to:

$$P(\theta, \sigma|\mathcal{D}_{\text{train}}) = \frac{P(\mathcal{D}_{\text{train}}|\theta, \sigma)P(\theta)P(\sigma)}{P(\mathcal{D}_{\text{train}})} \propto P(\mathcal{D}_{\text{train}}|\theta, \sigma)P(\theta)P(\sigma)$$

$$= P(\mathcal{D}_{\text{train}}|\theta, \sigma) \left(\frac{2\pi}{\lambda}\right)^{-\frac{|\theta|}{2}} \exp\left\{ -\frac{\lambda\|\theta\|_2^2}{2} \right\} \frac{1}{\sigma}. \tag{13}$$

In conclusion, the log of the posterior distribution is equal to the following equation, up to an additive constant not depending on the parameters

$$\log P(\theta, \sigma|\mathcal{D}_{\text{train}}) \approx$$

$$-(n+1)\log\sigma - \frac{1}{2\sigma^2} \sum_{r \in \mathcal{D}_{\text{train}}} (\log_{10} k_r - \log_{10} \hat{k}_r)^2 - \frac{\lambda}{2}\|\theta\|_2^2 \tag{14}$$

where the squared $L2$ norm in Eq. 14 is computed as $\|\theta\|_2^2 = \alpha^2 + |\ln k_{\text{uni}}|^2 + |\ln k_{\text{bi}}|^2$ for the Metropolis model and as $\|\theta\|_2^2 = \alpha^2 + \sum_{l \in C} |\ln A_l|^2 + \sum_{l \in C} |E_l|^2$ for the Arrhenius model. Note that $|\theta| = 2$ for the Metropolis model and $|\theta| = 15$ for the Arrhenius model.

Our MAP approach seeks a unique parameter set that maximizes the normalized log posterior of the dataset (Eq. 14). We use the Nelder-Mead optimization method [18], a gradient-free local optimizer. For MCMC, we use the *emcee* software package [10], that implements an affine invariant ensemble sampling algorithm.

6   Results

Table 2 shows the performance of the Metropolis and the Arrhenius models with the MAP and MCMC approaches. For details on computational settings for the approaches see Sect. 8. The Arrhenius model fits the training data better than the Metropolis model (for details see the online appendix, Figs. S3–S14), which is unsurprising when considering the increase of adjustable parameters in the Arrhenius model (2 vs. 15). However, the Arrhenius model also has better predictive qualities for the testing set, as evidenced by the MCMC ensemble mean standard deviation of $\sqrt{0.99} = 0.99$ for the Metropolis model and $\sqrt{0.42} = 0.64$ for the Arrhenius model. The improvement in the prediction of the testing set is apparent in Fig. 4, where both models predict the Machinek et al. study of toehold-mediated 3-way strand displacement with mismatches, and in predictions of opening and

Table 2. Performance of the Metropolis and the Arrhenius models on the training and testing sets. The Mean Squared Error (MSE) is the mean of $|\log_{10} k_r - \log_{10} \hat{k}_r|^2$ over $r \subset \mathcal{D}$. The Within Factor of Three metric shows the percentage of reactions for which $|\log_{10} k_r - \log_{10} \hat{k}_r| \leq \log_{10} 3$. Initial is the initial parameter set of the MAP approach (Sect. 8). MAP is the MAP inference method. Mode is the parameter set from the MCMC ensemble that has the highest posterior on $\mathcal{D}_{\text{train}}$. Ensemble is the MCMC ensemble method where the reaction rate constant $\hat{k}_r$ is averaged over all parameter sets.

		Mean Squared Error		Within Factor of Three	
		$\mathcal{D}_{\text{train}}$	$\mathcal{D}_{\text{test}}$	$\mathcal{D}_{\text{train}}$	$\mathcal{D}_{\text{test}}$
Metropolis	Initial	.55	1.3	69.3%	33.9%
	MAP	.33	.94	79.0%	41.0%
	Mode	.33	.95	79.0%	41.0%
	Ensemble	.33	.99	79.6%	37.5%
Arrhenius	Initial	.59	1.3	71.2%	33.9%
	MAP	.14	.47	92.1%	73.2%
	Mode	.12	.40	92.8%	78.5%
	Ensemble	.12	.42	93.4%	78.5%

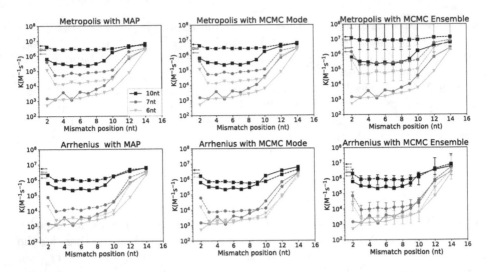

Fig. 4. Model predictions (dashed lines) of reaction rate constants (y axis) for toehold-mediated 3-way strand displacement with mismatches, experimental data (solid lines) from Fig. 2d of Machinek et al. [15]. For the MCMC ensemble method, error bars indicate the range (minimum to maximum) of 100 predictions (see Sect. 8). Arrows indicate no mismatch. The mismatch in the invading strand affects the reaction rate. The length of the toehold domain is ten, seven, and six nucleotides long for ■, ●, and ▼, respectively.

closing rates for hairpin with short stems (1–2 nt) (Figs. S15 and S16 in the online appendix). It is impressive that the models, when trained on a comprehensive training dataset, can predict the results of experiments not seen during training.

There are two reasons for the superior performance of the Arrhenius model. First, the presence of the temperature dependent activation energy allows the Arrhenius model to better calibrate to measurements at varying temperatures. On average, the reaction rate constants $k_{uni}(l, r)$ double in the Arrhenius model between $T = 25\,°C$ and $T = 60\,°C$ (this follows from the parameter values in which $\mathbb{E}[E_l + E_r] = 3.32$ kcal mol^{-1}). A second factor is the relation between the activation energy of a transition and the local context. In Fig. 5, the inferred distribution of $k_{uni}(l, r)$ is given for all local contexts that occur in the model. Strikingly, for many local contexts, the $k_{uni}(l, r)$ are narrowly distributed and often mutually exclusive, indicating that our model captures intrinsic qualitative differences in activation energy.

7   Discussion

A common problem for Arrhenius models in biophysics is that the limited range of temperatures in experimental data can result in ambiguous parameter inference, and this is indeed the case for our model with the current data set. Despite the generally narrow bands for the transition rates (Fig. 5a), the inferred A and E

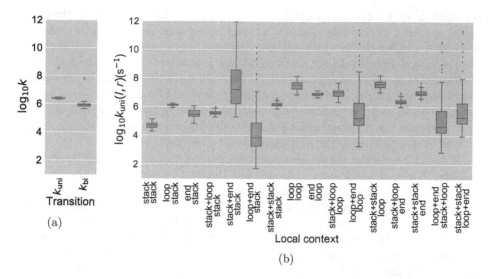

Fig. 5. Box plots of model features inferred by the MCMC ensemble method, using a sample of 100 parameter sets. Edges of the box correspond to the first and third quartile of the distribution. The whisker length is set to cover all parameter values in the sample, or is limited to at most 1.5 times the box height with the outliers plotted separately. (a) k_{uni} and k_{bi} for the Metropolis model. (b) $k_{uni}(l,r)$ at 25 °C for the Arrhenius model. Combinations that do not occur in the model are not shown.

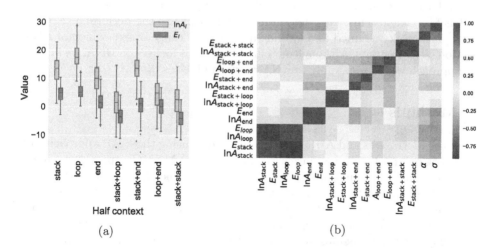

Fig. 6. The Arrhenius model parameters inferred by the MCMC ensemble method. (a) Box plots of the half context parameters. Edges of the box correspond to the first and third quartile of the distribution. The whisker length is set to cover all parameter values in the sample, or is limited to at most 1.5 times the box height with the outliers plotted separately. (b) The Pearson correlation coefficients $R_{ij} = \frac{\text{cov}(\theta_i, \theta_j)}{\sigma_{\theta_i} \sigma_{\theta_j}}$, where $\text{cov}(X, Y) = \mathbb{E}[(X - \mathbb{E}[X])(Y - \mathbb{E}[Y])]$ and $\sigma_X = \sqrt{\mathbb{E}[(X - \mathbb{E}[X])^2]}$. (Color figure available online)

parameters are poorly constrained, as is evident from the wide range in the parameter posterior probability distribution and correlation matrix (Fig. 6). Mathematically, measurements at a single temperature only restrict $\ln A_l + \frac{-E_l}{RT}$ rather than A_l and E_l independently, and a significant fraction of the measurements were performed at constant temperature. If further mining of the existing experimental literature does not resolve the issue, one solution would be to develop customized experiments to calibrate the model further. Interestingly, the relative lack of correlation between the parameters for different half contexts suggests that there could be benefit in subdividing the half context categories further.

We envision further improvements to the model by adjusting the state space and the thermodynamic energy model. For the state space, the requirement for hybridizing strands to only engage in perfectly aligned base pairing is not realistic, and we plan on using a state space generated directly from stochastic Multistrand simulations to avoid these problems. Our simulation depends on the model of thermal stability implemented in the NUPACK software [27] and adjustments to the thermodynamic model also could improve the quality of our predictions. For example, hairpin closing rates are known to depend on the loop sequence, as open poly(A) loops are more rigid than poly(T) loops [1]. The current thermodynamic model does not incorporate this effect, and we avoid comparing the model to measurements on poly(A) loop hairpins. Similarly, the initiation of branch migration is known to have a significant thermodynamic cost, with one study measuring a cost of 2.0 kcal mol^{-1} at room temperature [23]. This initialization cost is not yet incorporated in NUPACK.

We have reported the initial results of our effort to develop accurate kinetic models for nucleic acids. Our Arrhenius model surpasses the performance of the Metropolis model, trained and evaluated on a wide range of experimental DNA reaction rate constants. Although our current analysis focuses on DNA, we believe our approach would also apply to RNA reaction kinetics.

8 Methods

We fit the Metropolis and Arrhenius kinetic models using the MAP approach to a learn parameter set that maximizes Eq. 14. Using the MCMC approach, we maximize the same equation, but instead obtain an ensemble of parameter sets.

The MAP method is sensitive to the initial parameters, and for the Metropolis model, we use $k_{uni} = 8.2 \times 10^6$ s^{-1} and $k_{bi} = 3.3 \times 10^5$ M^{-1}s^{-1}, following known estimates for a one dimensional model of toehold-mediated strand displacement [23]. For the Arrhenius model, we initialize $E_r = 3$ kcal mol^{-1} for all $r \in \mathcal{C}$ and we initialize α and A_r such that, at $T = 23\,^{\circ}$C, equally $k_{uni}(l, r) = 8.2 \times 10^6$ s^{-1} and $k_{bi}(l, r) = 3.3 \times 10^5$ M^{-1}s^{-1} for all local contexts $l, r \in \mathcal{C}$. For both models, we initialize $\sigma = 1$.

Results for the MCMC should generally depend less on the initial value of the sets in the ensemble. To initialize the parameter assignment for each parameter set in the MCMC ensemble, we realize random variables

$$
\begin{aligned}
&E_r \sim U(0,6) \times \text{kcal mol}^{-1} && A_r \sim U(0,10^4) \times s^{-1/2} && \forall r \in \mathcal{C} \\
&k_{\text{uni}} \sim U(0,10^8) \times s^{-1} && k_{\text{bi}} \sim U(0,10^8) \times M^{-1}s^{-1} \\
&\alpha \sim U(0,10) \times M^{-1} && \sigma \sim U(0,1) && (15)
\end{aligned}
$$

where $U(a,b)$ is the uniform distribution over (a,b). During the inference, the parameters are not restricted to initialization bounds, and instead we only require $k_{\text{uni}}, k_{\text{bi}}, A_l, \alpha$ and σ to be positive.

In the emcee software [10], an ensemble of walkers each represents a set of parameters, which are updated through *stretch moves*. Given two walkers θ_1 and θ_2, a new parameter assignment θ_1' for the first walker is generated as

$$
\theta_1' = Z\theta_1 + (1-Z)\theta_2 \qquad g(Z=z) \propto \begin{cases} \frac{1}{\sqrt{z}} & \text{if } z \in \left[\frac{1}{a}, a\right] \\ 0 & \text{otherwise} \end{cases} \qquad (16)
$$

where $g(z)$ is the probability density of Z. We use $a = 2$ (default value) and an ensemble of 100 walkers. We only use the last step of each walker to make predictions, which results in an ensemble of 100 parameter sets for each model.

For the MAP approach, we continue the inference until an absolute tolerance of 10^{-4} is reached. For the MCMC approach, we continue the inference until 750 iterations are performed per walker.

We implemented our framework in Python. All experiments were run on a system with 16 2.93GHz Intel Xeon processors and 64GB RAM, running openSUSE 42.1. On this system, each iteration takes less than 6 s.

Our framework and dataset, as well as an online appendix that has a full description of the state space, more experimental plots and analysis, and algorithms that underlie our framework, are available at https://github.com/DNA-and-Natural-Algorithms-Group/ArrheniusInference.

Acknowledgements. We thank the U.S. National Science Foundation (awards 0832824, 1213127, 1317694, 1643606), the Gordon and Betty Moore Foundation's Programmable Molecular Technology Initiative, and the Natural Sciences and Engineering Research Council of Canada for support. We also thank the anonymous reviewers for their helpful comments and suggestions. XR's current address is Descartes Labs, Los Alamos, NM, USA.

References

1. Aalberts, D.P., Parman, J.M., Goddard, N.L.: Single-strand stacking free energy from DNA beacon kinetics. Biophys. J. **84**, 3212–3217 (2003)
2. Altan-Bonnet, G., Libchaber, A., Krichevsky, O.: Bubble dynamics in double-stranded DNA. Phys. Rev. Lett. **90**, 138101 (2003)

3. Andronescu, M., Aguirre-Hernandez, R., Condon, A., Hoos, H.H.: RNAsoft: a suite of RNA secondary structure prediction and design software tools. Nucleic Acids Res. **31**, 3416–3422 (2003)
4. Andronescu, M., Condon, A., Hoos, H.H., Mathews, D.H., Murphy, K.P.: Computational approaches for RNA energy parameter estimation. RNA **16**(12), 2304–2318 (2010)
5. Bonnet, G.: Dynamics of DNA breathing and folding for molecular recognition and computation. Ph.D. thesis, Rockefeller University (2000)
6. Bonnet, G., Krichevsky, O., Libchaber, A.: Kinetics of conformational fluctuations in DNA hairpin-loops. Proc. Natl. Acad. Sci. **95**(15), 8602–8606 (1998)
7. Chen, S.J.: RNA folding: conformational statistics, folding kinetics, and ion electrostatics. Annu. Rev. Biophys. **37**, 197–214 (2008)
8. Dabby, N.L.: Synthetic molecular machines for active self-assembly: prototype algorithms, designs, and experimental study. Ph.D. thesis, California Institute of Technology (2013)
9. Flamm, C., Fontana, W., Hofacker, I.L., Schuster, P.: RNA folding at elementary step resolution. RNA **6**, 325–338 (2000)
10. Foreman-Mackey, D., Hogg, D.W., Lang, D., Goodman, J.: emcee: the MCMC hammer. Publ. Astron. Soc. Pacific **125**, 306 (2013)
11. Gibbs, J., DiMarzio, E.: Statistical mechanics of helix-coil transitions in biological macromolecules. J. Chem. Phys. **30**, 271–282 (1959)
12. Hofacker, I.L.: Vienna RNA secondary structure server. Nucleic Acids Res. **31**, 3429–3431 (2003)
13. Jeffreys, H.: An invariant form for the prior probability in estimation problems. Proc. Roy. Soc. Lond. A Math. Phys. Eng. Sci. **186**, 453–461 (1946). The Royal Society
14. Kim, J., Doose, S., Neuweiler, H., Sauer, M.: The initial step of DNA hairpin folding: a kinetic analysis using fluorescence correlation spectroscopy. Nucleic Acids Res. **34**, 2516–2527 (2006)
15. Machinek, R.R., Ouldridge, T.E., Haley, N.E., Bath, J., Turberfield, A.J.: Programmable energy landscapes for kinetic control of DNA strand displacement. Nat. Commun. **5**, 5324 (2014)
16. Mathews, D.H., Sabina, J., Zuker, M., Turner, D.H.: Expanded sequence dependence of thermodynamic parameters improves prediction of RNA secondary structure. J. Mol. Biol. **288**(5), 911–940 (1999)
17. Morrison, L.E., Stols, L.M.: Sensitive fluorescence-based thermodynamic and kinetic measurements of DNA hybridization in solution. Biochemistry **32**, 3095–3104 (1993)
18. Nelder, J.A., Mead, R.: A simplex method for function minimization. Comput. J. **7**, 308–313 (1965)
19. Reynaldo, L.P., Vologodskii, A.V., Neri, B.P., Lyamichev, V.I.: The kinetics of oligonucleotide replacements. J. Mol. Biol. **297**, 511–520 (2000)
20. Schaeffer, J.M.: Stochastic simulation of the kinetics of multiple interacting nucleic acid strands. Ph.D. thesis, California Institute of Technology (2012)
21. Schaeffer, J.M., Thachuk, C., Winfree, E.: Stochastic simulation of the kinetics of multiple interacting nucleic acid strands. In: Proceedings of the 21st International Conference on DNA Computing and Molecular Programming, vol. 9211 (2015)
22. Schreck, J.S., Ouldridge, T.E., Romano, F., Šulc, P., Shaw, L.P., Louis, A.A., Doye, J.P.: DNA hairpins destabilize duplexes primarily by promoting melting rather than by inhibiting hybridization. Nucleic Acids Res. **43**(13), 6181–6190 (2015)
23. Srinivas, N., Ouldridge, T.E., Šulc, P., Schaeffer, J.M., Yurke, B., Louis, A.A., Doye, J.P., Winfree, E.: On the biophysics and kinetics of toehold-mediated DNA strand displacement. Nucleic Acids Res. **41**, 10641–10658 (2013)

24. Suhov, Y., Kelbert, M.: Probability and Statistics by Example. Volume 2: Markov Chains: A Primer in Random Processes and Their Applications, vol. 2. Cambridge University Press, Cambridge (2008)
25. Xayaphoummine, A., Bucher, T., Isambert, H.: Kinefold web server for RNA/DNA folding path and structure prediction including pseudoknots and knots. Nucleic Acids Res. **33**, W605–W610 (2005)
26. Xu, Z.Z., Mathews, D.H.: Experiment-assisted secondary structure prediction with RNAstructure. RNA Struct. Determ. Methods Protoc. **1490**, 163–176 (2016)
27. Zadeh, J.N., Steenberg, C.D., Bois, J.S., Wolfe, B.R., Pierce, M.B., Khan, A.R., Dirks, R.M., Pierce, N.A.: NUPACK: analysis and design of nucleic acid systems. J. Comput. Chem. **32**, 170–173 (2011)
28. Zhang, D.Y., Winfree, E.: Control of DNA strand displacement kinetics using toehold exchange. J. Am. Chem. Soc. **131**, 17303–17314 (2009)
29. Zuker, M.: Mfold web server for nucleic acid folding and hybridization prediction. Nucleic Acids Res. **31**, 3406–3415 (2003)

Simplifying Analyses of Chemical Reaction Networks for Approximate Majority

Anne Condon, Monir Hajiaghayi$^{(\boxtimes)}$, David Kirkpatrick, and Ján Maňuch

The Department of Computer Science, University of British Columbia,
Vancouver, Canada
{condon,monirh,kirk,jmanuch}@cs.ubc.ca

Abstract. Approximate Majority is a well-studied problem in the context of chemical reaction networks (CRNs) and their close relatives, population protocols: Given a mixture of two types of species with an initial gap between their counts, a CRN computation must reach consensus on the majority species. Angluin, Aspnes, and Eisenstat proposed a simple population protocol for Approximate Majority and proved correctness and $O(\log n)$ time efficiency with high probability, given an initial gap of size $\omega(\sqrt{n}\log n)$ when the total molecular count in the mixture is n. Motivated by their intriguing but complex proof, we provide simpler, and more intuitive proofs of correctness and efficiency for two bi-molecular CRNs for Approximate Majority, including that of Angluin et al. Key to our approach is to show how the bi-molecular CRNs essentially emulate a tri-molecular CRN with just two reactions and two species. Our results improve on those of Angluin et al. in that they hold even with an initial gap of $\Omega(\sqrt{n\log n})$. Our analysis approach, which leverages the simplicity of a tri-molecular CRN to ultimately reason about bi-molecular CRNs, may be useful in analyzing other CRNs too.

Keywords: Approximate Majority · Chemical reaction networks · Population protocols

1 Introduction

Stochastic chemical reaction networks (CRNs) and population protocols (PPs) model the dynamics of interacting molecules in a well-mixed solution [1] or of resource-limited agents that interact in distributed sensor networks [2]. CRNs are also a popular molecular programming language for computing in a test tube [3,4]. A central problem in these contexts is Approximate Majority [2,5]: in a mixture of two types of species where the gap between the counts of the majority and minority species is above some threshold, which species is in the majority? Angluin et al. [6] proposed and analyzed a PP for Approximate Majority, noting that "Unfortunately, while the protocol itself is simple, proving that it converges quickly appears to be very difficult". Here we provide a new, simpler analysis of CRNs for Approximate Majority.

© Springer International Publishing AG 2017
R. Brijder and L. Qian (Eds.): DNA 23 2017, LNCS 10467, pp. 188–209, 2017.
DOI: 10.1007/978-3-319-66799-7_13

1.1   CRNs and Population Protocols

A CRN is specified as a finite set of chemical reactions, such as those in Fig. 1. The underlying model describes how counts of molecular species evolve when molecules interact in a well-mixed solution. Any change in the molecular composition of the system is attributable to a sequence of one or more interaction events that trigger reactions from the specified set. The model is probabilistic at two levels. First, which interaction occurs next, as well as the time between interaction events, is stochastically determined, reflecting the dynamics of collisions in a well-mixed solution [7]. Second, an interaction can trigger more than one possible reaction, and rate constants associated with reactions determine the relative likelihood of each outcome. For example, reactions (0'x) and (0'y) of Fig. 1(c) are equally likely reactions triggered by an interaction involving one molecule of species X and one of species Y. Soloveichik et al. [8]'s method for simulating CRNs with DNA strand displacement cascades can support such probabilistic reactions.

Angluin et al. [2] introduced the closely related population protocol (PP) model, in which agents interact in a pairwise fashion and may change state upon interacting. Agents and states of a PP naturally correspond to molecules and species of a CRN. A scheduler specifies the order in which agents interact, e.g., by choosing two agents randomly and uniformly, somewhat analogous to stochastic collision kinetics of a CRN. The models differ in other ways. For example, PP interactions always involve two agents, and as such correspond to bi-molecular interactions, while the CRN model allows for interactions of other orders, including unimolecular and tri-molecular interactions. Unlike CRNs, PP interactions may be asymmetric: one agent is the designated initiator and the other is the responder, and their new states may depend not only on their current states but also on their designation. Also, while CRN reaction outcomes may be probabilistic, PP state transition function outcomes are deterministic. Nevertheless, probabilistic transitions can be implemented in PPs by leveraging both asymmetry and the randomness of interaction scheduling [6,9].

$$X+Y \xrightarrow{1/2} X+B \ (0\text{'x})$$

		$X+Y \to B+B \ (0\text{'})$	$X+Y \xrightarrow{1/2} Y+B \ (0\text{'y})$
$X+X+Y \to X+X+X \ (1)$		$X+B \to X+X \ (1\text{'})$	$X+B \to X+X \ (1\text{'})$
$X+Y+Y \to Y+Y+Y \ (2)$		$Y+B \to Y+Y \ (2\text{'})$	$Y+B \to Y+Y \ (2\text{'})$

(a) Tri-molecular CRN.   (b) Double-B CRN.   (c) Single-B CRN.

Fig. 1. A tri-molecular and two bi-molecular chemical reaction networks (CRNs) for Approximate Majority. Reactions (0'x) and (1'y) of Single-B have rate constant 1/2 while all other reactions have rate constant 1.

1.2   The Approximate Majority Problem

Consider a mixture with n molecules, some of species X and the rest of species Y. Here and throughout, we denote the number of copies of X and Y during a CRN computation by random variables x and y respectively. The *Approximate Majority* problem [6] is to reach *consensus* — a configuration in which all molecules are X ($x = n$) or all are Y ($y = n$), from an initial configuration in which $x + y = n$ and the gap $|x - y|$ is above some threshold. If initially $x > y$, the consensus should be X-majority ($x = n$), and if initially $y > x$ the consensus should be Y-majority. We focus on the case when initially $x > y$ since the CRNs that we analyze are symmetric with respect to X and Y.

Angluin et al. [10] proposed and analyzed the Single-B CRN of Fig. 1(c). Informally, reactions (0'x) and (0'y) are equally likely to produce B's (blanks) from X's or Y's respectively, while reactions (1') and (2') recruit B's to become X's and Y's respectively. (Angluin et al. described this as a population protocol, using asymmetry, that provides $1/2$ rates, and the randomness of the scheduler to implement the random reactions (0'x) and (0'y).) When X is initially in the majority ($x > y$ initially), a productive reaction event (i.e., resulting in some chemical changes) is more likely to be (1') than (2'), with the bias towards (1') increasing as x gets larger. Angluin et al. showed *correctness*: if initially $x - y = \omega(\sqrt{n} \log n)$, then with high probability Single-B reaches X-majority consensus. They also showed *efficiency*: with "high" probability $1 - n^{-\Omega(1)}$, for any initial gap value $x - y$, Single-B reaches consensus within $O(n \log n)$ interaction events. They also proved correctness and efficiency in more general settings, such as in the presence of $o(\sqrt{n})$ Byzantine agents.

Doerr et al.'s [11] "median rule" protocol for stabilizing consensus with two choices in a distributed setting involves rules that are identical to the interactions of our tri-molecular protocol of Fig. 1(a). Their model differs somewhat from that of CRNs in that interactions happen in rounds, in which each process (molecule) initiates exactly one interaction with two other processes chosen uniformly at random. They provide a simple and elegant analysis of the protocol, showing that it achieves consensus with high probability in their model within $O(\log n)$ rounds. They note that the consensus value agrees with that of the initial majority when the initial gap is $\omega(\sqrt{n \log n})$. Doerr et al. did not analyze protocols in which interactions involve just two processes.

Several others have subsequently and independently studied the problem; we'll return to related work after describing our own contributions.

1.3   Our Contributions

We analyze three CRNs for Approximate Majority: a simple tri-molecular CRN whose reactions involve just the two species X and Y that are present initially, and two bi-molecular CRNs, which we call Double-B and Single-B, that use an additional "blank" species B – see Fig. 1. As noted earlier, the Single-B CRN is the same as that of Angluin et al. The Double-B CRN is symmetric even in the PP setting, and was among the earliest CRN algorithms constructed with strand displacement chemistry, by Chen et al. [12].

Our primary motivation is to provide the simplest and most intuitive proofs of correctness and efficiency that we can, with the hope that simple techniques can be adapted to reason about CRNs for other problems. A bonus is that our results apply with high probability when the initial gap is $\Omega(\sqrt{n \log n})$, and thus are a factor of $\sqrt{\log n}$ stronger than Angluin et al.'s results in this situation. We do not concern ourselves with smaller initial gaps, but note that even with no initial gap we can still expect efficiency, since the expected number of interaction events until a gap of $\sqrt{n \log n}$ is reached is $O(n \log n)$. This would be true even if there were no bias in favour of reaction (1') as x, the majority species, increases. We suspect that the complexity of Angluin et al.'s proof stems from the case when the initial gap is small $(o(\sqrt{n \log n}))$, and the fact that they show efficiency with high probability, rather than expected efficiency for such an initial setup.

First, in Sect. 3 we analyze the tri-molecular CRN of Fig. 1(a). Intuitively, its reactions sample triples of molecules and amplify the majority species by exploiting the facts that (i) every triple must have a majority of either X or Y, and (ii) the ratio of the number of triples with two X-molecules and one Y-molecule to the number of triples with two Y-molecules and one X-molecule, is exactly the ratio of X-molecules to Y-molecules.

We analyze the CRN in three phases. In the first phase we model the evolution of the gap $x - y$ as a sequence of random walks with increasing bias of success (i.e., increase in $x - y$). Similarly, in the second phase we model the evolution of the count of y as a sequence of random walks with increasing bias of success (decrease in y). We use a simple biased random walk analysis to show that these walks make forward progress with high probability, thereby ensuring correctness. To show efficiency of each random walk, we model it as a sequence of independent trials, observe a natural lower bound on the probability of progress, and apply Chernoff bounds. In the third and last phase we model the "end game" as y decreases from $\Theta(\log n)$ to 0, and apply the random walk analysis and Chernoff bounds a final time to show correctness and efficiency, respectively.

Then in Sect. 4 we analyze the bi-molecular CRNs of Fig. 1 by relating them to the tri-molecular CRN. For the Double-B CRN, we show that with high probability, after a short initial start-up period and continuing almost until consensus is reached, the number of B's is at least proportional to y and is at most $n/2$, in which case reaction events are reactions (1') or (2') with probability $\Omega(1)$. Moreover, blanks are in a natural sense a proxy for $X + Y$ (an interaction between X and Y), and so reactions (1') and (2') behave exactly like the corresponding reactions of our tri-molecular CRN. Essentially the same argument applies to Single-B. We present empirical results in Sect. 5.

Our analysis of the tri-molecular protocol is quite similar to that of Doerr et al.'s median rule algorithm, although the models of interaction are different. We discuss the similarities in Sect. 6, as well as directions for future work.

1.4   Related Work

Perron et al. [13] analyze Single-B when $x + y = n$ and $y \leq \epsilon n$. They use a biased random walk argument to show that Single-B reaches consensus on X-majority with exponentially small error probability $1 - e^{-\Theta(n)}$. The results of Perron et al. do not apply to smaller initial gaps. Mertzios et al. [14] showed somewhat weaker results for Single-B when initially $x - y \geq \epsilon n$ (the main focus of their paper is when interactions are governed by a more general interaction network). Cruise and Ganesh [15] devise a family of protocols in network models where agents (nodes) can poll other agents in order to update their state. Their family of protocols provides a natural generalization of our tri-molecular CRN and their analysis uses connections between random walks and electrical networks.

Yet other work on Approximate Majority pertains to settings with different assumptions about the number of states per agent, the types of interaction scheduling rules, and possibly adversarial behaviour [9,11,14,16], or analyze more general multi-valued consensus problems [10,11,17,18].

2   Preliminaries

2.1   Chemical Reaction Networks

Let $\mathcal{X} = \{X_1, X_2, \ldots X_m\}$ be a finite set of *species*. A solution *configuration* $c = (x_1, x_2, \ldots, x_m)$, where the x_i's are non-negative integers, specifies the number of molecules of each species in the mixture. Molecules in close proximity are assumed to interact. We denote an *interaction* that simultaneously involves s_i copies of X_i, for $1 \leq i \leq m$, by a vector $s = (s_1, s_2, \ldots, s_m)$, and define the *order* of the interaction to be $s_1 + s_2 + \ldots + s_m$.

We model interacting molecules in a well-mixed solution, under fixed environmental conditions such as temperature. The well-mixed assumption has two important implications that allow us to draw on aspects of both CRN models [1,3,19] and also PP models [2], aiming to serve as a bridge between the two. The first, that all molecules are equally likely to reside in any location, supports a stochastic model of chemical kinetics, in which the time between molecular interactions of fixed order is a continuous random variable that depends only on the number of molecules and the volume of the solution. The second, that any fixed interaction is equally likely to involve any of the constituent molecules, and is therefore sensitive to the counts of different species, supports a discrete, essentially combinatorial, view of interactions reminiscent of, but more general than, those in standard PP models. In the Appendix we compare our model with that of Cook et al. [3].

In this paper we will only be interested in interactions of a single order (either two or three). According to a stochastic model of chemical kinetics [1], at any moment, the *time* until the next interaction of order o, what we refer to as an *interaction event*, occurs is exponentially distributed with parameter $\binom{n}{o} / v^{o-1}$, where n denotes the total number of molecules and v denotes the total volume of the solution. Accordingly, if n and v remain fixed, the expected time between interaction events of order o is $v^{o-1}/\binom{n}{o}$ and the variance is $(v^{o-1}/\binom{n}{o})^2$. It follows that, if $v = \Theta(n)$, the time T_n for n interaction events has expected value $E[T_n] = \Theta(n^o/\binom{n}{o}) = \Theta(1)$ and variance $\mathrm{Var}[T_n] = \Theta((n^o/\binom{n}{o})^2/n) = \Theta(1/n)$. By Chebyshev's inequality, we have that:

$$\mathbb{P}[|T_n - \mathrm{E}[T_n]|] \geq h\sqrt{\mathrm{Var}[T_n]}] = \mathbb{P}[|T_n - n^o/\binom{n}{o}| \geq h(n^o/\binom{n}{o}))/\sqrt{n}] \leq 1/h^2.$$

By setting $h = \sqrt{n}$ we see that the time for n interaction events is $O(1)$ with probability at least $1 - 1/n$. Thus we are led to use the number of interaction events, divided by n, as a proxy for time.

When the solution is in configuration $c = (x_1, x_2, \ldots, x_m)$ where $\sum_i x_i = n$, the well-mixed property dictates that the probability that a given interaction event of order o is the particular interaction $s = (s_1, s_2, \ldots, s_m)$ is $\lambda(c, s) = \left[\prod_{i=1}^m \binom{x_i}{s_i}\right]/\binom{n}{o}$.

Some interaction events lead to an immediate change in the configuration of the solution, while others do not. The change (possibly null) arising from an interaction can be described as a (possibly unproductive) reaction event. Formally, a *reaction* $r = (s, t) = ((s_1, s_2, \ldots, s_m), (t_1, t_2, \ldots, t_m))$ is a pair of non-negative integer vectors describing reactants and products, where, for *productive* reactions, at least one i, $s_i \neq t_i$. Reaction r is *applicable* in configuration $c = (x_1, x_2, \ldots, x_m)$ if $s_i \leq x_i$, for $1 \leq i \leq m$. If reaction r occurs in configuration c, the new configuration of the mixture is $c' = (x_1 - s_1 + t_1, x_2 - s_2 + t_2, \ldots, x_m - s_m + t_m)$. In this case we say that the transition from configuration c to configuration c' is *realized* by reaction r and we write $c \to^r c'$. Each reaction r has an associated rate constant $0 < k_r \leq 1$, specifying the probability that the reaction is consummated, given the interaction specified by the reactant vector is satisfied, so the probability that reaction $r = (s, t)$ occurs as the result of an interaction event in a configuration c is just $k_r \lambda(c, s)$.

A *chemical reaction network (CRN)* is a pair $(\mathcal{X}, \mathcal{R})$, where $\mathcal{X}$ is a finite set of species and $\mathcal{R}$ is a finite set of productive reactions, such that, for all reactant vectors s, if $\mathcal{R}(s)$ is the subset of $\mathcal{R}$ with reactant vector s, then $\sum_{r \in \mathcal{R}(s)} k_r \leq 1$. To ensure that all interactions have a fully specified outcome, we take as implicit in this formulation the existence, for every reactant vector s, including all possible interactions of order o, of a non-productive reaction with rate constant $1 - \sum_{r \in \mathcal{R}(s)} k_r$.

2.2   CRN Computations

Next we describe how the mixture of molecules evolves when reactions of a CRN $(\mathcal{X}, \mathcal{R})$ occur. For the CRNs that we analyze, there is some order o such that for every reaction (s, t) of $\mathcal{R}$, $s_1 + s_2 + \ldots s_m = t_1 + t_2 + \ldots t_m = o$. Thus the number n of molecules in the system does not change over time. We furthermore assume that the volume v of the solution is fixed and proportional to n.

A random sequence of interaction events triggers a sequence of (not necessarily productive) reaction events, reflected in a sequence of configurations that we interpret as a computation. More formally, a *computation* of the CRN $(\mathcal{X}, \mathcal{R})$, with respect to an initial configuration c_0, is a discrete Markov process whose states are configurations. The probability of a transition, via a reaction event, from configuration c to configuration c' is just the sum of the probabilities of all reactions r such that $c \rightarrow^r c'$.

2.3 Analysis Tools

We will use the following well-known property of random walks, Chernoff tail bounds on functions of independent random variables, and Azuma's inequality.

Lemma 1 (Asymmetric one-dimensional random walk [20] (XIV.2)).
If we run an arbitrarily long sequence of independent trials, each with success probability at least p, then the probability that the number of failures ever exceeds the number of successes by b is at most $(\frac{1-p}{p})^b$.

Lemma 2 (Chernoff tail bounds [21]). *If we run N independent trials, with success probability p, then S_N, the number of successes, has expected value $\mu = Np$ and, for $0 < \delta < 1$,*

(a) $\mathbb{P}[S_N \le (1-\delta)\mu] \le \exp(-\frac{\delta^2\mu}{2})$, and
(b) $\mathbb{P}[S_N \ge (1+\delta)\mu] \le \exp(-\frac{\delta^2\mu}{3})$.

Lemma 3 (Azuma's inequality [22]). *Let $Q_1, \ldots, Q_k$ be independent random variables, with Q_r taking values in a set A_r for each r. Suppose that the (measurable) function $f : \Pi A_r \to R$ satisfies $|f(x) - f(x')| \le c_r$ whenever the vectors x and x' differ only in the rth coordinate. Let Y be the random variable $f(Q_1, \ldots, Q_k)$. Then, for any $t > 0$,*

$$\mathbb{P}[|Y - E[Y]| \ge t] \le 2\exp\left(-2t^2 / \sum_{r=1}^{k} c_r^2\right).$$

3 Approximate Majority Using Tri-molecular Reactions

In this section we analyse the behaviour of the tri-molecular CRN of Fig. 1(a). We prove the following:

Theorem 1. *For any constant $\gamma > 0$, there exists a constant c_γ such that, provided the initial molecular count of X exceeds that of Y by at least $c_\gamma\sqrt{n\lg n}$, a computation of the tri-molecular CRN reaches a consensus of X-majority, with probability at least $1 - n^{-\gamma}$, in at most $c_\gamma n\lg n$ interaction events.*

Recall that we denote by x and y the random variables corresponding to the molecular count of X and Y respectively. We note that the probability that an interaction event triggers reaction (1) (respectively, reaction (2)) is just $\binom{x}{2}y/\binom{n}{3}$ (respectively, $\binom{y}{2}x/\binom{n}{3}$). Hence, the probability that an interaction even triggers one of these (a productive reaction event) is $xy(x + y - 2)/(2\binom{n}{3})$, and the probability that such a reaction event is reaction (1) is $(x - 1)/(x + y - 2) \geq x/(x + y)$, provided $x \geq y$.

We divide the computation into a sequence of three, slightly overlapping and possibly degenerate, phases, where c_γ, d_γ and e_γ are constants depending on γ:

phase 1 $c_\gamma/2\sqrt{n \lg n} < x - y \leq n(d_\gamma - 2)/d_\gamma$. It ends as soon as $y \leq n/d_\gamma$.
phase 2 $e_\gamma \lg n < y < 2n/d_\gamma$. It ends as soon as $y \leq e_\gamma \lg n$.
phase 3 $0 \leq y < 2e_\gamma \lg n$. It ends as soon as $y = 0$.

Of course the assertion that a computation can be partitioned in such a way that these phases occur in sequence holds only with sufficiently high probability. To facilitate this argument, as well as the subsequent efficiency analysis, we divide both phase 1 and phase 2 into $\Theta(\lg n)$ stages, defined by integral values of t and s, as follows:

- A typical stage in phase 1 starts with $x \geq y + 2^t\sqrt{n \lg n}$ and ends with $x \geq y + 2^{t+1}\sqrt{n \lg n}$, where $\lg c_\gamma \leq t \leq (\lg n - \lg \lg n)/2 + \lg((d_\gamma - 2)/(2d_\gamma))$.
- A typical stage in phase 2 starts with $y \leq n/2^s$ and ends with $y < n/2^{s+1}$, where $\lg d_\gamma \leq s \leq \lg n - \lg \lg n - \lg e_\gamma - 1$.

Our proof of correctness (the computation proceeds through the specified phases as intended) and our timing analysis (how many interaction events does it take to realize the required number of productive reaction events) exploit the simple and familiar tools set out in the previous section, taking advantage of bounds on the probability of reactions (1) and (2) that hold throughout a given phase/stage:

(a) [Low probability of unintended phase/stage completion]. The relative probability of reactions (1) and (2) is determined by the relative counts of X and Y. This allows us to argue, using a biased random walk analysis (Lemma 1 above), that, with high probability, there is no back-sliding; when the computation leaves a phase/stage it is always to a higher indexed phase/stage (cf. Corollaries 1, 2 and 3, below).
(b) [High probability of intended phase/stage completion within a small number of productive reaction events]. Within a fixed phase/stage the computation can be viewed as a sequence of independent trials (choice of reaction (1) or (2)) with a fixed lower bound on the probability of success (choice of reaction (1)). This allows us to establish, by a direct application of Chernoff's upper tail bound Lemma 2, an upper bound, for each phase/stage, on the probability that the phase/stage completes within a specified number of productive reaction events (cf. Corollaries 4, 5 and 6, below).
(c) [High probability that the productive reaction events occur within a small number of molecular interactions]. Within a fixed phase/stage the choice of productive reaction events, among interaction events, can be viewed as a sequence of independent trials with a fixed lower bound on the probability of success (the interaction corresponds to a productive reaction event). Thus our timing analysis (proof of efficiency) is another direct application of Chernoff's upper tail bound (Lemma 2) (cf. Corollary 7, below).

Lemma 4. *At any point in the computation, if $x - y = \Delta$ then the probability that $x - y \leq \Delta/2$ at some subsequent point in the computation is less than $(1/e)^{\Delta^2/(2n+2\Delta)}$.*

Proof. Since $x - y > \Delta/2$ up to the point when we first have $x - y \leq \Delta/2$, it follows that $x \geq n/2 + \Delta/4$ and $y \leq n/2 - \Delta/4$. We can view the change in $x - y$ resulting from productive reaction events as a random walk, starting at Δ, with success (an increase in $x - y$, following reaction (1)) probability p satisfying $p \geq 1/2 + \Delta/(4n)$.

It follows from Lemma 1 that reaching a configuration where $x - y \leq \Delta/2$ (which entails an excess of $\Delta/2$ failures to successes) is less than $(\frac{1}{1+\Delta/n})^{\Delta/2}$ which is at most $(1/e)^{\Delta^2/(2n+2\Delta)}$.

Corollary 1. *In stage t of phase 1, $x - y$ reduces to $2^{t-1}\sqrt{n \lg n}$ with probability less than $1/n^{2^{2t-2}}$.*

Lemma 5. *At any point in the computation, if $y = n/k$ then the probability that $y > 2n/k$ at some subsequent point in the computation is less than $(2/(k-2))^{n/k}$.*

Proof. Since $y \leq 2n/k$ up to the point when we first have $y > 2n/k$, we can view the change in y resulting from productive reaction events as a random walk, starting at n/k, with success (a decrease in y, following reaction (1)) probability p satisfying $p \geq 1 - 2/k$.

It follows from Lemma 1 that reaching a configuration where $y > 2n/k$ (which entails an excess of n/k failures to successes) is less than $(2/(k-2))^{n/k}$.

Corollary 2. *In stage s of phase 2, y increases to $n/2^{s-1}$ with probability less than $(2/(2^s - 2))^{n/2^s}$.*

Corollary 3. *In phase 3, y increases to $2e_\gamma \lg n$ with probability less than $(2e_\gamma \lg n/(n - 2e_\gamma \lg n))^{e_\gamma \lg n}$.*

Lemma 6. *At any point in the computation, if $x - y = \Delta \leq n/2$ then, assuming that $x - y$ never reduces to $\Delta/2$, the probability that $x - y$ increases to 2Δ within at most λn productive reaction events is at least $1 - \exp(-\frac{(\lambda-2)\Delta^2}{\lambda(2n+\Delta)})$.*

Proof. We view the choice of productive reaction as an independent trial with success corresponding to reaction (1), and failure to reaction (2). We start with $x - y = \Delta$ and run until either $x - y = \Delta/2$ or we have completed λn productive reactions. By Lemma 2, the probability that we complete λn productive reactions with fewer than $\lambda n/2 + \Delta/2$ successes, which is necessary under our assumptions if we finish with $x - y < 2\Delta$, is at most $\exp(-\frac{(\lambda-2)\Delta^2}{\lambda(2n+\Delta)})$.

Corollary 4. *In stage t of phase 1, assuming that $x - y$ never reduces to $2^{t-1}\sqrt{n \lg n}$, the probability that $x - y$ increases to $2^{t+1}\sqrt{n \lg n}$ within at most λn productive reaction events is at least $1 - \exp(-\frac{(\lambda-2)2^{2t} \lg n}{3\lambda})$.*

Lemma 7. *At any point in the computation, if* $y = n/k$ *then, assuming that* y *never increases to* $2n/k$, *the probability that* y *decreases to* $n/k - r$ *within* $f(n) > 2r$ *productive reaction events is at least* $1 - \exp(-\Theta(f(n))$.

Proof. We view the choice of productive reaction as an independent trial with success corresponding to reaction (1), and failure to reaction (2). We start with $y = n/k$ and run until either $y = n/k - r$ or we have completed $f(n)$ productive reaction events. (We assume, by Lemma 5, that $y < 2n/k$, and so $p > 1 - 2/k$, throughout.)

By Lemma 2, the probability that we complete $f(n)$ productive reactions with fewer than $(f(n) + r)/2$ successes, which is necessary under our assumptions if we finish with $y > \frac{n}{k-r}$, is at most

$$\exp(-\frac{f(n)(k-2)/2 - (f(n)+r)/2]^2}{2f(n)(k-2)/k}),$$

which is at most $\exp(-\Theta(f(n))$, when $f(n) > 2r$.

Corollary 5. *In stage* s *of phase 2, assuming that* y *never increases to* $n/2^{s-1}$, y *decreases to* $n/2^{s+1}$, *ending stage* s, *in at most* $\lambda n/2^s$ *productive reaction events, with probability at least* $1 - \exp(-\Theta(\lambda n/2^s))$.

Corollary 6. *In phase 3, assuming that* y *never increases to* $2e_\gamma \lg n$, y *decreases to 0, ending phase 3 (and the entire computation), in at most* $\lambda \lg n$ *productive reaction events, with probability at least* $1 - \exp(-\Theta(\lambda \lg n))$.

The following is an immediate consequence of Lemma 2:

Lemma 8. *If during some sequence of* m *interaction events the total probability of all productive reactions is at least* p *then the probability that the sequence gives rise to fewer than* $mp/2$ *productive reaction events is no more than* $\exp(-mp/8)$.

Corollary 7.

(i) *The* λn *productive reaction events of each stage of phase 1 occur within* $(8/3)d_\gamma \lambda n$ *interaction events, with probability at least* $1 - \exp(-\lambda n/4)$.
(ii) *The* $\lambda(n/2^s)$ *productive reaction events of stage* s *of phase 2 occur within* $(16/3)\lambda n$ *interaction events, with probability at least* $1 - \exp(-\lambda n/2^{s+2})$.
(iii) *The* $\lambda \lg n$ *productive reaction events of phase 3 occur within* $(8/3)\lambda n \lg n$ *interaction events, with probability at least* $1 - \exp(\lambda \lg n/4)$.

Proof. It suffices to observe the following lower bounds on the probability that an interaction event triggers reaction (1) in individual phases/stages:

(i) in phase 1, $x > y \geq n/d_\gamma$, so this probability is greater than $3/(4d_\gamma)$;
(ii) in stage s of phase 2, $x > n(1 - 2^{s-1})$ and $y \geq n/2^{s+1} \geq (\lg n)/2$, so this probability is at least $3/2^{s+3}$;
(iii) in phase 3, $x \geq n - \lg n$ and $y \geq 1$, so this probability is at least $3/(4n)$.

Finally, we prove Theorem 1 using the pieces proved until now.

Proof (of Theorem 1).

(i) [Correctness]. It follows directly from Corollaries 1 and 4 (respectively, 2 and 5, 3 and 6) that phase 1 (respectively phase 2, phase 3) completes in the intended fashion, within at most $\lambda n \lg n$ (respectively, λn, $\lambda \lg n$) productive reaction events, with probability at least $1 - \exp(-\Theta(c_\gamma \lg n))$ (respectively, $1 - \exp(-\Theta(\lambda n/d_\gamma))$, $1 - \exp(-\Theta(\lambda \lg n))$).

(ii) [Efficiency]. It is immediate from Corollary 7 that the required number of productive reaction events in phases 1, 2 and 3 occur within $\Theta(\lambda n \lg n)$ interaction events, with probability at least $1 - \exp(-\Theta(\lambda \lg n))$.

4   Approximate Majority Using Bi-molecular Reactions

Here we show correctness and efficiency of the Double-B and Single-B CRNs, essentially by showing that both CRNs respect the more abstract tri-molecular CRN of the previous section.

4.1   The Double-B CRN

Theorem 2. *For any constant $\gamma > 0$, there exists a constant c_γ such that, provided (i) the initial molecular count of X and Y together is at least $n/2$, and (ii) the count of X exceeds that of Y by at least $c_\gamma \sqrt{n \lg n}$, a computation of Double-B reaches a consensus of X-majority, with probability at least $1 - n^{-\gamma}$, in at most $c_\gamma n \lg n$ interaction events.*

Comparing with Theorem 1, it becomes clear that the role of the molecule B is simply to facilitate a bimolecular emulation of the tri-molecular CRN. The sense in which Double-B can be seen as emulating the earlier tri-molecular CRN is that we can analyse its behaviour using exactly the same three phases (and the same sub-phase stages) that we used in our tri-molecular analysis.

Correctness of the Emulation. We measure progress throughout in terms of the change in the molecular counts $\hat{x}$, defined as $x + b/2$, and $\hat{y}$, defined as $y + b/2$, noting that reaction (0') leaves these counts unchanged and reactions (1') and (2') change $\hat{x}$ and $\hat{y}$ at exactly half the rate that the corresponding tri-molecular reactions (1) and (2) change x and y. In each phase, we note that the relative probability of reaction (1') to that of (2'), equals or exceeds the relative probability of reaction (1) to that of (2) in the tri-molecular CRN, and we argue that the total probability of reactions (1') and (2') is at least some constant fraction of the total probability of reactions (1) and (2). This allows us to conclude that Double-B reaches the same conclusion as the tri-molecular CRN, using at most twice as many productive reaction events as the tri-molecular CRN to complete each corresponding phase/stage.

Efficiency of the Emulation. We argue that the productive reaction events needed to carry out the emulation of the tri-molecular CRN occur within a number of interaction events that is at most some constant multiple of the number of interaction events needed to realize the required productive reaction events in the tri-molecular CRN.

This argument is made most simply by setting out bounds on b, the molecular count of molecule B that, with high probability, hold after the first $\Theta(n)$ interaction events, and continue to hold thereafter.

Our bounds are summarized in Lemma 9 below. The proof, a straightforward application of Chernoff bounds, is in the Appendix. In the interests of simplicity, the bounds we provide here are not the tightest possible, but are sufficient for us to conclude immediately that the probability of reactions (1') and (2') of Double-B are each at most a constant factor smaller than those of reactions (1) and (2) in the corresponding phases/stages of the tri-molecular CRN.

Lemma 9. *Let I be any interval of $n/64$ interaction events of a computation of Double-B. Let x_0, x_e, $x_{\min}$ and $x_{\max}$, the initial, final, minimum and maximum values of x in the interval I (similarly, for y and b). Then for any constant $\gamma > 0$, there exists a constant f_γ such that, if $y_0 \geq f_\gamma \lg n$, the following bounds hold with probability at least $1 - 1/n^\gamma$:*

(a) *[Upper bounds] If $b_0 \leq 15n/32$ then $b_e \leq 15n/32$ and $b_{\max} \leq n/2$.*
(b) *[Lower bounds] Even if $b_0 = 0$, $b_e \geq y_e/265$. Furthermore, if $b_0 \geq y_0/265$ then $b_{\min} \geq y_{\max}/292$.*

The efficiency of Double-B follows similarly from the earlier analysis of the tri-molecular CRN presented in Corollary 7. There we observed that it sufficed to bound from below the probability of reaction (1). For the corresponding analysis of Double-B, we observe that in all corresponding phases/stages the probability of reaction (1') is up to a constant factor the same as that of reaction (1). This follows immediately from the upper bound $(n/2)$ on b, which ensures that the molecular count of X is at least $n/4$, and the lower bound $(y/292)$ on b, which ensures that the molecular count of B is at least a constant fraction of that of Y. The constant e_γ that is used in demarking the end of phase 2 and the start of phase 3 will now depend on the constant f_γ of Lemma 9, in order to ensure that this lower bound on b holds throughout phase 2 with high probability.

4.2   The Single-B CRN

Here, we study the behaviour of Single-B, originally proposed by Angluin et al. [10] and shown in Fig. 1(c):

Theorem 3. *For any constant $\gamma > 0$, there exists a constant $c_\gamma > \gamma$ such that, provided (i) the initial molecular count of X and Y together is at least $n/2$, and (ii) the count of X exceeds that of Y by at least $c_\gamma \sqrt{n \lg n}$, a computation of the Single-B CRN reaches a consensus of X-majority, with probability at least $1 - n^{-\gamma}$, in $c_\gamma n \log n$ interactions.*

Comparing the Double-B and Single-B CRNs, we notice that the only difference is that reaction (0') is replaced by probabilistic reactions (0'x) and (0'y) which are equally likely and thus on average, have no effect on $\hat{x}$ and $\hat{y}$. An advantage of Single-B is that B-majority consensus is never reached[1]. The analysis of Single-B proceeds in phases that are essentially the same as for Double-B, except for the need to account for drift in the gap $\hat{x} - \hat{y}$ caused by fluctuations in the number of (0'x) vs (0'y) reactions. For example, this drift may cause $\hat{x} - \hat{y}$ to initially dip lower when Single-B executes than it does when Double-B executes. To address this, we show in Lemma 10 that, despite the drift, the gap will remain at least $(c_\gamma - \gamma)/2\sqrt{n \lg n}$ with all but exponentially small probability, and accordingly we change the definition of phase 1 to be:

phase 1. $(c_\gamma - \gamma)/2\sqrt{n \lg n} < \hat{x} - \hat{y} \leq n(d_\gamma - 2)/d_\gamma$. It ends as soon as $\hat{y} \leq n/d_\gamma$.

Further minor adjustments, described in Appendix A.3, do not require any further changes to the definitions of phases and stages.

Lemma 10. *Starting from* $\hat{x} - \hat{y} \geq c_\gamma \sqrt{n \lg n}$, *where* $c_\gamma > \gamma$, $\hat{x} - \hat{y}$ *reduces to* $(c_\gamma - \gamma)\sqrt{n \log n}$ *within n reaction events with probability less than* $1/n^{(\gamma^2)}$.

Proof. Starting from $\hat{x} - \hat{y} \geq c_\gamma \sqrt{n \lg n}$, the probability that $\hat{x} - \hat{y}$ increases is at least as much as the probability that it decreases. As a worst case scenario, we can view the changes in $\hat{x} - \hat{y}$ as an unbiased random walk which starts at $c_\gamma \sqrt{n \lg n}$. Let $Q_1, \ldots, Q_r$ denote independent random variables where $0 \leq r \leq n$ taking values in set $A_r = [1, -1]$. The Q_r satisfy the conditions of Azuma's inequality (Lemma 3) with $c_r = 2$, the expected change $\sqrt{n}$ (assuming an unbiased random walk), and function $Y = f(Q_1, \ldots, Q_n) = \max_{1 \leq r \leq n} |\sum_{i=1}^{r} Q_i|$ which gives us the maximum translation distance over n reaction events. Now, using Azuma's inequality, we can infer that $\mathbb{P}[|Y - \sqrt{n}| \geq \gamma\sqrt{n \lg n}] \leq 1/n^{\gamma^2}$. Thus in our unbiased random walk the maximum distance from the origin is at most $\gamma\sqrt{n \lg n}$ with high probability.

5   Empirical Results

Figure 2 illustrates the progress of computations of each of our CRNs in each of the three phases, on a single run. In the first phase, the gap $x - y$ (red line) increases steadily. Once the gap is sufficiently high, phase 2 starts and the count of y for the tri-molecular CRN, and $\hat{y}$ for the bi-molecular CRNs, decrease steadily. In the last phase, as the counts of y and $\hat{y}$ are small, there is more noise in the evolution of y and $\hat{y}$, but they do reach 0. Figure 3 compares time (efficiency) and success rates (probability of correctness) of the three CRNs to reach consensus, as a function of the log of the initial count n of molecules, or the log of the volume. The plots show that time grows linearly with the log of the

---

[1] We note that although the B-majority consensus is reachable in the Double-B CRN, the probability of such an event is easily shown to be very small (i.e., $n^{\Omega(-\lg(n))}$).

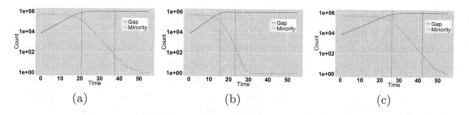

Fig. 2. The gap $x - y$ (red line) and minority (count y for tri-molecular CRN and $\hat{y}$ for bi-molecular CRNs) (blue line), as a function of time, of sample runs of the (a) tri-molecular, (b) Double-B, and (c) Single-B CRNs. The initial count is $n = 10^6$, the initial gap $x - y$ is $2\sqrt{n \lg n}$ and parameters c_γ, d_γ and e_γ are set to 2, 8, and 2 respectively. The vertical dotted lines demark transitions between phases 1, 2 and 3. (Color figure online)

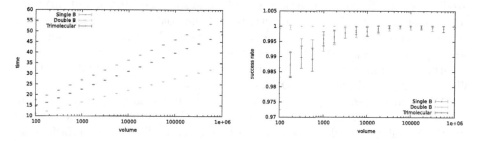

Fig. 3. Comparison of the time (left) and success rate, i.e., probability of correctness (right) of Single-B, Double-B and the tri-molecular CRN for Approximate Majority. Each point in the plot is an average over 5,000 trials. The initial configuration has no B's and the imbalance between X's and Y's is $\sqrt{n \ln n}$. Plots show confidence intervals at 99% confidence level.

molecular count, and the success rate is close to 1 for large n. A fit to the data of that figure shows that the expected times of the tri-molecular, Double-B and Single-B CRNs grow as $3.4 \ln n$, $2.4 \ln n$, and $4.0 \ln n$ respectively. For $n \geq 100$, the tri-molecular CRN has at least 99% probability of correctness and the bi-molecular CRNs have at least 97% probability of correctness. These probabilities all tend to 1 as n gets larger.

6   Discussion

As noted earlier, Doerr et al. [11] analyse what they call the median rule consensus protocol, which bears strong resemblance to our tri-molecular CRN for approximate majority. The median rule protocol assumes rounds of n concurrent interactions, with each of n participating processes initiating one interaction that involves two additional processes chosen uniformly at random. The result of each such round is very similar to what is accomplished in one time unit of the CRN or PP models, in which a sequence of n random interactions occur. Accordingly there are strong similarities between our analysis and theirs. For example, our analysis is staged in a way that allows us to assume that interactions within each stage are driven by essentially the same population sizes. Note however that in our CRN model, unlike the Doerr et al. model, there may be molecules that participate in no interaction within a given unit of time. This difference becomes evident in our end game analysis, which requires $\Theta(n \log n)$ time units to ensure that, with high probability, the few remaining Y interact and thus are converted to X's. In contrast, the end game is completed in $O(1)$ rounds with high probability in the Doerr et al. model. More significant differences between the Doerr et al. model versus the CRN and PP models arise when the initial gap $x - y$ is small, a case that we do not analyze and that appears to be significantly harder to handle in the CRN model.

There are several ways in which we can extend our results. Angluin et al. [10] analyze settings in which (i) some agents (molecules) have Byzantine, i.e., adversarial, behaviour upon interactions with others, (ii) some molecules are "activated" (become eligible for reaction) by epidemic spread of signal, and (iii) there are three or more species present initially and the goal is to reach consensus on the most populous species (multi-valued consensus). We believe that our techniques can be generalized to these settings.

Other generalizations are motivated by practicalities of molecular systems. When a CRN is "compiled" to a DNA strand displacement system, it may be that the DNA strand displacement reaction rate constants closely approximate, but are not exactly equal to, the CRN reaction rates. It could be helpful to describe how the initial gap needed to guarantee correct and efficient computations for Approximate Majority with high probability depends on the uncertainty in the rate constants. Also, our techniques may be useful for proving correctness of the Chen et al. strand displacement implementation of Double-B [12], which involves so-called fuel species and waste products in addition to molecules that represent the species of the CRN. Third, it could be useful to analyze variants of the CRNs analyzed here, or other CRNs, in which some or all of the reactions are reversible. For example, if the blank-producing reaction (0') of Double-B is made reversible, the modified CRN is still both correct and efficient, while having the additional nice property that a stable state with neither X-consensus nor Y-consensus cannot be reached, even with very low probability. On the other hand, some caution needs to be applied when reversing reactions. For instance, making reactions (0'x) and (0'y) of Single-B reversible can lead to a system that fluctuates around a state with an equal number of Xs and Ys, and some ratio of Bs. This would happen when the rate of reversed reactions (0'x) and (0'y) is greater or equal to the rate of reactions (1') and (2'). Again, we believe that our analyses can easily generalize to these scenarios.

A   Appendix

In this appendix we (1) relate our CRN model to that of Cook et al. [3], (2) prove our lower and upper bounds on the number of B molecules in the Double-B CRN, and (3) prove the lemmas which make the analysis of Single-B parallel to that of the tri-molecular CRN.

A.1   Relationship Between Our CRN Model and that of Cook et al.

Other CRN models define reaction probabilities and computation time somewhat differently than we do, but these differences can easily be reconciled. For example, in the model of Cook et al. [3], if k_r' is the rate constant associated with reaction $r = (s, t)$ of order o and the system is in configuration $c = (x_1, x_2, \ldots, x_m)$, then the propensity, or rate, of r is

$$\rho_r(c) = k_r' [\prod_{i=1}^{m} (x_i!/(x_i - s_i)!)]/v^{o-1}.$$

If $\rho^{tot}(c) = \sum_r \rho_r(c)$ for all reactions r of order o, then the probability that a reaction event is reaction r is $\rho_r(c)/\rho^{tot}(c)$, and the expected time until a reaction event occurs is $1/\rho^{tot}(c)$. (In this model, reaction rate constants can be greater than 1, and may depend not only on the number of reactants of each species, but also on other properties of a species such as its shape, capturing the fact that the likelihood of different types of interactions may not all be the same.)

If in our model we set $k_r = k_r' \prod_{i=1}^{m} s_i!$ for each productive reaction, and normalize by $\sum_r k_r$ if necessary to ensure that $\sum_{r \in \mathcal{R}(s)} k_r \leq 1$ (adjusting the underlying time unit accordingly), a straightforward calculation shows that, when in a given configuration c, the probability that a reaction event is a given reaction r is the same in our model and that of Cook et al.[2] See the example of Fig. 4. Also,

---

[2] Here is the calculation for the probability conversion.

$$\rho_r(c) = k_r'.[\prod_{i=1}^{m} (x_i!/(x_i - s_i)!)]/v^{o-1}$$

$$= k_r'.[\prod_{i=1}^{m} s_i!].[\prod_{i=1}^{m} \binom{x_i}{s_i}]/v^{o-1}$$

$$= [\binom{n}{o}/v^{o-1}]k_r'[\prod_{i=1}^{m} s_i!].[\prod_{i=1}^{m} \binom{x_i}{s_i}]/\binom{n}{o}$$

$$= [\binom{n}{o}/v^{o-1}]k_r[\prod_{i=1}^{m} \binom{x_i}{s_i}]/\binom{n}{o},$$

where

$$k_r = [k_r'.[\prod_{i=1}^{m} s_i!]. \tag{1}$$

We can interpret the last of these expressions for $\rho_r(c)$ as the product of three terms. The first term, namely $\binom{n}{o}/v^{o-1}$, corresponds to the (normalized) average rate of an interaction of order o. The last term, namely $[\prod_{i=1}^{m} \binom{x_i}{s_i}]/\binom{n}{o}$, is the probability that the reaction of order o has exactly the reactants of r. The middle term k_r depends on the s_i's, but could also model situations where different types of interactions have different rates, e.g., if some molecular species are larger than others. Normalizing the k_r's by $\sum k_r$ yields rate constants for our model.

$$X + X + Y \rightarrow^1 X + X + X \ (r_1) \qquad X + X + Y \rightarrow^{2/14} X + X + X \ (r_1)$$
$$Y + Y + Y \rightarrow^2 X + X + Y \ (r_2) \qquad Y + Y + Y \rightarrow^{12/14} X + X + Y \ (r_2)$$
$$\text{(a)} \qquad\qquad\qquad\qquad\qquad \text{(b)}$$

Fig. 4. (a) A CRN specified with respect to the Cook et al. model. The reaction rates when the system is in configuration (3, 3) are $k'_{r_1} = 18/v^2$ and $k'_{r_2} = 12/v^2$. The reaction probabilities are $\rho_{r_1}((3,3)) = 3/5$ and $\rho_{r_2}((3,3)) = 2/5$. (b) The mapping of the CRN of part (a) to our model by changing the rate constants (using Eq. 1 of footnote 2) and normalizing by $\sum k_r$. The probability that a reaction event is r_1 is $(18/14)/(30/14) = 18/30$, and the probability of r_2 is $12/30$. Thus, reaction probabilities are preserved exactly.

the expected time until the next reaction event differs between the models by a constant factor that is independent of c. Conversely, to convert from our model to that of Cook et al., divide our rate constant k_r by $[\prod_{i=1}^m s_i!]$ (and multiply all rate constants by the same constant factor in order to adjust time units as needed).

A.2   Bounds on b, the Molecular Count of B, in the Double-B CRN

Here we provide a proof of Lemma 9, omitted from Sect. 3. We note that the probability that an interaction event in the interval I triggers reaction (0') (respectively, reaction (1'), reaction (2')) is just $xy/\binom{n}{2}$ (respectively, $xb/\binom{n}{2}$, $yb/\binom{n}{2}$). In the following, we simplify calculations by replacing $\binom{n}{2}$ with $n^2/2$.

Upper Bounds on b. Note that reaction (0') has probability at most $(n/2)(n/2)/(n^2/2) = 1/2$, so at most $n/64$ new B molecules are produced by reaction (0') over interval I, in expectation, and at most $n/32$ are produced, with probability $1 - \exp(\Theta(n))$. Thus, $b_{\max} \le b_{\min} + n/32$.

Given this, we can clearly assume that $b_{\min} \ge 14n/32$, since otherwise $b_{\max}$ (and, of course b_e) is less than $15n/32$. Thus $x + y \le 18n/32$ throughout the interval, and so reaction (0') has probability at most $(18n/64)^2/(n^2/2)$ which is less than $1/6$. Hence fewer than $2(n/64)/6 = n/192$ new B molecules are produced by reaction (0') over interval I, in expectation, and fewer than $n/175$ are produced, with probability $1 - \exp(\Theta(n))$ (here we use a Chernoff upper tail bound). Assuming $b_0 \le 15n/32$, it follows that $b_{\max} < b_0 + n/175 < n/2$, and so $x + y > n/2$ throughout interval I.

It follows that the total probability of reactions (1') and (2') is at least $(n/2)(14n/32)/(n^2/2) = 14/32$ throughout interval I, which means that at least $(14/32)(n/64) > n/148$ B molecules are consumed by these reactions, in expectation, and at least $n/160$ are consumed, with probability $1 - \exp(\Theta(n))$, over the course of interval I (here we use a Chernoff lower tail bound). Thus, with probability $1 - \exp(\Theta(n))$, the net change in b is less than $n/175 - n/160 < 0$, and so $b_e < b_0 \le 15n/32$. We note that this upper bound holds with probability $1 - \exp(-\Theta(n))$, which is stronger than in the statement of the lemma.

Lower Bounds on b. Note that $x - y$ is not changed by reaction (0'), and by Lemma 4, it never reaches $(x_0 - y_0)/2$ through reactions (1') and (2'). Therefore, $b_{\max} \leq n/2$, it follows that $x + y \geq n/2$ and hence $x \geq n/4$. We will use this fact throughout.

We first show that even if $b_0 = 0$, $b_e \geq y_e/292$. Since $b_{\max} \leq n/2$ it follows that reaction (2') has probability at most $y_{\max} b_{\max}/(n^2/2) \leq y_{\max}/n$. Thus reaction (2') increases y from its minimum value $y_{\min}$ by at most $y_{\max}/64$, in expectation, and by at most $y_{\max}/32$, with high probability, over the course of interval I. Here, the high probability follows from the fact that $y_{\max} \geq y_0 \geq f_\gamma \lg n = \Omega(\log n)$, and application of a Chernoff tail bound. Thus, $y_e \leq y_{\max} \leq y_{\min} + y_{\max}/32$ and so

$$y_{\min} \geq (31/32)y_{\max} = \Omega(\log n). \tag{*}$$

Now suppose that

$$b_{\max} > (1/16)y_{\min}. \tag{**}$$

Thus we also have that $b_{\max} = \Omega(\log n)$, by (*). Since $x + y \leq n$, reactions (1') and (2') together have probability at most $nb_{\max}/(n^2/2)$, and so these reactions reduce b from its maximum value $b_{\max}$ by at most $b_{\max}/32$, in expectation, and by at most $b_{\max}/16$, with high probability, over the course of interval I. Here, the high probability follows from the fact that $b_{\max} = \Omega(\log n)$, and application of a Chernoff tail bound. Thus, with high probability,

$$b_e \geq (15/16)b_{\max}. \tag{***}$$

Then,

$$
\begin{aligned}
b_e &\geq (15/16)b_{\max} &&\text{by (***)} \\
&> (15/16)(1/16)y_{\min} &&\text{by (**)} \\
&\geq (15/16)(1/16)(31/32)y_{\max} &&\text{by (*)} \\
&> y_{\max}/18 \geq y_e/18.
\end{aligned}
$$

On the other hand, suppose that

$$b_{\max} \leq (1/16)y_{\min}. \tag{****}$$

Since reaction (0') has probability at least $x_{\min}y_{\min}/(n^2/2) \geq y_{\min}/(2n)$, reaction (0') increases b by at least $y_{\min}/64$, in expectation, and at least $y_{\min}/128$, with high probability, over the course of interval I. Since reactions (1') and (2') together have probability at most $nb_{\max}/(n^2/2) \leq n(y_{\min}/16)/(n^2/2)$ by (****), we know that together they decrease b by at most $y_{\min}/512$, in expectation, and at most $y_{\min}/256$, with high probability, over the course of interval I. Here, the high probability follows from the fact that $y_{\min} = \Omega(\log n)$ by (*), and application of a Chernoff tail bound. Thus the net change in b is at least $y_{min}/128 - y_{min}/256$, with high probability. Also,

$$y_{min}/128 - y_{min}/256 = y_{min}/256$$
$$\geq (31/32)(1/256)y_{max} \text{ by } (*)$$
$$> y_{max}/265.$$

So, $b_e > y_{max}/265 \geq y_e'/265$, even if $b_0 = 0$.

Finally, assume that $b_0 \geq y_0/265$. Let b_{max}' be the maximum value of b between b_0 and b_{min} in the course of interval I. By an argument similar to the one used for equation $(***)$, with high probability, we get

$$b_{min} \geq (15/16)b_{max}' \geq (15/16)b_0 \qquad\qquad (*****)$$

Therefore, we have

$$b_{min} \geq (15/16)b_0 \qquad\qquad \text{by } (*****)$$
$$\geq (15/16)y_0/265$$
$$\geq (15/16)y_{min}/265$$
$$> (15/16)(31/32)y_{max}/265. \text{ by } (*)$$

and so $b > y/292$ throughout interval I.

A.3   Adjustments Required for the Proof of Single-B

Here we describe additional adjustments to the proof of correctness and efficiency of the tri-molecular CRN that are needed to account for changes to random variables $\hat{x}$ and $\hat{y}$ due to reactions (0'x) and (0'y). Note that reactions (0'x) and (1') increase $\hat{x}$ by $1/2$ and decrease $\hat{y}$ by $1/2$, while reactions (0'y) and (2') decrease $\hat{x}$ by $1/2$ and increase $\hat{y}$ by $1/2$.

First, in the proof of the upper $(n/2)$ and lower $(y/292)$ bounds on b in Lemma 9, we simply adjust the probabilities of a change in $\hat{x}$ or $\hat{y}$ to account for reactions (0'x) and (0'y). (We remark that we are able to provide tighter lower and upper bounds on b with respect to variable y, i.e., $\frac{y}{2\alpha} \leq b \leq 2\alpha y$, where $\alpha \geq 20$, and $b = \Omega(\log n)$, for the Single-B CRN - details omitted.) Then, utilizing the lower bound on b, Lemma 11 shows that the ratio of total probability of reactions (0'x) and (1') to that of reactions (0'y) and (2') is lower than the ratio of the probability of reaction (1) to that of reaction (2) in the tri-molecular CRN by at most a small constant. Therefore, the analysis of phase 1 of Single-B parallels that of the tri-molecular CRN.

Lemma 11. *At any point in the computations, assuming that $\hat{x} - \hat{y} \geq \Delta/2$, the probability that $\hat{x} - \hat{y}$ increases is at least $1/2 + \Theta(\Delta/n)$.*

Proof. Let p denote the probability of a success ($\hat{x} - \hat{y}$ increases) and q denote the probability of a failure ($\hat{x} - \hat{y}$ increases). So, given that $x \leq n$, and $y/292 < b$, we have that

(1) $\dfrac{q}{p} = \dfrac{1/2xy + yb}{1/2xy + xb} \leq 1 - \dfrac{(\hat{x} - \hat{y})b}{1/2xy + xb} \leq 1 - \dfrac{(\Delta/2)b}{x(1/2y + b)} \leq 1 - \Theta(\Delta/n),$

(2) $q + p = 1$.

It follows from Eqs. 1 and 2 that $p \geq 1/2 + \Theta(\Delta/4n)$.

Similarly, we can revise Lemmas 5 and 7 (and their related corollaries) to make the analysis of phases 2 and 3 of Single-B also parallel to those of the tri-molecular CRN–see Lemmas 12 and 13.

Lemma 12. *At any point in the computation, if $\hat{y} = n/k$ then the probability that $\hat{y} > 2n/k$ at some subsequent point in the computation is less than $(1 - \Theta(1))^{n/k}$.*

Proof. Let p denote the probability of a success ($\hat{y}$ decreases) and q denote the probability of a failure ($\hat{y}$ increases). So, assuming that $x \leq n$, $\hat{x} - \hat{y} \geq n - n/4k$, and $y < 292b$, we can compute the ratio q/p on a reaction event as follows.

$$\frac{q}{p} = \frac{1/2xy + yb}{1/2xy + xb} \leq 1 - \frac{(\hat{x} - \hat{y})b}{1/2xy + xb} \leq 1 - \frac{(n - 4n/k)b}{n(1/2y + b)} \leq 1 - \Theta(1).$$

By Lemma 1, we conclude that reaching a configuration where $y > 2n/k$ (which entails an excess of n/k failures to successes) is less than $(1 - \Theta(1))^{n/k}$.

Lemma 13. *At any point in the computation, if $\hat{y} = n/k$ then, assuming that $\hat{y}$ never increases to $2n/k$, the probability that $\hat{y}$ decreases to $n/k - r$ within $f(n) > \Theta(r)$ reaction events is at least $1 - exp(-\Theta(f(n)))$.*

Proof. The proof is completely parallel to the proof of Lemma 7. We only need to compute the probability of a success ($\hat{y}$ decrease). By Lemma 12, $q/p = 1 - \Theta(1)$. So, considering $p + q = 1$, it's straightforward to obtain $p \geq \frac{1}{2} + \Theta(1)$.

Finally, we employ Lemma 8 to complete the proof of efficiency. Using the upper bound on b, which confirms that $x \geq n/4$ and the lower bound on b, which confirms $b \geq y/292$, we can conclude that the total probability of reactions (0'x), (0'y), (1'), and (2') is at least some constant fraction of the total probability of reactions (1) and (2) in tri-molecular CRN. Therefore, the total number of interactions in Single-B is at most some constant multiple times the required number of interactions in the tri-molecular CRN.

References

1. Gillespie, D.T.: Exact stochastic simulation of coupled chemical reactions. J. Phys. Chem. **81**, 2340–2361 (1977)
2. Angluin, D., Aspnes, J., Diamadi, Z., Fischer, M.J., Peralta, R.: Computation in networks of passively mobile finite-state sensors. Distrib. Comput. **18**(4), 235–253 (2006)
3. Cook, M., Soloveichik, D., Winfree, E., Bruck, J.: Programmability of chemical reaction networks. In: Condon, A., Harel, D., Kok, J., Salomaa, A., Winfree, E. (eds.) Algorithmic Bioprocesses, pp. 543–584. Springer, Heidelberg (2009). doi:10.1007/978-3-540-88869-7_27
4. Soloveichik, D., Cook, M., Winfree, E., Bruck, J.: Computation with finite stochastic chemical reaction networks. Nat. Comput. **7**, 615–633 (2008)
5. Cardelli, L., Csikász-Nagy, A.: The cell cycle switch computes approximate majority. Nat. Sci. Rep. **2**, 656 (2012)
6. Angluin, D., Aspnes, J., Eisenstat, D.: Fast computation by population protocols with a leader. In: Dolev, S. (ed.) DISC 2006. LNCS, vol. 4167, pp. 61–75. Springer, Heidelberg (2006). doi:10.1007/11864219_5
7. Cardelli, L., Kwiatkowska, M., Laurenti, L.: Programming discrete distributions with chemical reaction networks. In: Rondelez, Y., Woods, D. (eds.) DNA 2016. LNCS, vol. 9818, pp. 35–51. Springer, Cham (2016). doi:10.1007/978-3-319-43994-5_3
8. Soloveichik, D., Seelig, G., Winfree, E.: DNA as a universal substrate for chemical kinetics. PNAS **107**(12), 5393–5398 (2010)
9. Alistarh, D., Aspnes, J., Eisenstat, D., Gelashvili, R., Rivest, R.L.: Time-space trade-offs in population protocols. In: Proceedings of the Twenty-Eighth Annual ACM-SIAM Symposium on Discrete Algorithms, pp. 2560–2579 (2017)
10. Angluin, D., Aspnes, J., Eisenstat, D.: A simple population protocol for fast robust approximate majority. Distrib. Comput. **21**(2), 87–102 (2008)
11. Doerr, B., Goldberg, L.A., Minder, L., Sauerwald, T., Scheideler, C.: Stabilizing consensus with the power of two choices. In: Proceedings of the Twenty-third Annual ACM Symposium on Parallelism in Algorithms and Architectures, SPAA 2011, pp. 149–158. ACM, New York (2011)
12. Chen, Y.-J., Dalchau, N., Srinivas, N., Phillips, A., Cardelli, L., Soloveichik, D., Seelig, G.: Programmable chemical controllers made from DNA. Nat. Nanotechnol. **8**(10), 755–762 (2013)
13. Perron, E., Vasudevan, D., Vojnovic, M.: Using three states for binary consensus on complete graphs. In: Proceedings of the 28th IEEE Conference on Computer Communications (INFOCOM), pp. 2527–2535 (2009)
14. Mertzios, G.B., Nikoletseas, S.E., Raptopoulos, C.L., Spirakis, P.G.: Determining majority in networks with local interactions and very small local memory. Distrib. Comput. **30**(1), 1–16 (2017)
15. Cruise, J., Ganesh, A.: Probabilistic consensus via polling and majority rules. Queueing Syst. **78**(2), 99–120 (2014)
16. Draief, M., Vojnovic, M.: Convergence speed of binary interval consensus. SIAM J. Control Optim. **50**(3), 1087–1109 (2012)
17. Becchetti, L., Clementi, A.E.F., Natale, E., Pasquale, F., Trevisan, L.: Stabilizing consensus with many opinions. In: Proceedings of the Twenty-Seventh Annual ACM-SIAM Symposium on Discrete Algorithms, pp. 620–635 (2016)

18. Becchetti, L., Clementi, A., Natale, E., Pasquale, F., Silvestri, R., Trevisan, L.: Simple dynamics for plurality consensus. Distrib. Comput. **30**, 1–14 (2016)
19. van Kampen, N.: Stochastic Processes in Physics and Chemistry (1997). (revised edition)
20. Bruguière, C., Tiberghien, A., Clément, P.: Introduction. In: Bruguière, C., Tiberghien, A., Clément, P. (eds.) Topics and Trends in Current Science Education. CSER, vol. 1, pp. 3–18. Springer, Dordrecht (2014). doi:10.1007/978-94-007-7281-6_1
21. Chernoff, H.: A measure of asymptotic efficiency for tests of a hypothesis based on the sum of observations. Ann. Math. Stat. **23**, 493–507 (1952)
22. McDiarmid, C.: On the method of bounded differences. Lond. Soc. Lect. Note Ser. **141**, 148–188 (1989)

Chemical Boltzmann Machines

William Poole[1], Andrés Ortiz-Muñoz[1], Abhishek Behera[2], Nick S. Jones[3],
Thomas E. Ouldridge[3], Erik Winfree[1], and Manoj Gopalkrishnan[2(✉)]

[1] California Institute of Technology, Pasadena, CA, USA
{wpoole,winfree}@caltech.edu
[2] India Institute of Technology Bombay, Mumbai, India
manoj.gopalkrishnan@gmail.com
[3] Imperial College London, London, UK
t.ouldridge@imperial.ac.uk

Abstract. How smart can a micron-sized bag of chemicals be? How can an artificial or real cell make inferences about its environment? From which kinds of probability distributions can chemical reaction networks sample? We begin tackling these questions by showing three ways in which a stochastic chemical reaction network can implement a Boltzmann machine, a stochastic neural network model that can generate a wide range of probability distributions and compute conditional probabilities. The resulting models, and the associated theorems, provide a road map for constructing chemical reaction networks that exploit their native stochasticity as a computational resource. Finally, to show the potential of our models, we simulate a chemical Boltzmann machine to classify and generate MNIST digits in-silico.

1 Introduction

To carry out complex tasks such as finding and exploiting food sources, avoiding toxins and predators, and transitioning through critical life-cycle stages, single-celled organisms and future cell-like artificial systems must make sensible decisions based on information about their environment [1,2]. The small volumes of cells makes this enterprise inherently probabilistic: environmental signals and the biochemical networks within the cell are noisy, due to the stochasticity inherent in the interactions of small, diffusing molecules [3–5]. The small volumes of cells also raises questions not only about how stochasticity influences circuit function, but also about how much computational sophistication can be packed into the limited available space.

Perhaps surprisingly, neural network models provide an attractive architecture for the types of computation, inference, and information processing that cells must do. Neural networks can perform deterministic computation using circuits that are smaller and faster than boolean circuits composed of AND, OR, and NOT gates [6], can robustly perform tasks such as associative recall [7], and can

---

W. Poole, A. Ortiz-Muñoz and A. Behera—Contributed Equally.

© Springer International Publishing AG 2017
R. Brijder and L. Qian (Eds.): DNA 23 2017, LNCS 10467, pp. 210–231, 2017.
DOI: 10.1007/978-3-319-66799-7_14

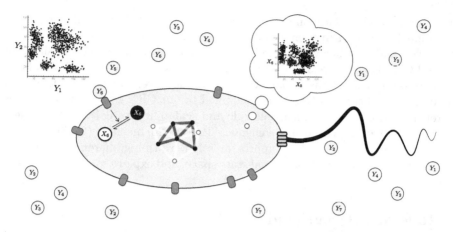

Fig. 1. In a micron-scale environment, molecular counts are low and a real (or synthetic) cell will have to respond to internal and environmental cues. Probabilistic inference using chemical Boltzmann machines provides a framework for how this may be achieved.

naturally perform Bayesian inference [8]. Furthermore, the structure of biochemical networks, such as signal transduction cascades [1,9,10] and genetic regulatory networks [11–15], can map surprisingly well onto neural network architectures. Chemical implementations of neural networks and related machine learning models have also been proposed [16–20], and limited examples demonstrated [21–24], for synthetic biochemical systems.

Most previous work on biochemical neural networks and biochemical inference invoked models based on continuous concentrations of species representing neural activities. Such models are limited in their ability to address questions of biochemical computation in small volumes, where discrete stochastic chemical reaction network models must be used to account for the low molecular counts. The nature of biochemical computation changes qualitatively in this context. In particular, stochasticity has been widely studied in genetic regulatory networks [25], signaling cascades [26], population level bet hedging in bacteria [27], and other areas [28,29] – where the stochasticity is usually seen as a challenge limiting correct function, but is occasionally also viewed as a useful resource [30]. Our work falls squarely in the latter camp: we attempt to exploit the intrinsic stochastic fluctuations of a formal chemical reaction network (CRN) to build natively stochastic samplers by implementing a stochastic neural network. This links to efforts to build natively stochastic hardware for Bayesian inference [31,32] and to the substantial literature attempting to model, and find evidence for, stochastic neural systems capable of Bayesian inference [33,34].

Specifically, we propose CRNs that implement Boltzmann machines (BMs), a flexible class of Markov random fields capable of generating diverse distributions and for which conditioning on data has straightforward physical interpretations [8,35]. BMs are an established model of probabilistic neural networks due to their analytic tractability and connections to spin systems in statistical physics [36] and Hopfield networks in computer science [7]. These networks have been studied extensively and used in a wide range of applications including image classification [37] and video generation [38]. We prove that CRNs can implement BMs and that this is possible using detailed balanced CRNs. Moreover, we show that many of the attractive features of BMs can be applied to our CRN constructions such as inference, a straightforward learning rule and scalability to real-world data sets. We thereby introduce the idea of a chemical Boltzmann machine (CBM), a chemical system capable of exactly or approximately performing inference using a stochastically sampled high-dimensional state space, and explore some of its possible forms.

2   Relevant Background

2.1   Boltzmann Machines (BMs)

Boltzmann machines are a class of binary stochastic neural networks, meaning that each node randomly switches between the values 0 and 1 according to a specified distribution. They are widely used for unsupervised machine learning because they can compactly represent and manipulate high-dimensional probability distributions. Boltzmann machines provide a flexible machine learning architecture because, as generative models, they can be used for a diverse set of tasks including data classification, data generation, and data reconstruction. Additionally, the simplicity of the model makes them analytically tractable. The use of hidden units (described below) allows Boltzmann machines to represent high order correlations in data. Together, these features make Boltzmann machines an excellent starting point for implementing stochastic chemical computers.

Fix a positive integer $N \in \mathbb{Z}_{>0}$. An N-node **Boltzmann machine** (BM) is specified by a quadratic **energy** function $E : \{0,1\}^N \to \mathbb{R}$

$$E(x_1, x_2, \ldots, x_N) = -\sum_{i<j} w_{ij} x_i x_j - \sum_i \theta_i x_i \qquad (1)$$

where $\theta_i \in \mathbb{R}$ is the **bias** of node i, and $w_{ij} = w_{ji} \in \mathbb{R}$ is the **weight** of the unordered pair (i,j) of nodes, with $w_{ii} = 0$. One may specify a BM **architecture**, or graph topology, by choosing additional weights w_{ij} that are to be set to 0. In this paper, we will use $\mathcal{N}(i) = \{j \text{ s.t. } w_{ij} \neq 0\}$ to denote the neighborhood of i. From a physical point of view, we are implicitly using *temperature units* $k_B T$ for energy, which we will continue to do throughout this paper. A BM describes a distribution $P(x)$ over **state vectors** $x = (x_1, \ldots, x_N) \in \{0,1\}^N$,

$$P(x) = \frac{1}{Z} e^{-E(x)} \quad \text{with} \quad Z = \sum_{x' \in \{0,1\}^N} e^{-E(x')}. \qquad (2)$$

Nodes of a BM are often partitioned into sets V and H of **visible** and **hidden**, respectively. Nodes in V represent data, and auxiliary nodes in H allow more complex distributions to be represented in the visible nodes. An **implementation** of a BM is a stationary stochastic process that samples from this distribution in the steady state. A BM can be implemented *in silico* using the **Gibbs sampling** algorithm [39], which induces a discrete time Markov chain (DTMC) on the state space $\{0,1\}^N$ in such a way that the stationary distribution of this Markov chain corresponds to the distribution $P(x)$. In each round, one node $i \in \{1, \ldots, N\}$ is chosen at random for update. For any two adjacent configurations x and x' which differ only at node i—i.e., $x_i = 1 - x_i'$ and $x_j = x_j'$ for all $j \neq i$—we set the transition probabilities $T_{x \to x'}$ of the DTMC so that

$$\frac{T_{x' \to x}}{T_{x \to x'}} = \frac{P(x)}{P(x')} = \frac{e^{-E(x)}}{e^{-E(x')}} = e^{\left(\theta_i + \sum_{j \in \mathcal{N}(i)} w_{ij} x_j\right)(x_i - x_i')}. \tag{3}$$

Any function $T_{x \to x'}$ can be chosen so long as (3) is satisfied. One common choice is $T_{x \to x'} = 1/(N(1 + e^{E(x') - E(x)}))$, where the factor $1/N$ represents the probability of choosing node i.

A Boltzmann machine is also an inference engine. One can do inference on $P(x)$ by conditioning on the values of a subset of the nodes. Suppose nonempty node subsets U and Y form a partition of the nodes $\{1, 2, \ldots, N\}$, and fix $u \in \{0,1\}^U$. To obtain samples from $P(y \mid u)$ where $y \in \{0,1\}^Y$, one **clamps** every node $i \in U$ to the state u_i while running Gibbs sampling, i.e., one does not allow these nodes to update. Clamping nodes to an input state is the same as specifying the input data for a statistical model. Steady state samples $y \in \{0,1\}^Y$ of this procedure are draws from the distribution $P(y \mid u)$.

Boltzmann machines can be used to learn a generative model from unlabeled data. After specifying the architecture, one then proceeds to find the weights, w_{ij}, and biases, θ_i, that maximize the likelihood of the observed data according to the model, using gradient descent from a random initial parameterization. This reduces to a very simple a two-phase Hebbian learning rule where weights on active edges are strengthened in a "wake phase" during which the network is clamped to observed data and are weakened in a "sleep phase" during which the network runs free [8,35]. Given a target distribution $Q(x)$, this gradient descent corresponds to calculating the gradient of the Kullback-Leibler divergence from P to Q, $D_{KL}(Q \parallel P) = \sum_x Q(x) \log \frac{Q(x)}{P(x)}$, with respect to the parameters θ_i and w_{ij}:

$$\frac{d\theta_i}{dt} = -\frac{\partial D_{KL}}{\partial \theta_i} = \langle x_i \rangle_Q - \langle x_i \rangle_P \quad \text{and} \quad \frac{dw_{ij}}{dt} = -\frac{\partial D_{KL}}{\partial w_{ij}} = \langle x_i x_j \rangle_Q - \langle x_i x_j \rangle_P \tag{4}$$

where $\langle \cdot \rangle_P$ and $\langle \cdot \rangle_Q$ denote expected values with respect to the distributions P and Q respectively. When hidden units are present, the distribution Q (which is defined on visible units only) is extended to hidden units based on clamping the visible units according to Q and using the conditional distribution $P(y|u)$ to determine the hidden units.

2.2   Chemical Reaction Networks (CRNs)

Fix a finite set $\mathcal{S} = (S_1, S_2, \ldots, S_M)$ of M **species**. A **reaction** r is a formal chemical equation

$$\sum_{i=1}^{M} \mu_r^i S_i \rightarrow \sum_{i=1}^{M} \nu_r^i S_i, \tag{5}$$

abbreviated as $r = \mu_r \rightarrow \nu_r$ where $\mu_r, \nu_r \in \mathbb{N}^{\mathcal{S}}$ are the stoichiometric coefficient vectors for the reactant and product species respectively, and $\mathbb{N} = \mathbb{Z}_{\geq 0}$. A **reaction rate constant**, $k_r \in \mathbb{R}_{>0}$, is associated with each reaction. In this paper, we define a **chemical reaction network** (CRN) as a triple $\mathcal{C} = (\mathcal{S}, \mathcal{R}, k)$ where $\mathcal{S}$ is a finite set of species, and $\mathcal{R}$ is a set of reactions, and k is the associated set of reaction rate constants.

We will denote chemical species by capital letters, and their counts by lower case letters; e.g., s_1 denotes the number of species S_1. Thus the state of a stochastic CRN is described by a vector on a discrete lattice, $s = (s_1, s_2 \ldots s_M) \in \mathbb{N}^{\mathcal{S}}$. The dynamics of a stochastic CRN are as follows [40]. The probability that a given reaction occurs per unit time, called its **propensity**, is given by

$$\rho_r(s) = k_r \prod_{i=1}^{M} \frac{s_i!}{(s_i - \mu_r^i)!} \quad \text{if} \quad s_i \geq \mu_r^i \quad \text{and 0 otherwise.} \tag{6}$$

Each time a reaction fires, state s changes to state $s + \Delta_r$, where $\Delta_r = \nu_r - \mu_r$ is called the reaction vector, and the propensity of each reaction may change. Viewed from a state space perspective, stochastic CRNs are continuous time Markov chains (CTMCs) with transition rates

$$R_{s \rightarrow s'} = \sum_{r \text{ s.t. } s' = s + \Delta_r} \rho_r(s) \tag{7}$$

and thus their dynamics follow

$$\frac{dP(s,t)}{dt} = \sum_{s' \neq s} R_{s' \rightarrow s} P(s', t) - R_{s \rightarrow s'} P(s, t), \tag{8}$$

where $P(s,t)$ is the probability of a state with counts s at time t. Equivalently, they are governed by the **chemical master equation**,

$$\frac{dP(s,t)}{dt} = \sum_{r \in \mathcal{R}} P(s - \Delta_r, t) \rho_r(s - \Delta_r) - P(s, t) \rho_r(s). \tag{9}$$

A stationary distribution $\pi(s)$ may be found by solving $\frac{dP(s,t)}{dt} = 0$ simultaneously for all s; in general, it need not be unique, and even may not exist. Given an initial state s_0, $\pi(s) = P(s, \infty)$ is unique if it exists. For that initial state, the **reachability class** $\Omega_{s_0} \subseteq \mathbb{N}^M$ is the maximal subset of the integer lattice accessible to the CRN via some sequence of reactions in $\mathcal{R}$. We will specify a CRN and a reachability class given an initial state as a shorthand for specifying a CRN and a set of initial states with identical reachability classes.

2.3   Detailed Balanced Chemical Reaction Networks

A CTMC is said to satisfy **detailed balance** if there exists a well-defined function of state s, $E(s) \in \mathbb{R}$, such that for every pair of states s and s', the transition rates $R_{s \to s'}$ and $R_{s' \to s}$ are either both zero or

$$\frac{R_{s \to s'}}{R_{s' \to s}} = e^{E(s) - E(s')}. \tag{10}$$

If the state space Ω is connected and the partition function $Z = \sum_{s \in \Omega} e^{-E(s)}$ is finite, then the steady state distribution $\pi(s) = \frac{1}{Z} e^{-E(s)}$ is unique, and the net flux between all states is zero in that steady state.

There is a related but distinct notion of detailed balance for a CRN. An equilibrium chemical system is constrained by physics to obey detailed balance at the level of each reaction. In particular, for a dilute equilibrium system, each species $S_i \in \mathcal{S}$ has an energy $G[S_i] \in \mathbb{R}$ that relates to its intrinsic stability, and

$$\frac{k_{r+}}{k_{r-}} = e^{-\sum_{i=1}^{M} \Delta_{r+}^i G[S_i]} = e^{-\Delta G_{r+}}, \tag{11}$$

where Δ_{r+}^i is the ith component of $\Lambda_{r+} = \nu_{r+} - \mu_{r+}$, and we have defined $\Delta G_{r+} = \sum_{i=1}^{N} \Delta_{r+}^i G[S_i]$. Any CRN for which there exists a function G satisfying (11) is called a detailed balanced CRN. To see that the CTMC for a detailed balanced CRN also itself satisfies detailed balance in the sense of (10), let $s' = s + \Delta_{r+}$ and note that (6) and (11) imply that

$$\frac{\rho_{r+}(s)}{\rho_{r-}(s')} = e^{\mathcal{G}(s) - \mathcal{G}(s')} \quad \text{with} \quad \mathcal{G}(s) = \sum_{i=1}^{M} s_i G[S_i] + \log(s_i!), \tag{12}$$

for all reactions r^+. Here, $\mathcal{G}(s)$ is a well-defined function of state s (the free energy) that can play the role of E in (10). If there are multiple reactions that bring s to s', they all satisfy (12), and therefore the ratio $R_{s \to s'}/R_{s' \to s}$ satisfies (10) and the CTMC satisfies detailed balance.

It is possible to consider non-equilibrium CRNs that violate (11). Such systems must be coupled to implicit reservoirs of fuel molecules that can drive the species of interest into a non-equilibrium steady state [41–43]. Usually – but not always [44,45] – the resultant Markov chain violates detailed balance. In Sect. 3.1, we shall consider a system that exhibits detailed balance at the level of the Markov chain, but is necessarily non-equilibrium and violates detailed balance at the detailed chemical level.

Given an initial condition s_0, a detailed balanced CRN will be confined to a single reachability class Ω_{s_0}. Moreover, from the form of $\mathcal{G}(s)$, the stationary distribution $\pi(s)$ on Ω_{s_0} of any detailed balanced CRN exists, is unique, and is a product of Poisson distributions restricted to the reachability class [46],

$$\pi(s) = \frac{1}{Z} e^{-\mathcal{G}(s)} = \frac{1}{Z} \prod_{i=1}^{M} \frac{e^{-s_i G[S_i]}}{s_i!}, \tag{13}$$

with the partition function $Z = \sum_{s' \in \Omega_{s_0}} e^{-\mathcal{G}(s')}$ dependent on the reachability class. Note that this implies that the partition function is always finite, even for an infinite reachability class.

3   Exact Constructions and Theorems

3.1   Clamping and Conditioning with Detailed Balanced CRNs

In a Boltzmann machine that has been trained to generate a desired probability distribution when run, inference can be performed by freezing, also known as clamping, the values of known variables, and running the rest of the network to obtain a sample; this turns out to exactly generate the conditional probability. A similar result holds for a subclass of detailed balanced CRNs that generate a distribution, for an appropriate notion of clamping in a CRN. Imagine a "demon" that, whenever a reaction results in a change in the counts of one of the clamped species, will instantaneously change it back to its previous value. If every reaction is such that either no clamped species change, or else every species that changes is clamped, then the demon is effectively simply "turning off" those reactions. More precisely, consider a CRN, $\mathcal{C} = (\mathcal{S}, \mathcal{R}, k)$. We will partition the species into two disjoint groups $Y = \mathcal{S}_{free}$ and $U = \mathcal{S}_{clamped}$, where $\mathcal{S}_{free}$ will be allowed to vary and $\mathcal{S}_{clamped}$ will be held fixed. We will define free reactions, $\mathcal{R}_{free}$, as reactions which result in neither a net production nor a net consumption of any clamped species. Similarly, clamped reactions, $\mathcal{R}_{clamped}$ are reactions which change the counts of any clamped species. The clamped CRN will be denoted $\mathcal{C}|_{U=u}$ to indicate the the species $U_i \in U$ have been clamped to the values u_i. The clamped CRN is defined by $\mathcal{C}|_{U=u} = (\mathcal{S}, \mathcal{R}_{free}, k_{free})$, that is, the entire set of species along with the reduced set of reactions and their rate constants. In the clamped CRN it is apparent that the clamped species will not change from their initial conditions because no reaction in $\mathcal{R}_{free}$ can change their count. However, these clamped species may still affect the free species catalytically. If the removed reactions, $\mathcal{R}_{clamped}$, never change counts of non-clamped species, then $\mathcal{C}|_{U=u}$ is equivalent to the action of the "demon" imagined above.

We use Eq. 13 to prove that clamping a detailed balanced CRN is equivalent to calculating a conditional distribution, and to show when the conditional distributions of a detailed balanced CRN will be independent. Together, these theorems provide guidelines for devising detailed balanced CRNs with interesting (non-independent) conditional distributions and for obtaining samples from these distributions via clamping.

We will need one more definition. Let $\mathcal{C}$ be a detailed balanced CRN with reachability class Ω_{s_0} for some initial condition $s_0 = (u_0, y_0)$. Let Γ_{s_0} be the reachability class of the clamped CRN $\mathcal{C}|_{U=u_0}$ with species U clamped to u_0 and species Y free. We say clamping **preserves reachability** if $\Omega_{s_0|U=u_0}^{Y} = \Gamma_{s_0}^{Y}$ where $\Omega_{s_0|U=u_0}^{Y} = \{y \text{ s.t. } (u_0, y) \in \Omega_{s_0}\}$ and $\Gamma_{s_0}^{Y} = \{y \text{ s.t. } (u_0, y) \in \Gamma_{s_0}\}$. In other words, clamping preserves reachability if, whenever a state $s = (u_0, y)$ is reachable from s_0 by any path in $\mathcal{C}$, then it is also reachable from s_0 by some path in $\mathcal{C}|_{U=u_0}$ that never changes u.

Theorem 1. *Consider a detailed balanced CRN $\mathcal{C} = (\mathcal{S}, \mathcal{R}, k)$ with reachability class Ω_{s_0} from initial state s_0. Partition the species into two disjoint sets $U = \{U_1, \ldots, U_{M_u}\} \subset \mathcal{S}$ and $Y = \{Y_1, \ldots, Y_{M_y}\} \subset \mathcal{S}$ with $M_u + M_y = M = |\mathcal{S}|$. Let the projection of s_0 onto U and Y be u_0 and y_0. The conditional distribution $P(y \mid u)$ implied by the stationary distribution π of $\mathcal{C}$ is equivalent to the stationary distribution of a clamped CRN, $\mathcal{C}|_{U=u}$ starting from initial state s_0 with $u_0 = u$, provided that clamping preserves reachability.*

Proof. We have $\mathcal{G}(u, y) = \sum_{i=1}^{M_u} u_i G[U_i] + \log(u_i!) + \sum_{i=1}^{M_y} y_i G[Y_i] + \log(y_i!)$. Let the reachability class of $\mathcal{C}|_{U=u}$ be Γ_{s_0}, its projection onto Y be $\Gamma_{s_0}^Y$, and $\Omega_{s_0|U=u_0}^Y = \{y \text{ s.t. } (u_0, y) \in \Omega_{s_0}\}$ with $\Omega_{s_0|U=u_0}^Y = \Gamma_{s_0}^Y$. Then, the conditional probability distribution of the unclamped CRN is given by

$$P(y \mid u) = \frac{\pi(u, y)}{\sum_{y' \in \Gamma_{s_0}^Y} \pi(u, y')} = \frac{e^{-\mathcal{G}(u,y)}}{\sum_{y' \in \Gamma_{s_0}^Y} e^{-\mathcal{G}(u,y')}}. \tag{14}$$

Simply removing pairs of forward and backward reactions will preserve detailed balance for unaffected transitions, and hence the clamped system remains a detailed balanced CRN with the same free energy function. We then readily see that the clamped CRN's stationary distribution, $\pi_c(y|u)$ is given by

$$\pi_c(y|u) = \frac{e^{-\mathcal{G}(u,y)}}{Z_c(u)} \quad \text{with} \quad Z_c(u) = \sum_{y' \in \Gamma_{s_0}^Y} e^{-\mathcal{G}(u,y')}. \quad \blacksquare \tag{15}$$

The original CRN and the clamped CRN do not need to have the same initial conditions as long as the initial conditions have the same reachability classes. However, even if the two CRNs have the same initial conditions, it is possible that the clamping process will make some part of $\Omega_{s_0|U=u_0}^Y$ inaccessible to $\mathcal{C}|_{U=u}$, in which case this theorem will not hold.

Theorem 2. *Assume the reachability class of a detailed balanced CRN can be expressed as the product of subspaces, $\Omega_{s_0} = \prod_{j=1}^L \Omega_{s_0}^j$. Then the steady-state distributions of each subspace will be independent for each product space: $\pi(s) = \prod_{j=1}^L \pi^j(s^j)$, where $s = (s^1, \ldots, s^L)$ and π^j is the distribution over $\Omega_{s_0}^j$.*

Proof. If Ω_{s_0} is decomposable into a product of subspaces $\Omega_{s_0}^j$, with $j = 1 \ldots L$, then each subspace involves disjoint sets of species $Y^j = \{S_1^j, \ldots, S_{M_j}^j\}$. In this case the steady-state distribution of a detailed balanced CRN can be factorized due to the simple nature of $\mathcal{G}(s)$ given by Eq. (12):

$$\pi(s) = \frac{e^{-\mathcal{G}(s)}}{Z} = \frac{\prod_{j=1}^L e^{-\mathcal{G}(s^j)}}{\prod_{j=1}^L \sum_{s^{j'} \in \Omega_{s_0}^j} e^{-\mathcal{G}(s^{j'})}} = \prod_{j=1}^L \frac{e^{-\mathcal{G}(s^j)}}{\sum_{s^{j'} \in \Omega_{s_0}^j} e^{-\mathcal{G}(s^{j'})}} = \prod_{j=1}^L \frac{e^{-\mathcal{G}(s^j)}}{Z^j}, \tag{16}$$

where $s^j = (s_1^j, s_2^j, \ldots, s_{M_j}^j)$ is the state of the set of species within subspace j. $\blacksquare$

The product form $\pi(s)$ means that species from separate subspaces $\Omega_{s_0}^j$ are statistically independent. To develop non-trivial conditional probabilities for the states of different species, therefore, it is necessary either to use a non-detailed balanced CRN by driving the system out of equilibrium, or to generate complex interdependencies through conservation laws that constrain reachability classes and "entangle" the state spaces for different species. We explore both of these possibilities in the following sections.

3.2   Direct Implementation of a Chemical Boltzmann Machine (DCBM)

We first consider the most direct way to implement an N-node Boltzmann machine with a chemical system. Recall that a BM has a state space $\Omega_{BM} = \{0,1\}^N$ and an energy function $E(x_1, x_2, \ldots, x_N) = -\sum_{i<j} w_{ij} x_i x_j - \sum_i \theta_i x_i$. We use a dual rail representation of each node i by two CRN species X_i^{ON} and X_i^{OFF} and reactions that respect a conservation law, $x_i^{ON} + x_i^{OFF} = 1$. The species X_i^{ON} and X_i^{OFF} could represent activation states of an enzyme. The CRN has $M = 2N$ species and states $s = (x_1^{ON}, x_1^{OFF}, \ldots, x_N^{ON}, x_N^{OFF})$. Although there are 2^{2N} states in which each species has a count of at most one, only $1/2^N$ of these states are reachable due to the conservation laws. Let Ω_{DCBM} be the states reachable from a valid initial state. There exists a one-to-one invertible mapping $\mathcal{F} : \Omega_{BM} \to \Omega_{DCBM}$ which maps the states $x \in \Omega_{BM}$ of a BM to states $s = \mathcal{F}(x) \in \Omega_{DCBM}$ of the CBM, according to $x_i^{ON} = x_i$ and $x_i^{OFF} = 1 - x_i$.

Reactions are intended to provide a continuous-time analog of the typical BM implementations, such as the Gibbs sampling method discussed in Sect. 2.1. In each reaction r, only the species X_i^{ON} and X_i^{OFF}, corresponding to a single node i, change ($\nu_r - \mu_r$ has two non-zero components). To reproduce the stationary distribution of a Boltzmann machine with energy function $E(x)$, it is sufficient to require that the CTMC for the CRN satisfies

$$s \rightleftharpoons s' \quad \text{with} \quad \frac{R_{s \to s'}}{R_{s' \to s}} = \frac{e^{-E(s')}}{e^{-E(s)}} = e^{\theta_i + \sum_{j \in \mathcal{N}(i)} w_{ij} x_j^{ON}} \tag{17}$$

where s is any reachable state with $x_i^{OFF} = 1$, and s' has $x_i^{ON} = 1$ but is otherwise the same. Such a choice would enforce detailed balance of the CTMC, with the desired steady-state distribution

$$\pi(s) = \frac{1}{Z} e^{-E(s)} = \frac{1}{Z} e^{-\sum_{i<j} w_{ij} x_i^{ON} x_j^{ON} - \sum_i \theta_i x_i^{ON}}. \tag{18}$$

To implement such a CRN, we define a reaction set $\mathcal{R}$ that contains a distinct pair of reactions for each possible state of the neighbors of i for which $w_{ij} \neq 0$. Let $\alpha^i \in \{ON, OFF\}^{|\mathcal{N}(i)|}$ denote a state of neighboring species. Then, the necessary reactions and rate constants are

$$X_i^{ON} + \sum_{j \in \mathcal{N}(i)} X_j^{\alpha_j^i} \underset{k_{i+|\alpha_i}}{\overset{k_{i-|\alpha_i}}{\rightleftharpoons}} X_i^{OFF} + \sum_{j \in \mathcal{N}(i)} X_j^{\alpha_j^i}, \quad \frac{k_{i+|\alpha^i}}{k_{i-|\alpha^i}} = e^{\theta_i + \sum_{j \in \mathcal{N}(i)} w_{ij} x_j^{ON}}, \tag{19}$$

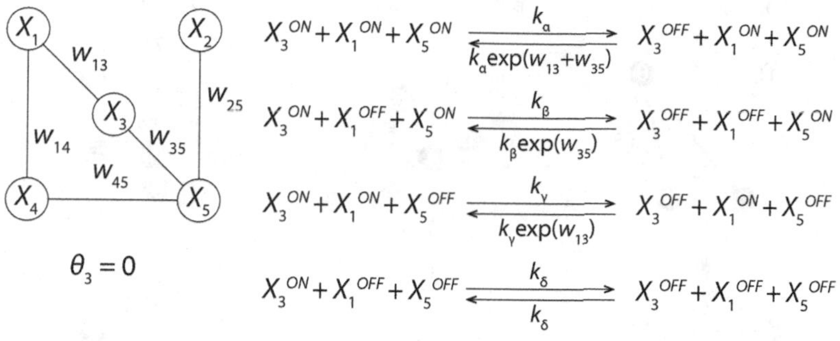

$$X_3^{ON} + X_1^{ON} + X_5^{ON} \xrightleftharpoons[k_\alpha \exp(w_{13}+w_{35})]{k_\alpha} X_3^{OFF} + X_1^{ON} + X_5^{ON}$$

$$X_3^{ON} + X_1^{OFF} + X_5^{ON} \xrightleftharpoons[k_\beta \exp(w_{35})]{k_\beta} X_3^{OFF} + X_1^{OFF} + X_5^{ON}$$

$$X_3^{ON} + X_1^{ON} + X_5^{OFF} \xrightleftharpoons[k_\gamma \exp(w_{13})]{k_\gamma} X_3^{OFF} + X_1^{ON} + X_5^{OFF}$$

$$X_3^{ON} + X_1^{OFF} + X_5^{OFF} \xrightleftharpoons[k_\delta]{k_\delta} X_3^{OFF} + X_1^{OFF} + X_5^{OFF}$$

$$\theta_3 = 0$$

Fig. 2. The reactions required by the dynamics of a single node using the direct CBM implementation. We consider a simple network with the illustrated topology, and display the required reactions for node 3. Since node 3 has degree 2, there are 4 possible states of its neighbors, and hence four distinct pairs of reactions for the species of node 3. The relative rates of each pair of reactions is set by w_{ij} as indicated (where, for simplicity, we have assumed $\theta_3 = 0$).

for each i and every possible state α. In physical terms, the species representing the neighbors of node i collectively catalyze $X_i^{OFF} \rightleftharpoons X_i^{ON}$, with a separate pair of reactions for every possible α^i. While this entails a large number of reactions ($2|\mathcal{N}(i)|+1$ for each node i), it allows the rate constants for each configuration of neighbors to be distinct, and thus to satisfy the ratio of rate constants given in (19). For CRN states that satisfy the conservation laws $x_i^{ON} + x_i^{OFF} = 1$, there will be a unique reaction that can flip any given bit, and thus the CTMC detailed balance (17) also holds, yielding the correct $\pi(s)$. The construction is illustrated by example in Fig. 2 and compared to other constructions in Fig. 3.

The distribution $\pi(s)$ is identical to that of the BM, both with and without clamping. Reachability is preserved by clamping, as all states satisfying the conservation laws and clamping can be reached in the clamped CRN. All results derived for traditional BMs therefore apply, including conditional inference and the Hebbian learning rule. The construction can be generalized to any graphical model and indeed to any finite Markov chain defined on a positive integer lattice.

With the DCBM, we have shown that CRNs can express the same distributions as BMs, and are thus very expressive. However, since each possible state α^i of $\mathcal{N}(i)$ is associated with two reactions, the number of reactions of the CRN is exponentially large in the typical node degree d in the original BM. Moreover, the scheme requires high molecularity reactions in which multiple catalysts effect a single transition (the molecularity grows linearly with d). Physical implementations are therefore likely to be challenging. We further note that as a consequence of Theorem 2, the DCBM cannot be detailed balanced at the level of the underlying chemistry, due to its simple conservation laws. Physically, this means that the DCBM must use a fuel species to drive each reaction. Details of this argument are given in the Appendix (A.1).

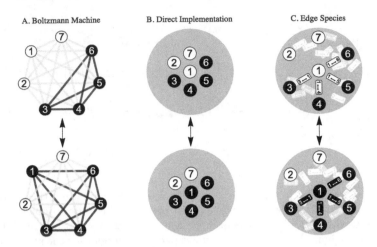

Fig. 3. Comparison of the switching of a node in exact constructions for fully-connected topologies. Black circles indicate ON species (or nodes), and white circles indicate OFF species. Similarly, black/white rectangles indicate ON/OFF edge species. Species not involved in the reaction have been grayed out. A. A Boltzmann machine. Black edges contribute to the energy function. B. The direct implementation of a chemical Boltzmann machine. All species jointly catalyze the conversion of X_1^{OFF} to X_1^{ON}. C. The edge species chemical Boltzmann machine. X_1^{OFF} is converted to ON simultaneously with W_{14}^{OFF}, W_{15}^{OFF}, W_{16}^{OFF} and W_{17}^{OFF}; all other node species involved act as catalysts.

3.3   The Edge Species CBM Construction (ECBM)

Can a detailed balanced CRN also implement a Boltzmann machine, or is it necessary to break detailed balance at the level of the CRN reactions, as in the DCBM? Here we show that it is not necessary by introducing a detailed balanced CRN that uses species to represent both the nodes and edges of a BM. The N nodes of a BM are converted into N pairs of species, X_i^{ON} and X_i^{OFF}, via a dual rail implementation identical to that used in the DCBM. Similarly, the edges w_{ij} are represented by dual rail edge species W_{ij}^{ON} and W_{ij}^{OFF} with the conservation law $w_{ij}^{ON} + w_{ij}^{OFF} = 1$ for $1 \le j < i \le N$. Note that we may slightly abuse notation and let $W_{ij}^{\alpha_{ij}}$ and $W_{ji}^{\alpha_{ij}}$, with $\alpha_{ij} \in \{ON, OFF\}$, represent the same chemical species.

To have detailed balance, we associate energies to each node species determined by the bias in a BM, $G[X_i^{ON}] = -\theta_i$ and $G[X_i^{OFF}] = 0$. Similarly, each edge species has an energy determined by the corresponding edge weight $G[W_{ij}^{ON}] = -w_{ij}$ and $G[W_{ij}^{OFF}] = 0$. Finally, we define a set of catalytic reactions that ensure that the states of edge and node species are consistent, meaning $w_{ij}^{ON} = 1$ if and only if $x_i^{ON} = 1$ and $x_j^{ON} = 1$. To achieve this coupling, the reactions that switch node i are always catalyzed by the species corresponding to the set of neighboring nodes $\mathcal{N}(i)$. Simultaneously, these reactions switch edge

ij if $j \in \mathcal{N}(i)$ and $x_j^{ON} = 1$, maintaining $x_i^{ON} x_j^{ON} = w_{ij}^{ON}$. The set of reactions that result from this scheme are

$$X_i^{OFF} + \sum_{j \in \mathcal{N}(i)} X_j^{\alpha_j^i} + \sum_{j \in \mathcal{N}(i), x_j^{ON}=1} W_{ij}^{OFF} \rightleftharpoons X_i^{ON} + \sum_{j \in \mathcal{N}(i)} X_j^{\alpha_j^i} + \sum_{j \in \mathcal{N}(i), x_j^{ON}=1} W_{ij}^{ON}.$$

(20)

This reaction scheme is visualized in Fig. 3. Just like in the DCBM, there is a separate pair of reactions for each node i for each state of its neighbors α^i. In this case, however, the backward reaction in (20) does represent a transition that is a true chemical inversion of the forward reaction. So the rate constants can be set to agree with detailed balance (11). Further, given a valid initial state, clamping any subset of the $X_i^{ON/OFF}$ species preserves reachability.

Theorem 3. *The stationary distribution $\pi(x^{ON}, x^{OFF}, w^{ON}, w^{OFF})$ of the ECBM is equivalent to the stationary distribution of a Boltzmann machine, $P(x)$, provided that the ECBM begins in a valid state obeying $w_{ij}^{ON} = x_i^{ON} x_j^{ON}$ and one applies a one-to-one invertible mapping $\mathcal{F}$ between BM and ECBM states, as described below.*

Proof. If this CRN begins in a consistent state, then every subsequent reaction will conserve this condition. The combined conservation laws $x_i^{ON} + x_i^{OFF} = 1$, $w_{ij}^{ON} + w_{ij}^{OFF} = 1$, and $w_{ij}^{ON} = x_i^{ON} x_j^{ON}$ ensure that the set of values x_i^{ON} uniquely determine the CRN state for the ECBM, and thus—similar to how the BM and DCBM states were related—we can define a one-to-one invertible mapping $\mathcal{F}$ that sets $x_i^{ON} = x_i$ and obeys the conservation laws.

The ECBM is detailed balanced and therefore its stationary distribution has the form (13). Substituting the conservation law $w_{ij} = x_i x_j$ and omitting species with 0 energy results in

$$\pi(x^{ON}, x^{OFF}, w^{ON}, w^{OFF}) = \frac{1}{Z_\pi} e^{-\sum_{i \neq j} G[W_{ij}^{ON}] x_i^{ON} x_j^{ON} - \sum_i G[X_i^{ON}] x_i^{ON}}$$

(21)

Comparing this expression to the distribution of a BM, Eq. (2), the above expressions are equivalent provided that their partition functions are equivalent. To see this is the case, notice that: (1) the partition function is just a sum over the Gibbs factors across the entire state space. (2) The Gibbs factors take the same form between the ECBM and BM (as shown above). And (3) the reachable state spaces spaces are equivalent. Thus a sum over all possible Gibbs factors will be equal. Therefore, $Z_{BM} = Z_\pi$ and the theorem is proven. ∎

Via the ECBM, we have shown that even detailed balanced CRNs can represent rich distributions and are able to calculate conditional distributions through clamping as proven in Theorem 1. Due to being detailed balanced, this construction requires no fuel molecules and performs sampling via the intrinsic equilibrium fluctuations of the CRN. Moreover, it is only necessary to tune molecular energies in this construction, since appropriate relative rate constants follow by definition. This construction is possible due to the complex set of conservation laws that ensure that the reachability classes of all the $X_i^{ON/OFF}$ species are tightly coupled via the $W_{ij}^{ON/OFF}$ species. One implication is that this construction does not generalize easily to non-binary species counts. Additionally, issues related to high molecularity reactions and large number of reactions remain.

4   Approximate Bimolecular Implementations

The DCBM and the ECBM both require reactions of high molecularity. High molecularity reactions and systems involving many species are physically challenging to implement and also potentially suffer from long mixing times. In this section, we discuss an approximation scheme to create CBMs with lower molecularity reactions and thus overcome these issues.

4.1   Taylor Series Chemical Boltzmann Machine (TCBM)

Here, we demonstrate a compact CBM that approximates a BM. It is not detailed balanced on either the Markov chain or the CRN level, but uses only $2N$ species and $\mathcal{O}(N^2)$ unimolecular and bimolecular reactions. The TCBM is a non-equilibrium CBM of the kind discussed in Sect. 3.1 that uses a dual-rail representation and single-node transitions to approximately implement a BM. The reactions are given by:

$$X_i^{OFF} \underset{k}{\overset{k}{\rightleftharpoons}} X_i^{ON}$$

$$X_j^{ON} + X_i^{OFF} \xrightarrow{ka_{ij}} X_j^{ON} + X_i^{ON}$$

$$X_j^{ON} + X_i^{ON} \xrightarrow{kb_{ij}} X_j^{ON} + X_i^{OFF} \tag{22}$$

which, with appropriate initial conditions, preserve the conservation law that $x_i^{ON} + x_i^{OFF} = 1$.

This model's parameters can be taken directly from the weights of a BM, w_{ij}. First, define a symmetric matrix W. Decompose this matrix into the difference of two positive matrices, $W = A - B$, where $a_{ij} \in A$ are all $w_{ij} > 0$ and $b_{ij} \in B$ are the absolute values of all $w_{ij} < 0$. Finally, k is an arbitrary overall rate. This construction can be understood as an approximation of Eq. (17), which dictates that for two states s and s' that differ only in bit i with $x_i^{ON} = 1$ in state s', the CTMC transition rates must satisfy

$$\frac{R_{s \to s'}}{R_{s' \to s}} = \frac{e^{\sum_{j \neq i} a_{ij} x_j^{ON}}}{e^{\sum_{j \neq i} b_{ij} x_j^{ON}}} = \frac{1 + \sum_{j \neq i} a_{ij} x_j^{ON} + \mathcal{O}((\sum_{j \neq i} a_{ij} x_j^{ON}))^2}{1 + \sum_{j \neq i} b_{ij} x_j^{ON} + \mathcal{O}((\sum_{j \neq i} b_{ij} x_j^{ON}))^2}, \tag{23}$$

The bias θ_i has been absorbed into w_{ij} for notational clarity by assuming there is some $x_0^{ON} = 1$ whose weights act as biases. The TCBM is a bimolecular CRN obeying the same conservation laws as the DCBM in which each species j acts as an independent catalyst for transitions in i with reaction rates determined by a_{ij} and b_{ij}. The relative propensities of this network are exactly equal to the linear expansion of the relative propensities shown in the last term in (23). Specifically, the numerator is the sum of the reaction propensities for reactions that convert or catalyze $X_i^{OFF} \rightarrow X_i^{ON}$ and the denominator is the sum of the reaction propensities for $X_i^{ON} \rightarrow X_i^{OFF}$, in each case plus a constant term due to the unimolecular reactions. We thus propose the simple scheme in (22) as an approximate CBM; Fig. 4A depicts this TCBM schematically. This model bears some resemblance to protein phosphorylation networks where adding or removing a phosphate group is analogous to turning a species on or off; both are driven, catalytic processes capable of diverse computation.

4.2   Approximate BCRN Inference

Remarkably, this simple approximate CBM can reasonably approximate the inferential capabilities of a BM. We demonstrate this by using (22) to convert a BM trained on the MNIST dataset [47] to a TCBM (Fig. 4). We then compare the BM and the TCBM side by side. Digit classification is shown in

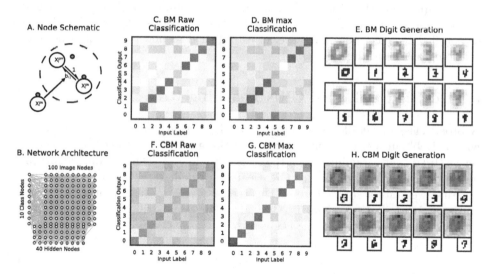

Fig. 4. A. CRN underlying an individual node of the TCBM approximation. In this case a negative weight, $w_{ij} < 0$ is shown because X_i^{ON} catalyzes $X_j^{ON} \rightarrow X_j^{OFF}$. B. Network architecture used for simulations is fully connected but only 10 percent of edges are shown for clarity. C. Average raw classification output of a BM running with clamped MNIST digits. D. Average max classification output of a BM running with clamped MNIST digits. E. Digits generated by a BM by clamping individual class nodes. Small sub-boxes in the bottom corners are plots of the top 85th percentile of pixels. F. Average raw classification output of a TCBM running with clamped MNIST digits. G. Average max classification output of a TCBM running with clamped MNIST digits. H. Digits generated by a TCBM by clamping individual class nodes. Small sub-boxes in the bottom corners are plots of the top 85th percentile of pixels.

Fig. 4 panels C and D for a BM and in Fig. 4 panels F and G for a CBM as confusion heatmaps. Classification is carried out by clamping the image nodes to MNIST images and averaging the values of the classification nodes. As is apparent from these plots, the BM does a fairly reasonable job classifying these digits, but struggles on the number 5. The CBM functions as a very noisy version of the BM with nodes in general much more likely to be on. The CBM has also faithfully inherited the capabilities and limitations of the BM and similarly struggles to classify the digit 5. Digit generation is shown in Fig. 4E for a BM and 4H for a CBM. Generation was carried out by clamping a single class node to 1 and all other class nodes to 0, then averaging the output of the image nodes after the network had equilibrated. For each generated image, we show the raw output and the top 85th percentile of nodes, a thresholding which helps visualize the noisy output. As is apparent from the raw output, the CBM approximation scheme does not generate images nearly as distinctly as the BM. However, this approximation does faithfully reproduce plausible digits when filtered for the top 85th percentile. Additional training and simulation details can be found in the appendix (A.2).

The overall performance of the CRN is reasonable, given the fact that weights were simply imported from a BM without re-optimization. The TCBM only approximates the distribution implied by these weights and, in the absence of detailed balance, does not have an established formal relationship between clamping and conditioning.

5   Detailed Balanced CRN Learning Rule

A broad class of detailed balanced chemical reaction networks can be trained with a Hebbian learning rule between a waking phase (clamped) and sleeping phase (free) that is reminiscent of the classic gradient descent learning algorithm for a BM [8,35]. We present the CRN learning rule here.

First we state a simple case of Theorem 4 where we just want a CRN with stationary distribution π over Ω_{s_0} to learn a target distribution Q also defined on Ω_{s_0}. Then, the learning rule is given by

$$\frac{dg_i}{dt} = -\frac{\partial D_{KL}}{\partial g_i} = \langle s_i \rangle_Q - \langle s_i \rangle_\pi. \tag{24}$$

Here, $\langle s_i \rangle_\pi$ and $\langle s_i \rangle_Q$ denote the expected count of the species S_i with respect to the probability distributions π and Q, respectively, and $g_i = G[S_i]$ is the energy of species S_i. Theorem 4 generalizes this procedure to cases with hidden species.

Theorem 4. *Let $\mathcal{C} = (\mathcal{S}, \mathcal{R}, k)$ be a detailed balanced chemical reaction network with stationary distribution $\pi(s)$ on Ω_{s_0}. Consider a partition (V, H) of the set $\mathcal{S}$ of species into visible and hidden species such that $\pi(s) = \pi(v, h)$. Require that for all visible states v, the clamped CRN $\mathcal{C}|_{V=v}$ preserves reachability. Let $Q(v) > 0$ for all $v \in \Omega_{s_0}^V = \{v \text{ s.t. } (v, h) \in \Omega_{s_0}\}$ be a target distribution defined on the projection of Ω_{s_0} onto V. Furthermore, let $\pi_Q(v, h) = Q(v)\pi(h \mid v)$*

be the weighted mixture of stationary distributions of the clamped CRNs $C|_{V=v}$ with v drawn from the distribution Q. Then, the gradient of the Kullback-Leibler divergence from π_V to Q with respect to the energy, $g_i = G[S_i]$, of the species S_i is given by

$$\frac{\partial D_{KL}(Q\|\pi_V)}{\partial g_i} = \langle s_i \rangle_\pi - \langle s_i \rangle_{\pi_Q} \tag{25}$$

where $\pi_V(v) = \sum_{h \in \Omega_{s_0}^H} \pi(v, h)$ is the marginal $\pi(v, h)$ over hidden species H.

Proof. Applying Theorem 1, the clamped CRN ensemble $\pi_Q(s)$ may be written

$$\pi_Q(s) = \pi_Q(v, h) = Q(v)\pi(h \mid v) = Q(v)\frac{\pi(v, h)}{\sum_{h \in \Omega_{s_0}^H} \pi(v, h)} = Q(v)\frac{\pi(v, h)}{\pi_V(v)}. \tag{26}$$

Additionally we will need the partial derivative of a Gibbs factor and the partition function with respect to g_i,

$$\frac{\partial e^{-\mathcal{G}(s)}}{\partial g_i} = -s_i e^{-\mathcal{G}(s)} \quad \text{and} \quad \frac{\partial Z}{\partial g_i} = -Z\langle s_i \rangle_\pi. \tag{27}$$

Using these results, the partial derivative of any detailed balanced CRN's distribution at a particular state s, with respect to an energy g_i, is

$$\frac{\partial \pi(s)}{\partial g_i} = \frac{\partial}{\partial g_i} \frac{1}{Z} e^{-\mathcal{G}(s)} = \langle s_i \rangle_\pi \pi(s) - s_i \pi(s). \tag{28}$$

Noting that Q has no dependence on g_i, the gradient of the Kullback-Leibler divergence can then be written,

$$\frac{\partial D_{KL}(Q\|\pi_V)}{\partial g_i} = \frac{\partial}{\partial g_i} \sum_{v \in \Omega_{s_0}^V} Q(v) \log \frac{Q(v)}{\pi_V(v)} = \sum_{v \in \Omega_{s_0}^V} -\frac{Q(v)}{\pi_V(v)} \frac{\partial \pi_V(v)}{\partial g_i}$$

$$= -\sum_{v \in \Omega_{s_0}^V} \sum_{h \in \Omega_{s_0}^H} \frac{Q(v)}{\pi_V(v)} \pi(v, h)\langle s_i \rangle_\pi - \frac{Q(v)}{\pi_V(v)} \pi(v, h)s_i$$

$$= -\sum_{(v,h) \in \Omega_{s_0}} \pi_Q(v, h)\langle s_i \rangle_\pi - \pi_Q(v, h)s_i = -\langle s_i \rangle_\pi + \langle s_i \rangle_{\pi_Q} \qquad \blacksquare$$

In the special case where there are no hidden species, which is to say the target distribution Q is defined over the whole reachability class Ω_{s_0}, then $\pi_V(v) = \pi(s)$ and $\pi_Q(s) = Q(s)$ and the gradient has the simple form shown in Eq. (24).

Applying gradient descent via $\frac{dg_i}{dt} = -\frac{\partial D_{KL}}{\partial g_i}$, we thus have a simple *in silico* training algorithm to train any detailed balanced CRN such that it minimizes the Kullback-Leibler divergence from π_V to Q. If $H = \emptyset$, simulate the CRN freely to estimate the average counts $\langle s_i \rangle$ under $\pi(s)$. Then compare to the average counts under the target $Q(s)$ and update the species' energies accordingly. If $H \neq \emptyset$, clamp the visible species to some $v \in \Omega_{s_0}^V$ with probability $Q(v)$ and simulate the dynamics of the hidden units. Repeat to sample an ensemble of clamped CRNs $\mathcal{C}|_{V=q}$. Because clamping v preserves reachability, Gillespie simulations of the CRN with the V species clamped to the data values v will sample appropriately. This gives the average counts under π_Q.

This CBM learning rule is more general than the classical Boltzmann machine learning rule, as it applies to arbitrary detailed balanced CRNs, including those with arbitrary conservation laws and arbitrarily large species counts (but still subject to the constraint that reachability under clamping must be preserved). That said, at first glance the CBM learning rule appears weaker than the classical Boltzmann machine learning rule, as it depends exclusively on mean values $\langle s_i \rangle$, whereas the Boltzmann machine learning rule relies primarily on second-order correlations $\langle x_i x_j \rangle$. In fact, though, conservation laws within the CRN can effectively transform mean values into higher-order correlations. A case in point would be to apply the CBM learning rule to the ECBM network: For $g_i = G[X_i^{ON}] = -\theta_i$, $\frac{d\theta_i}{dt} = -\frac{dg_i}{dt} = \langle x_i^{ON} \rangle_{\pi_Q} - \langle x_i^{ON} \rangle_\pi$, and for $g_i = G[W_{ij}^{ON}] = -w_{ij}$, $\frac{dw_{ij}}{dt} = -\frac{dg_i}{dt} = \langle w_{ij}^{ON} \rangle_{\pi_Q} - \langle w_{ij}^{ON} \rangle_\pi = \langle x_i^{ON} x_j^{ON} \rangle_{\pi_Q} - \langle x_i^{ON} x_j^{ON} \rangle_\pi$, which exactly matches the classical Boltzmann machine learning rule if we assert that the energies of OFF species are fixed at zero.

6   Discussion

We have given one approximate and two exact constructions that allow CRNs to function as Boltzmann machines. BMs are a "gold standard" generative model capable of performing numerous computational tasks and approximating a wide range of distributions. Our constructions demonstrate that CRNs have the same computational power as a BM. In particular, CRNs can produce the same class of distributions and can compute conditional probabilities via the clamping process. Moreover, the TCBM construction appears similar in architecture to protein phosphorylation networks. Both models are non-equilibrium (i.e., require a fuel source) and make use of molecules that have an on/off (e.g., phosphorylated/unphosphorylated) state. Additionally, there are clear similarities between our exact schemes and combinatorial regulation of genetic networks by transcription factors. In this case, both models make use of combinatoric networks of detailed-balanced interactions (e.g., binding/unbinding) to catalyze a state change in a molecule (e.g., by turning a gene on/off). We note that our constructions differ from some biological counterparts in requiring binary molecular counts. However, in some cases we believe that biology may make use of conservation laws (such as having only a single copy of a gene) to allow for chemical networks to perform low-cost computations. In the future, we plan to examine these cases in a biological setting as well as generalize our models to higher counts.

Developing these CBMs leads us to an important distinction between equilibrium, detailed-balanced CRNs with steady state distributions determined by molecular energies, and CRNs that do not obey detailed balance in the underlying chemistry. The second category includes those that nonetheless appear detailed balanced at the Markov chain level. Physically, this distinction is especially important: a non-detailed balanced CRN will always require some kind of implicit fuel molecule (maintained by a chemostat) to run and the steady state will not be an equilibrium steady state due to the continuous driving from the fuel molecules. A detailed balanced CRN (at the chemical level) requires no fuel molecules: and thus *the chemical circuit can act as a sampler without fuel cost*. Despite this advantage, working with detailed balanced CRNs presents additional challenges: to ensure that chemical species do not have independent distributions, species counts must be carefully coupled via conservation laws.

Table 1. The complexity and underlying properties of our constructions for reproducing a BM with N nodes of degree d. Detailed balance describes whether the construction is detailed balanced at the CRN level, at the CTMC level, or neither.

Model	Species	Reactions	Molecularity	Detailed balance
Direct CBM	$2N$	$N2^{d+1}$	$d+1$	CTMC
Edge CBM	$2N + dN$	$N2^{d+1}$	$\leq 2d+1$	CRN and CTMC
Taylor CBM	$2N$	$2N + 2dN$	≤ 2	Neither

Our constructions also highlight important complexity issues underlying CBM design. The number of species, the number of reactions, and the reaction molecularity needed to implement a particular BM are relevant. Trade-offs appear to arise between these different factors and the thermodynamic requirements of a given design. A breakdown of the main features of each CBM is given in Table 1. Summarizing, the TCBM is by far the simplest construction, using $\mathcal{O}(N)$ species, at most $\mathcal{O}(N^2)$ reactions, with molecularity ≤ 2. However, this happens at the expense of not being an exact recreation of a BM, and the requirement of a continuous consumption of fuel molecules. The DCBM is the next simplest in complexity terms, using $\mathcal{O}(N)$ species, $\mathcal{O}(N2^N)$ reactions, and molecularity of at most N. Like the TCBM, the DCBM requires fuel molecules because it is not detailed balanced at the CRN level. The ECBM is considerably more complex than the DCBM, using quadratically more species, $\mathcal{O}(N^2)$, the same number of reactions, $\mathcal{O}(N2^N)$ and double the reaction molecularity. The ECBM makes up for this increased complexity by being detailed balanced at the CRN level, meaning that it functions in equilibrium without implicit fuel species.

Finally, we have shown that a broad class of detailed balanced CRNs can be trained using a Hebbian learning rule between a waking phase (clamped) and sleeping phase (free) reminiscent of the gradient descent algorithm for a BM. This exciting finding allows for straightforward optimization of detailed balanced CRNs' distributions.

This work provides a foundation for future investigations of probabilistic molecular computation. In particular, how more general restrictions on reachability classes can generate other "interesting" distributions in detailed balanced CRNs remains an exciting question. We also wonder if the learning rule algorithm can be generalized to certain classes non-detailed balanced CRNs, and whether our exact CBM constructions can be generalized to non-binary molecular counts. From a physical standpoint, plausible implementations of the clamping process and the energetic and thermodynamic constraints require investigation. Indeed, a more realistic understanding of how a CBM might be implemented physically will help us identify when these kinds of inferential computations are being performed in real biological systems and could lead to building a synthetic CBM.

Acknowledgements. This work was supported in part by U.S. National Science Foundation (NSF) graduate fellowships to WP and to AOM, by NSF grant CCF-1317694 to EW, and by the Gordon and Betty Moore Foundation through Grant GBMF2809 to the Caltech Programmable Molecular Technology Initiative (PMTI), by a Royal Society University Research Fellowship to TEO, and by a Bharti Centre for Communication in IIT Bombay award to AB.

A   Appendix

A.1   Application of Theorem 2: The Direct CBM Must Use Implicit Fuel Species

Here, we use Theorem 2 to analyze the direct implementation of a CBM and show that it cannot be detailed balanced and thereby requires implicit fuel molecules. First, notice that the the conservation laws used in this construction are of a simple form. The states accessible by (X_i^{ON}, X_i^{OFF}) are independent of (X_j^{ON}, X_j^{OFF}) for $i \neq j$, and therefore the reachability class is a product over the subspaces of each individual node. As a consequence, by Theorem 2, the system must be out of equilibrium and violate detailed balance at the level of the CRN because, by construction, this system is equivalent to a BM and has correlations between nodes i and j whenever $w_{ij} \neq 0$. In physical terms, the presence of catalysts cannot influence the equilibrium yield of a species, and therefore a circuit which uses catalysis to bias distributions of species must be powered by a supply of chemical fuel molecules [41–43]. It is also worth noting that, as a consequence, this scheme cannot be implemented by tuning of (free) energies; it is fundamentally necessary to carefully tune all of the rate constants individually (via implicit fuel molecules) to ensure that detailed balance is maintained at the level of the Markov chain for the species of interest.

A.2   BM Training and TCBM Simulation Details

We trained a BM using stochastic gradient descent on the MNIST dataset, down sampled to be 10 pixels by 10 pixels [47]. The BM has 100 visible image units (representing a 10×10 image), 10 visible class nodes, and 40 hidden nodes as depicted in Fig. 4B. Our training data consisted of the concatenation of down sampled MNIST images and their classes projected onto the 10 class nodes. The weights and biases of the trained BM were converted to reaction rates for a CBM using the Taylor series approximation. This CBM consists of 300 species, 300 unimolecular reactions and 22350 bimolecular reactions. The resulting CBM was then compared side-by-side with the trained BM on image classification and generation. The BM was simulated using custom Gibbs sampling written in Python. The CRN was simulated on a custom Stochastic Simulation Algorithm (SSA) [40] algorithm written in Cython. All simulations, including network training, were run locally on a notebook or on a single high performance Amazon Cloud server.

Classification was carried out on all 10000 MNIST validation images using both the BM and the CBM. Each 10 by 10 gray-scale image was converted to a binary sample image by comparing the gray-scale image's pixels (which are represented as real numbers between 0 and 1) to a uniform distribution over the same range. The network's image units were then clamped to the binary sample and the hidden units and class units were allowed to reach steady state. This process was carried out 3 times for each MNIST validation image, resulting in 30000 sample images being classified. Raw classification scores were computed by averaging the class nodes' outputs for 20000 simulation steps after 20000 steps of burn-in (Gibbs sampling for the BM, SSA for the CBM). Max classification was computed by taking the most probable class from the raw classification output. Raw classification and max classification confusion heatmaps, showing the average classification across all sample images as a function of the true label are shown in Fig. 4 panels C and D for a BM and in Fig. 4 panels F and G for a CBM.

Image generation was carried out by clamping the class nodes with a single class, 0...9, taking the value of 1 and all other classes being 0, and then allowing the network to reach steady state. Generated images were computed by averaging the image nodes over 50000 simulation steps (Gibbs sampling for the BM, SSA for the CBM) after 25000 steps of burn-in. Generation results are shown in Fig. 4E for a BM and Fig. 4H for a CBM.

References

1. Bray, D.: Protein molecules as computational elements in living cells. Nature **376**(6538), 307 (1995)
2. Bray, D.: Wetware: A Computer in Every Living Cell. Yale University Press, New Haven (2009)
3. McAdams, H.H., Arkin, A.: Stochastic mechanisms in gene expression. Proc. Natl. Acad. Sci. **94**(3), 814–819 (1997)
4. Elowitz, M.B., Levine, A.J., Siggia, E.D., Swain, P.S.: Stochastic gene expression in a single cell. Science **297**(5584), 1183–1186 (2002)
5. Perkins, T.J., Swain, P.S.: Strategies for cellular decision making. Mol. Syst. Biol. **5**(1), 326 (2009)
6. Muroga, S.: Threshold Logic and Its Applications. Wiley Interscience, New York (1971)
7. Hopfield, J.J.: Neural networks and physical systems with emergent collective computational abilities. Proc. Natl. Acad. Sci. **79**(8), 2554–2558 (1982)
8. Hinton, G.E., Sejnowski, T.J., Ackley, D.H.: Boltzmann Machines: Constraint Satisfaction Networks that Learn. Department of Computer Science, Carnegie-Mellon University, Pittsburgh (1984)

9. Bray, D.: Intracellular signalling as a parallel distributed process. J. Theor. Biol. **143**(2), 215–231 (1990)

10. Hellingwerf, K.J., Postma, P.W., Tommassen, J., Westerhoff, H.V.: Signal transduction in bacteria: phospho-neural network(s) in Escherichia coli. FEMS Microbiol. Rev. **16**(4), 309–321 (1995)

11. Mjolsness, E., Sharp, D.H., Reinitz, J.: A connectionist model of development. J. Theor. Biol. **152**(4), 429–453 (1991)

12. Mestl, T., Lemay, C., Glass, L.: Chaos in high-dimensional neural and gene networks. Physica D: Nonlin. Phenom. **98**(1), 33–52 (1996)

13. Buchler, N.E., Gerland, U., Hwa, T.: On schemes of combinatorial transcription logic. Proc. Natl. Acad. Sci. **100**(9), 5136–5141 (2003)

14. Deutsch, J.M.: Collective regulation by non-coding RNA. arXiv preprint arXiv:1409.1899 (2014)

15. Deutsch, J.M.: Associative memory by collective regulation of non-coding RNA. arXiv preprint arXiv:1608.05494 (2016)

16. Hjelmfelt, A., Weinberger, E.D., Ross, J.: Chemical implementation of neural networks and turing machines. Proc. Natl. Acad. Sci. **88**(24), 10983–10987 (1991)

17. Hjelmfelt, A., Ross, J.: Chemical implementation and thermodynamics of collective neural networks. Proc. Natl. Acad. Sci. **89**(1), 388–391 (1992)

18. Kim, J., Hopfield, J.J., Winfree, E.: Neural network computation by in vitro transcriptional circuits. In: Advances in Neural Information Processing Systems (NIPS), pp. 681–688 (2004)

19. Napp, N.E., Adams, R.P.: Message passing inference with chemical reaction networks. In: Advances in Neural Information Processing Systems (NIPS), pp. 2247–2255 (2013)

20. Gopalkrishnan, M.: A scheme for molecular computation of maximum likelihood estimators for log-linear models. In: Rondelez, Y., Woods, D. (eds.) DNA 2016. LNCS, vol. 9818, pp. 3–18. Springer, Cham (2016). doi:10.1007/978-3-319-43994-5_1

21. Hjelmfelt, A., Schneider, F.W., Ross, J.: Pattern recognition in coupled chemical kinetic systems. Science **260**, 335–335 (1993)

22. Kim, J., White, K.S., Winfree, E.: Construction of an in vitro bistable circuit from synthetic transcriptional switches. Mol. Syst. Biol. **2**, 68 (2006)

23. Kim, J., Winfree, E.: Synthetic in vitro transcriptional oscillators. Mol. Syst. Biol. **7**, 465 (2011)

24. Qian, L., Winfree, E., Bruck, J.: Neural network computation with DNA strand displacement cascades. Nature **475**(7356), 368–372 (2011)

25. Lestas, I., Paulsson, J., Ross, N.E., Vinnicombe, G.: Noise in gene regulatory networks. IEEE Trans. Autom. Control **53**, 189–200 (2008)

26. Lestas, I., Vinnicombe, G., Paulsson, J.: Fundamental limits on the suppression of molecular fluctuations. Nature **467**(7312), 174–178 (2010)

27. Veening, J.W., Smits, W.K., Kuipers, O.P.: Bistability, epigenetics, and bethedging in bacteria. Annu. Rev. Microbiol. **62**, 193–210 (2008)

28. Balázsi, G., van Oudenaarden, A., Collins, J.J.: Cellular decision making and biological noise: from microbes to mammals. Cell **144**(6), 910–925 (2011)

29. Tsimring, L.S.: Noise in biology. Rep. Prog. Phys. **77**(2), 26601 (2014)

30. Eldar, A., Elowitz, M.B.: Functional roles for noise in genetic circuits. Nature **467**(7312), 167–173 (2010)

31. Mansinghka, V.K.: Natively probabilistic computation. Ph.D. thesis, Massachusetts Institute of Technology (2009)

32. Wang, S., Zhang, X., Li, Y., Bashizade, R., Yang, S., Dwyer, C., Lebeck, A.R.: Accelerating Markov random field inference using molecular optical Gibbs sampling units. In: Proceedings of the 43rd International Symposium on Computer Architecture, pp. 558–569. IEEE Press (2016)
33. Fiser, J., Berkes, P., Orbán, G., Lengyel, M.: Statistically optimal perception and learning: from behavior to neural representations. Trends Cogn. Sci. **14**(3), 119–130 (2010)
34. Pouget, A., Beck, J.M., Ma, W.J., Latham, P.E.: Probabilistic brains: knowns and unknowns. Nat. Neurosci. **16**(9), 1170–1178 (2013)
35. Ackley, D.H., Hinton, G.E., Sejnowski, T.J.: A learning algorithm for Boltzmann machines. Cogn. Sci. **9**(1), 147–169 (1985)
36. Tanaka, T.: Mean-field theory of Boltzmann machine learning. Phys. Rev. E **58**(2), 2302 (1998)
37. Tang, Y., Sutskever, I.: Data normalization in the learning of restricted Boltzmann machines. Department of Computer Science, University of Toronto, Technical report UTML-TR-11-2 (2011)
38. Taylor, G.W., Hinton, G.E.: Factored conditional restricted Boltzmann machines for modeling motion style. In: Proceedings of the 26th Annual International Conference on Machine Learning (ICML), pp. 1025–1032. ACM (2009)
39. Casella, G., George, E.I.: Explaining the Gibbs sampler. Am. Stat. **46**(3), 167–174 (1992)
40. Gillespie, D.T.: Stochastic simulation of chemical kinetics. Annu. Rev. Phys. Chem. **58**, 35–55 (2007)
41. Qian, H.: Phosphorylation energy hypothesis: open chemical systems and their biological functions. Annu. Rev. Phys. Chem. **58**, 113–142 (2007)
42. Beard, D.A., Qian, H.: Chemical Biophysics: Quantitative Analysis of Cellular Systems. Cambridge University Press, Cambridge (2008)
43. Ouldridge, T.E.: The importance of thermodynamics for molecular systems, the importance of molecular systems for thermodynamics. arXiv preprint arXiv:1702.00360 (2017)
44. Joshi, B.: A detailed balanced reaction network is sufficient but not necessary for its Markov chain to be detailed balanced. arXiv preprint arXiv:1312.4196 (2013)
45. Erez, A., Byrd, T.A., Vogel, R.M., Altan-Bonnet, G., Mugler, A.: Criticality of biochemical feedback. arXiv preprint arXiv:1703.04194 (2017)
46. Anderson, D.F., Craciun, G., Kurtz, T.G.: Product-form stationary distributions for deficiency zero chemical reaction networks. Bull. Math. Biol. **72**(8), 1947–1970 (2010)
47. LeCun, Y., Cortes, C., Burges, C.J.C.: The MNIST database of handwritten digits (1998)

A General-Purpose CRN-to-DSD Compiler with Formal Verification, Optimization, and Simulation Capabilities

Stefan Badelt[1]([✉]), Seung Woo Shin[1], Robert F. Johnson[1], Qing Dong[2], Chris Thachuk[1], and Erik Winfree[1]([✉])

[1] California Institute of Technology, Pasadena, USA
{badelt,winfree}@caltech.edu
[2] Stony Brook University, New York, USA

Abstract. The mathematical formalism of mass-action chemical reaction networks (CRNs) has been proposed as a mid-level programming language for dynamic molecular systems. Several systematic methods for translating CRNs into domain-level strand displacement (DSD) systems have been developed theoretically, and in some cases demonstrated experimentally. Software that facilitates the simulation of CRNs and DSDs, and that helps automate the construction of DSDs from CRNs, has been instrumental in advancing the field, but as yet has not incorporated the fundamental enabling concept for programming languages and compilers: a rigorous abstraction hierarchy with well-defined semantics at each level, and rigorous correctness proofs establishing the correctness of compilation from a higher level to a lower level. Here, we present a CRN-to-DSD compiler, Nuskell, that makes a first step in this direction. To support the wide range of translation schemes that have already been proposed in the literature, as well as potential new ones that are yet to be proposed, Nuskell provides a domain-specific programming language for translation schemes. A notion of correctness is established on a case-by-case basis using the rate-independent stochastic-level theories of pathway decomposition equivalence and/or CRN bisimulation. The "best" DSD implementation for a given CRN can be found by comparing the molecule size, network size, or simulation behavior for a variety of translation schemes. These features are illustrated with a 3-reaction oscillator CRN and a 32-reaction feedforward boolean circuit CRN.

1 Introduction

Toehold-mediated nucleic acid strand displacement has become a widely used technology to control and fine-tune the interactions of DNA and RNA molecules [8,22]. This contribution focuses on automated construction – compilation – of nucleic acid networks, to realize larger, dynamically more complex and precise algorithms using DSD. We use the abbreviation DSD for "domain-level strand displacement" as opposed to the more common notion of "DNA strand displacement", because all results presented in this work are using the *domain-level* abstraction and we do not analyse any *sequence-level* details.

© Springer International Publishing AG 2017
R. Brijder and L. Qian (Eds.): DNA 23 2017, LNCS 10467, pp. 232–248, 2017.
DOI: 10.1007/978-3-319-66799-7_15

Domains are segments of a molecule with well defined properties. In the simplest case, we distinguish two types of domains: toehold domains and branch-migration domains. A DSD system contains many possible instances of those domains, where two domains can only bind if they are of the same type and complementary to each other. Toehold domains (short) bind reversibly, while branch-migration domains (long) bind irreversibly. Ensuring that these properties are fulfilled is something attributed to the *sequence-level*. This abstraction enables us to study nucleic acid reaction networks on a different level of detail, including rigorous proofs to guarantee the correctness of a domain-level compilation and simulations of DSD systems based on "typical" sequence-independent reaction rates. A correct domain-level network can then be compiled to either DNA or RNA sequences or combinations of different nucleic acids, as well as, hypothetically, other artificial polymers such as PNA sequences, or even proteins [1].

Formal CRNs are a natural language to formulate the intended dynamics of a nucleic acid network and therefore serve as the ideal input for a DSD compiler. We demonstrate automated translation of CRNs into DSD systems, as well as the formal verification and simulation using our compiler Nuskell (see Sect. 2). We show that there are many formally correct translations of particular CRNs, but that some types of CRNs are more efficiently implemented with different translation schemes [2–4,12,13,18,20]. Starting from CRNs highlights a fundamental difference from other existing compilers, e.g. VisualDSD [12], the most used software for DNA strand displacement design, that takes hand-crafted DSD modules as input in order to predict and verify their dynamics [10]. The DNA strand displacement compilers Seesaw [21] and Piperine [20] have each been developed for one experimentally tested/optimized translation scheme and translate digital circuits or bimolecular reactions respectively.

Formal CRNs themselves might be derived from higher-level languages such as digital-circuits, Turing machines, etc. [5,17,18]. A demonstration is shown in Sect. 3.3, where we present a translation from a logic circuit into a formal CRN, and then use Nuskell to compile this CRN into a DSD circuit. The DSD implementation is pathway decomposition equivalent [16] (see Sect. 2.3) with the input CRN.

2 Nuskell

The CRN-to-DSD compiler Nuskell is an open-source Python package[1] for the design, verification and analysis of DSD systems. Figure 1 provides an overview of the Nuskell compiler project. The translation from CRNs to DSD systems is described in Sect. 2.1, the domain-level reaction enumeration using the peppercornenumerator[2] library [7] in Sect. 2.2 and the two notions of stochastic trajectory-type CRN equivalence (pathway equivalence [16] and CRN bisimulation equivalence [9]) in Sect. 2.3.

---

[1] www.github.com/DNA-and-Natural-Algorithms-Group/nuskell.
[2] www.github.com/DNA-and-Natural-Algorithms-Group/peppercornenumerator.

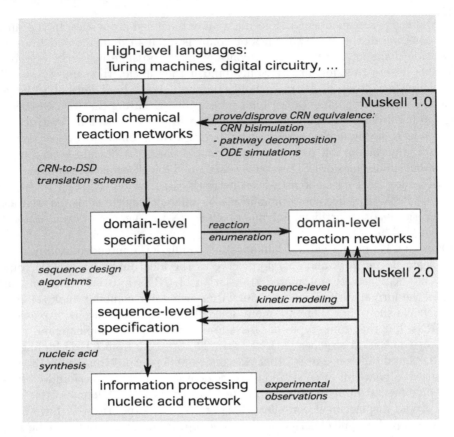

Fig. 1. The Nuskell compiler. **The current version** "Nuskell 1.0" translates formal CRNs into a set of domain-level nucleic acid complexes. The algorithm for translation can be chosen from multiple different CRN-to-DSD translation schemes (see Sect. 2.1). Domain-level complexes are input for a DSD reaction network enumerator (see Sect. 2.2). Two CRN equivalence notions (see Sect. 2.3) can be used to formally verify the equivalence between the domain-level reaction network and the formal CRN. Alternatively, initial complex concentrations can be specified to simulate the formal and/or enumerated CRN using ODEs. Domain-level specifications may be imported or exported to a plain-text format (*.pil) for manual adjustments or as an alternative to translation schemes. **The next version** "Nuskell 2.0" will translate *correct* domain-level specifications into sequence-level specifications and use sequence-level kinetic models to verify the correct implementation of domain-level reaction networks. **Eventually,** the Nuskell project may incorporate experimental feedback to train domain-level and sequence-level biophysics.

2.1   Translation: From a CRN to DSD Species

A multitude of translations from formal reactions into DSD systems have been shown previously [2–4,11–13,18,20], and there are many more possible variations. Nuskell's CRN-to-DSD translation is a *top-down* approach. First, one has to conceptualize a domain-level design in terms of an algorithm and then apply it to a formal input CRN. From this point on, we call a *translation scheme* a script written in the Nuskell programming language, which takes an input CRN and produces a DSD system. The language provides an "easy" formulation of translation schemes, and, more importantly, a standardized format for comparison, evaluation and debugging. This approach is in contrast to the *bottom-up* language of VisualDSD, where the user prototypes domain-level complexes as individual modules and combines them into a DSD system. In the bottom-up approach, it is not obvious whether a particular DSD implementation or its components can be generalized to implement different algorithms or whether conceptually new modules are required.

```
# Translate formal reactions with two reactants and two products.
# Lakin et al. (2012) "Abstractions for DNA circuit design." [Fig. 5]

# Define a global short toehold domain
global toehold = short();

# Define domains and structure of signal species
class formal(s) = "? t f" | ". . ."
  where { t = toehold; f = long() };

# Define fuel complexes for bimolecular reactions
class bimol_fuels(r, p) =
  [ "a t i + b t k + ch t c + dh t d + t* dh* t* ch* t* b* t* a* t*"
  | "( ( . + ( ( . + ( ( . + ( ( . +) ) ) ) ) ) ) ) .",
    "a t i" | ". . .", "t ch t dh t" | ". . . . ." ]
  where {
    a = r[0].f;
    b = r[1].f;
    c = p[0].f; ch = long();
    d = p[1].f; dh = long();
    i = long(); k = long();
    t = toehold };

# Module *rxn* applies the fuel production to every bimolecular reaction
module rxn(r) = sum(map(infty, fuels))
  where fuels =
    if len(r.reactants) != 2 or len(r.products) != 2 then
      abort('Reaction type not implemented')
    else
      bimol_fuels(r.reactants, r.products);

# Module *main* applies *rxn* to the crn
module main(crn) = sum(map(rxn, crn))
  where crn = irrev_reactions(crn);
```

Listing 1.1. An example of a translation scheme. The **formal** class defines a signal species for every formal species, here, consisting of three unpaired domains: a history domain, a global short domain and a unique long domain. The **main** module translates a CRN into a set of fuel complexes: the CRN is converted to irreversible reactions, every reaction translated into a set of fuel complexes and the sum over all sets returned by the main function.

A drawback of CRN-to-DSD translation schemes is that they require a particular DSD architecture. There are always two types of species involved: *signal species* and *fuel species*. Signal species are at low concentrations and they represent the dynamical information, e.g. input and output species. Fuel species are at high (ideally constant) concentrations and they mediate the information transfer by consuming and/or releasing signal species. All signal species must have the same domain-level constitution and structure, and they must be independent of each other. After compilation, every signal species corresponds to one species in the formal CRN.

We provide a continuously growing translation scheme library online[3], and a basic example in Listing 1.1 (translating Fig. 5 of [12]). The Nuskell programming language is inspired by the functional programming language Haskell and provides DSD specific classes, functions and macros to generalize translations for arbitrary CRNs [16]. There are two required parts: (i) the *formal* class defines sequence and structure of signal complexes, (ii) the *main* module produces a set of fuel species from the input CRN.

In some translation schemes, multiple signal species can correspond to the same formal species [2,12,18,20]. These schemes make use of so-called *history domains*. A history domain is considered to be an inert branch-migration domain of a signal species, but it is unique to the reaction that has produced the signal species. Hence, multiple species that differ only by their history domains map to the same formal species. When writing a translation scheme, a history domain is a wildcard: '?'. Together with the remainder of the molecule, a species with a wildcard forms a regular-expression, matching every other species in the system that differs only by a single domain in place of '?'. Nuskell automatically replaces history domains *after* domain-level enumeration, when it is known which species actually got produced. If there exists a species matching the regular expression, then the species with the wildcard domain and every enumerated reaction involving that species is removed from the system, otherwise, the wildcard domain is replaced by a new branch-migration domain.

A given translation schemes may be particularly efficient for certain types of formal reactions but inefficient or incorrect for other types, or it can be correct for every possible formal CRN, typically at the cost of being less efficient. For example, some translation schemes are particularly efficient for reversible reactions, while others implement reversible reactions using two irreversible reactions. To bypass the strict DSD system architectures that are imposed by translation schemes, Nuskell also provides an import/export file format (*.pil) to modify DSD systems, add extra modules or analyse bottom-up, hand-crafted, or alternatively compiled domain-level designs. In order to verify CRN equivalence, users have to ensure that names of signal species in the DSD implementation match the formal species in the input CRN, potentially using wildcards to indicate history domains. Also, Nuskell can export DSD systems to the VisualDSD file format (*.dna), which enables convenient access to the functionality of VisualDSD, including visualization, alternative reaction enumeration semantics and verification using probabilistic model checking [10].

---

[3] http://www.github.com/DNA-and-Natural-Algorithms-Group/nuskell/schemes.

2.2 Reaction Enumeration and Reaction Rate Calculation

The domain-level representation provides a more coarse-grained perspective on nucleic acid folding than the single-nucleotide level. At the nucleotide level every step is a base pair opening or closing reaction and the corresponding rate can be calculated from the free energy change of a reaction combined with inherent kinetic rate constants [6,15]. On the domain level, we consider a more diverse set of reactions in order to account for the fine-grained details that can happen on the sequence level. Nuskell uses the domain-level reaction enumerator Peppercorn [7] to predict desired and, potentially, undesired reactions emerging from interactions between previously compiled signal and fuel species.

Detailed Enumeration. The general types of reactions are summarized in Fig. 2. In particular: spontaneous binding and unbinding of domains, 3-way branch migration, 4-way branch migration and remote toehold branch migration. These reactions have been identified as most relevant in DSD systems. A typical DSD reaction pathway (also shown in Fig. 2) is a sequence of these detailed reaction steps. Peppercorn's enumeration semantics are justified based on the assumption that

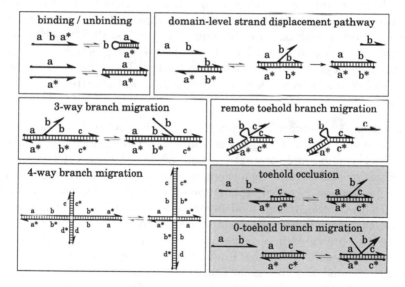

Fig. 2. Reaction semantics of the DSD enumerator Peppercorn [7]. Four generally supported detailed forms of reactions: intermolecular and intramolecular **binding/ unbinding** of domains, **3-way branch migration**, **4-way branch migration** and **remote toehold branch migration**. A typical detailed **domain-level strand displacement pathway**. The condensed reaction network notion removes the intermediate (transient) complex and calculates one irreversible rate (see main text). **Toehold occlusion** describes the effect of toehold binding to a complementary domain that does not have the correct adjacent branch-migration domain. **0-toehold branch migration** is an invalid reaction in the Peppercorn semantics, but it is a well-known unintended *leak* reaction.

the DSD system is operated at sufficiently low concentrations, such that unimolecular reactions always go to completion before the next bimolecular interaction takes place. A number of enumeration constraints are implemented to avoid combinatoric explosions [7].

Condensed Enumeration. Under the assumptions of low concentrations, a *condensed* CRN can be calculated, with reactions that indicate just the eventual results after all unimolecular reactions complete, and with rate constants systematically derived from the detailed reaction network rate constants. Reaction condensation can drastically reduce the size of an enumerated network by removing reactions that do not result in stable resting states. A particular example, *toehold occlusion* is shown in Fig. 2. The reversible binding of a single toehold domain without the prospect of initiating branch migration is captured in the detailed reaction network, but not in its condensed form. For more details and subtleties, see [7].

Limitations. There are other forms of interactions which cannot be modelled using the presented set of reactions. Most prominently, every conformation in the DSD system has to be free of pseudoknots. That means every bound domain dissects the structure into an independent left and a right part, such that there are no base pairs connecting them. Also, initiation of 3-way branch-migration reactions requires at least one already bound domain. So-called 0-toehold branch-migration reactions (see Fig. 2) have been observed in practice due to partial unbinding at helix ends, but cannot be enumerated. They belong to the broader category of *leak* reactions which we faithfully ignore in the current version of the compiler.

Reaction rates. Peppercorn uses empirical domain-level reaction rates derived from DNA strand displacement and general DNA biophysics experiments. The domain-level reaction rate constants assume perfect Watson-Crick complementary of domains and "typical" sequences, as they only depend on the length and the type of a reaction. Detailed explanation on rates, as well as their justification compared with thermodynamic models can be found in [7], but it is important to emphasize that domain-level designs may choose from a range of realistic rates, which are here presented in a discrete form as typical for certain toehold and branch-migration domain lengths. Finding sequences that confirm these chosen rates constants and verifying them using stochastic sequence-level simulations is the responsibility of a sequence-level compiler. There are many mechanisms, such as small variations in toehold sequence composition, single-nucleotide mismatches, wobble base pairs, and non-canonical base pairs that can be exploited to fine-tune reaction dynamics.

2.3   Verification of DSD Reaction Networks

The most fundamental requirement towards compilation of large scale DSD systems is verification. Every *formal reaction* is translated into multiple *implementation reactions*. Thus, there are many possibilities for introducing "bugs", i.e. unwanted side reactions that alter the implemented algorithm. We present two case-by-case verification strategies that compare formal CRNs with their implementations. As intended, our approach does not verify the general correctness of a particular scheme, but the correctness of a particular implementation.

Pathway Decomposition Equivalence. This notion was introduced in [16] together with an early version of the Nuskell compiler. The core idea is to represent each implementation trajectory as a combination of independent pathways of reactions between formal species. Pathway decomposition yields a set of pathways which are indivisible (or *prime*) and are called the *formal basis* of a CRN. The formal basis is unique for any valid implementation. Any two CRNs are said to be equivalent if they have the same formal basis. Conveniently, a CRN without intermediate species has itself as the formal basis, and it is worth pointing out that this equivalence relation allows for the comparison of one implementation with another implementation.

A common artifact of incorrect CRN-to-DSD translations is that intermediate species accumulate. That means the implementation network produces intermediate species, but they do not get cleaned up after a formal reaction goes to completion. In the notion of pathway equivalence, a given implementation is *tidy* if all intermediate species are cleaned up after a formal reaction goes to completion, and not tidy otherwise. The pathway decomposition verification method removes fuel species and inert waste products before equivalence testing, the compiler distinguishes formal from intermediate species.

CRN Bisimulation Verification. A CRN bisimulation [9] is an *interpretation* of the implementation CRN, where every implementation species is mapped to a multiset of formal species. This often yields so-called *trivial* reactions, where reactants and products do not change according to the interpretation. An interpretation is only a bisimulation if three conditions are fulfilled: (i) *atomic condition* – for every formal species there exists an implementation species that interprets to it, (ii) *delimiting condition* any reaction in the implementation is either trivial or a valid formal reaction, and (iii) *permissive condition* – for any initial condition in the implementation CRN, the set of possible next non-trivial reactions is exactly the same as it would be in the formal CRN. CRNs are said to be bisimulation equivalent, if the translation can be interpreted as an implementation of that formal CRN.

Bisimulation does not require any upfront information of which signal species correspond to formal species. In fact an implementation can be bisimulation equivalent without the intended correspondence between signal and formal species. For this reason, Nuskell provides a mapping from signal to formal species as a *partial interpretation* upfront, guaranteeing that the species are interpreted as intended, and also guaranteeing that the atomic condition is fulfilled.

Differences of Equivalence Notions. In most cases of practical interest, pathway decomposition and CRN bisimulation agree. However, it is worth pointing out examples where pathway decomposition and CRN bisimulation disagree.

First, note that pathway decomposition theory was intended to be applied to translation schemes that implement reversible reactions as two independent irreversible reaction pathways; it generally does not handle schemes that provide a single reversible implementation of each reversible reaction. For example, consider the following implementations using the scheme presented in [13]:

$A + B \rightleftharpoons C + D$	$A + B \rightleftharpoons B + C$	$A + B \rightleftharpoons C + B$
$A \rightleftharpoons i1$	$A \rightleftharpoons i1$	$A \rightleftharpoons i1$
$B + i1 \rightleftharpoons i2$	$B + i1 \rightleftharpoons i2$	$B + i1 \rightleftharpoons i2$
$i2 \rightleftharpoons C + i3$	$i2 \rightleftharpoons B + i3$	$i2 \rightleftharpoons C + i3$
$i3 \rightleftharpoons D$	$i3 \rightleftharpoons C$	$i3 \rightleftharpoons B$
not pathway equivalent	pathway equivalent	not pathway equivalent
bisimulation equivalent	bisimulation equivalent	bisimulation equivalent

It is easy to see that all CRNs are bisimulation equivalent, e.g. the interpretation $\{i1 = \{A\}; i2 = \{C, D\}; i3 = \{D\}\}$ is a valid bisimulation of $A + B \rightleftharpoons C + D$. However, the first example is not pathway equivalent, because the species C can be produced and then reverse without producing D. This form of *prematurely generated or consumed species* is forbidden in pathway equivalence, because it is problematic for implementations of irreversible reactions. In the second example of a catalytic reaction, this effect is not present, because the catalyst last and producing it first. Changing this order of reactants makes the two CRNs pathway inequivalent. On the other hand, bisimulation demands an interpretation of every species in terms of a formal species. A particularly relevant example is the *delayed choice* phenomenon [16]. Consider the formal CRN $\{A \rightarrow B; A \rightarrow C\}$ and its implementation $\{A \rightarrow i; i \rightarrow B; i \rightarrow C\}$. The two CRNs are clearly pathway equivalent, but bisimulation cannot interpret i such that both formal reactions are possible. Taken together, although both pathway decomposition and bisimulation capture the majority of intuitive equivalence relations, particular forms of very efficient implementations, or shortcuts might result in differences between the notions.

3   Case Studies

This section provides a glimpse into the future of automated DSD circuit design. We discuss potential problems of translation schemes, optimization strategies, and compare different schemes for a small oscillating CRN. Last but not least, we demonstrate the correct compilation of a large CRN implementing a digital circuit.

3.1   The Effects of Network Condensation (and Toehold Occlusion)

The intention behind network condensation is primarily to reduce the size of enumerated reaction networks. This makes verification methods, which often scale poorly with CRN size, more likely to be computationally tractable. However, here we use network condensation to study the effects of *toehold occlusion*, i.e. an effect where complementary toeholds bind "unintentionally" without actually triggering strand displacement reactions (see Fig. 2). Toehold occlusion is believed to influence the dynamics of a DSD system [14,20], especially in schemes where the consumption of fuels results in accumulation of waste species with accessible toeholds.

We start with compiling an oscillator CRN with a translation scheme that has recently been able to confirm DNA oscillations experimentally [20]. The CRN is composed of three autocatalytic reactions:

$$A + B \rightarrow 2B$$
$$B + C \rightarrow 2C \tag{1}$$
$$C + A \rightarrow 2A$$

As the scheme uses history domains that are unique to each reaction output and every formal species (A, B, C) is produced twice in the formal CRN, signal species exist in two versions: $A_1, A_2, B_1, B_2, C_1, C_2$. We simulate the system at the domain level using fuel concentrations at initially 100 nM and signal concentrations at $[A_1] = 2$ nM, $[B_1] = 10$ nM, $[C_1] = 15$ nM. Keeping the species at very low concentrations extends the number of oscillations, as fuel species get depleted more slowly.

Figure 3 shows data from multiple simulations, analysing the influence of reaction network condensation as a function of toehold length. We observed a roughly constant number of oscillations ranging from 9 to 11 peaks in total, across toehold lengths between 3–9 nt. The differences come from minor fluctuations when fuel species get depleted, i.e. the first 9 oscillation peaks are present across all examples. Hence, neither toehold length, nor reaction network condensation has a strong effect on the number of oscillation peaks.

The period of oscillations, however, changes drastically for chosen toehold lengths. At the typical lengths of 5–7 nt we observe the fastest oscillations according to the detailed reaction network. For this range of toehold lengths and concentations, binding and unbinding of toeholds occurs at a similar rate, which means (a) toeholds frequently bind to complementary sites and have enough time to initiate-branch migration and (b) toeholds bound to sites where branch migration cannot be initiated dissociate quickly. For shorter toeholds, both detailed and condensed enumerations agree, because toeholds dissociate at a high rate and the effects of toehold occlusion are insignificant. However, the fraction of toeholds completing branch migration is low, slowing down the oscillation period. For longer toeholds, detailed and condensed enumeration disagree. In the detailed network, toehold occlusion slows down the system, such that species cannot bind to their intended complementary regions. The condensed network does not simulate toehold-occlusion effects and therefore these networks oscillate faster with increased toehold length.

This result is particularly interesting because some translation schemes use a mechanism called *garbage collection* [2,3]. The intention is to collect waste

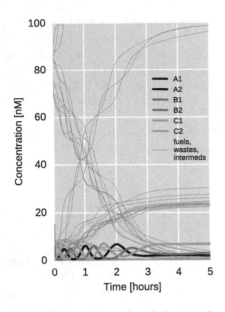
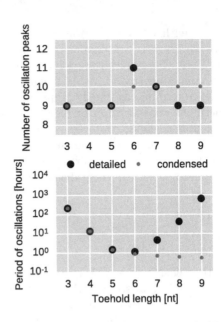

(a) ODE simulation: detailed network, 6 nt toeholds

(b) Simulations as a function of toehold length.

Fig. 3. Controlling the dynamics of a DSD oscillator via toehold length. Data compares an oscillator implemented using Srinivas' translation scheme [19], with initial conditions: $A_1 = 2$ nM, $B_1 = 10$ nM, $C_1 = 15$ nM. **(a)** Simulation of a detailed enumerated DSD reaction network. Black, red and blue lines correspond to the formal species A, B and C respectively. Note that there are two molecules with distinct history domains for each formal species. 9 oscillation peaks are clearly visible, starting with C (blue) and ending with B (red). These peaks are also present in all other simulations, independent of toehold length. However, there are actually two more hardly visible peaks for species A and C just before they reach equilibrium. **(b)** Number of oscillation peaks (top) and the period of oscillations (bottom) as a function of toehold length. Oscillation peaks are counted after the species with distinct history domains have been added (e.g. $A = A_1 + A_2$). The oscillation period for n oscillation peaks is calculated as $3(t_n - t_1)/(n - 1)$, where t_n is the time point of the last oscillation peak. (Color figure online)

species with available toehold domains into inert complexes. Therefore, garbage collection introduces additional complexes and reactions to keep the computation speed of a DSD system constant. However, in practice, this makes systems larger and harder to verify. In a physical realization, it increases the synthesis cost as well as the possibilities for *leak* reactions, such that experimental realizations have so far refrained from these additional complexes [4]. Studying the differences of detailed vs. condensed reaction networks shows that, if one chooses the rates for toehold binding appropriately and with respect to intended concentrations, toehold occlusion is not a limiting issue for the presented system.

3.2   Comparing DSD Oscillator Translations

In Fig. 4 we compare implementations of the above oscillator CRN using 13 different translation schemes. All schemes compared here verified correct for a single autocatalytic reaction $A + B \rightarrow 2B$, according to at least one of the two equivalence notions. Figure 4b shows verification results of the full system, including potential cross reactions between the three autocatalytic reactions. The schemes are *generalized* versions to support n-arity of reactions, but use exclusively reaction mechanisms shown (or described) in the original publication. A *variant* differs from the originally published version, either to correct the original version, to generalize it in a form that was not obvious from the publication, or to make a modification that enhances the performance of the scheme.

Figure 4 compares the size of the condensed enumerated network and the number of nucleotides in a system. The number of nucleotides is an indicator for the synthesis cost of nucleic acid sequences, calculated as the combined length of all distinct strands. The size of the implementation network is an indicator of computation efficiency, calculated as number of irreversible implementation reactions (i.e. reversible reactions are two irreversible reactions) in the condensed reaction network.

The implementations range from 27 to 108 reactions in the condensed enumerated CRN and from 693 to 1557 nucleotides. Obviously, removing garbage collection complexes reduces the total number of nucleotides as well as the number of reactions. The largest system in terms of reactions is `lakin2012_3D.ts`. A simple modification in `lakin2012_3D_var1.ts`: removing inert domains of strands that reverse the consumption of input strands, makes them independent of the implemented formal reaction and reduces the numbers to 48 reactions and 693 nucleotides. The 2-domain scheme `cardelli2013_2D_3I.ts` is implemented with the 3-domain irreversible step as suggested in the publication [3]. This scheme is particularly optimized for autocatalytic reactions such that they do not require extra garbage collection complexes and reactions. Hence, `cardelli2013_2D_3I.ts` and `cardelli2013_2D_3I_noGC.ts` both return the same set of fuel species for this CRN.

3.3   Towards Compilation of Large CRNs

We now demonstrate the domain-level implementation of larger systems. Our test case, adapted from [14], is a dual-rail implementation of a logic circuit computing the floor of the square root of a 4-bit binary number:

$$y_2 y_1 = \lfloor \sqrt{x_4 x_3 x_2 x_1} \rfloor \tag{2}$$

First, the logic circuit was translated into a CRN that consists of 32 uni- and bimolecular reactions (see Fig. 5a), second, the CRN was compiled using `Nuskell` with the scheme `soloveichik2010.ts` [18]. The condensed enumerated reaction network has 316 species (52 signal species, 92 fuel species, 172

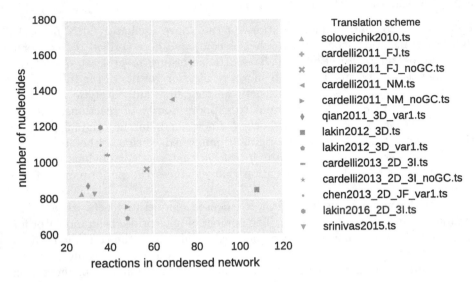

(a) Efficiency of translations schemes

Translation scheme	pathway equivalent	bisimulation equivalent
soloveichik2010.ts	True	True
cardelli2011_FJ.ts	False	-
cardelli2011_FJ_noGC.ts	False	-
cardelli2011_NM.ts	-	-
cardelli2011_NM_noGC.ts	False	True
qian2011_3D_var1.ts	True	True
lakin2012_3D.ts	False	-
lakin2012_3D_var1.ts	False	True
cardelli2013_2D_3I.ts	False	-
cardelli2013_2D_3I_noGC.ts	-	-
chen2013_2D_JF_var1.ts	-	-
lakin2016_2D_3I.ts	-	-
srinivas2015.ts	True	True

(b) Verification of translation schemes

Fig. 4. DSD oscillator implemented using 13 different translation schemes. A "_noGC" indicates that the scheme differs from the published version in that it does not implement garbage collection reactions. Other variants are indicated by "_var". lakin2012_3D.ts produces identical complexes with the scheme presented in Listing 1.1 for this input CRN. **(a)** The plot shows the total length of all distinct strands in the circuit as an indicator of synthesis cost, and the size of the enumerated reaction network as indication of computation speed. The number of nucleotides is calculated assuming 6 nt toeholds and 15 nt branch-migration domains. **(b)** A table summarizing the results of verification. None of the schemes was shown incorrect by both equivalence notions, but many reaction networks are too complicated, such that equivalence testing did not terminate within 1 hour.

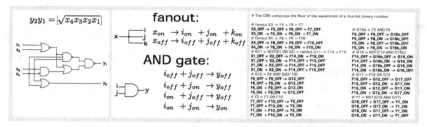

(a) from digital circuit to a feedforward CRN

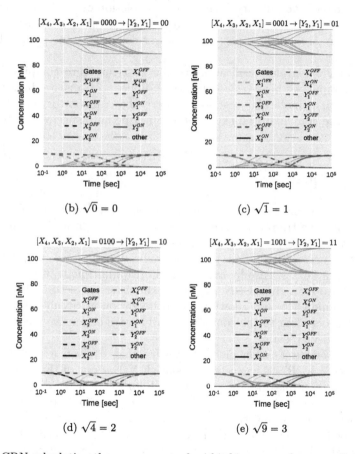

(b) $\sqrt{0} = 0$

(c) $\sqrt{1} = 1$

(d) $\sqrt{4} = 2$

(e) $\sqrt{9} = 3$

Fig. 5. A CRN calculating the square root of a 4-bit binary number compiled to a DSD system using the translation scheme presented in [18]. **(a)** A digital circuit taken from [14] is translated following the rules shown for fanouts and AND gates. NOT and OR gates follow the same principle, and the three-input AND gate is translated using two two-input AND gates. **(b-e)** The four simulations show the results for inputs $0, 1, 4, 9$. Input signal species are called X, output signal species called Y, all other signal species are "Gates", which are only transiently produced. All signal species exist in an ON and OFF version which is either initially present (10 nM) or absent (0 nM). Fuel species are initially at 100 nM.

intermediate species), 180 reactions, and it verifies as correct according to the pathway equivalence notion.

Figure 5 shows the simulations for four calculations: $0000, 0001, 0100, 1001$. Every input and output digit is represented by one ON and one OFF species, which are either initially present (10 nM) or absent (0 nM). The remaining 40 signal species (with unique history domains) represent the 12 transient formal species in the formal CRN, also called "Gates" in Fig. 5. The fuel species are initially at 100 nM; some of them are consumed during the DSD calculations, while others become more abundant. The reaction rates as calculated by the `peppercornenumerator` library suggest the completion of the DSD circuit after approximately 27 h, which is comparable to the computation time using the Seesaw architecture [14], with respect to the lower concentrations of initial species in this example.

Both enumeration and verification can be bottlenecks to compile large systems. For example, we have tested other techniques to translate digital circuits into CRNs with trimolecular reactions, where the enumerator had difficulties to deal with the combinatorial explosion of intermediate species due to history domains.

4   Conclusions

The strength of `Nuskell` comes from three features: First, formulating DSD design principles as translation schemes makes the design and optimization of complex networks easily accessible to a broad scientific community. Second, rigorous proofs of correctness guarantee a successful domain-level compilation, and are applied on a flexible case-by-case basis. Third, multiple translation schemes can be compared for a given CRN, exploiting the diversity of DSD circuits implementing the same CRN and allowing for optimization of circuits at the domain level, before proceeding to the computationally more expensive DNA sequence-level design and verification.

The discussed verification methods ensure that – given a particular reaction semantics – a CRN is correctly translated. Variations of these enumeration semantics are sometimes necessary and can help to identify problems. For example, some schemes are only correct if remote toehold branch migration or 4-way branch-migration reactions are disabled, which reveals clues for identifying unintended side reactions and making schemes more robust. On the other hand, ODE simulations of domain-level systems can be used to compare the performance of schemes, e.g. in terms of oscillation periods or fuel consumption; some schemes can be technically incorrect but with a probability of error that decreases with molecular counts in the stochastic regime and disappears entirely in the large-volume deterministic regime. Future versions of the compiler might calculate leak reaction rates to fine-tune the length of particular domains and to more efficiently combat leak during nucleic acid sequence design.

There are many open questions about the limitations of the *algorithmic behavior* that can be programmed into nucleic acid systems. How complex can

DSD systems get? Does efficiency decrease with a larger number of reactions? Can particularly efficient translation schemes be combined? Compilers can be used to study and optimize DSD systems in order to reveal their full potential.

Acknowledgements. We thank the U.S. National Science Foundation for support: NSF Grant CCF-1213127 and NSF Grant CCF-1317694 ("The Molecular Programming Project"). The Gordon and Betty Moore Foundation's Programmable Molecular Technology Initiative (PMTI). SB is funded by the Caltech Biology and Biological Engineering Division Fellowship. SWS's current address is Google, Mountain View, California. QD's current address is Epic Systems, Madison, Wisconsin.

References

1. Boyken, S.E., Chen, Z., Groves, B., Langan, R.A., Oberdorfer, G., Ford, A., Gilmore, J.M., Xu, C., DiMaio, F., Pereira, J.H., et al.: De novo design of protein homo-oligomers with modular hydrogen-bond network-mediated specificity. Science **352**(6286), 680–687 (2016)
2. Cardelli, L.: Strand algebras for DNA computing. Nat. Comput. **10**(1), 407–428 (2011)
3. Cardelli, L.: Two-domain DNA strand displacement. Math. Struct. Comput. Sci. **23**(02), 247–271 (2013)
4. Chen, Y.J., Dalchau, N., Srinivas, N., Phillips, A., Cardelli, L., Soloveichik, D., Seelig, G.: Programmable chemical controllers made from DNA. Nat. Nanotechnol. **8**(10), 755–762 (2013)
5. Cook, M., Soloveichik, D., Winfree, E., Bruck, J.: Programmability of chemical reaction networks. In: Condon, A., Harel, D., Kok, J., Salomaa, A., Winfree, E. (eds.) Algorithmic Bioprocesses. Natural Computing Series, pp. 543–584. Springer, Heidelberg (2009)
6. Flamm, C., Fontana, W., Hofacker, I.L., Schuster, P.: RNA folding at elementary step resolution. RNA **6**, 325–338 (2000)
7. Grun, C., Sarma, K., Wolfe, B., Shin, S.W., Winfree, E.: A domain-level DNA strand displacement reaction enumerator allowing arbitrary non-pseudoknotted secondary structures. arXiv:1505.03738 (2014)
8. Hochrein, L.M., Schwarzkopf, M., Shahgholi, M., Yin, P., Pierce, N.A.: Conditional dicer substrate formation via shape and sequence transduction with small conditional RNAs. J. Am. Chem. Soc. **135**(46), 17322–17330 (2013)
9. Johnson, R.F., Dong, Q., Winfree, E.: Verifying chemical reaction network implementations: a bisimulation approach. In: Rondelez, Y., Woods, D. (eds.) DNA 2016. LNCS, vol. 9818, pp. 114–134. Springer, Cham (2016). doi:10.1007/978-3-319-43994-5_8
10. Lakin, M.R., Parker, D., Cardelli, L., Kwiatkowska, M., Phillips, A.: Design and analysis of DNA strand displacement devices using probabilistic model checking. J. Roy. Soc. Interface **9**, 1470–1485 (2012)
11. Lakin, M.R., Stefanovic, D., Phillips, A.: Modular verification of chemical reaction network encodings via serializability analysis. Theoret. Comput. Sci. **632**, 21–42 (2016)
12. Lakin, M.R., Youssef, S., Cardelli, L., Phillips, A.: Abstractions for DNA circuit design. J. Roy. Soc. Interface **9**(68), 470–486 (2012)

13. Qian, L., Soloveichik, D., Winfree, E.: Efficient Turing-universal computation with DNA polymers. In: Sakakibara, Y., Mi, Y. (eds.) DNA 2010. LNCS, vol. 6518, pp. 123–140. Springer, Heidelberg (2011). doi:10.1007/978-3-642-18305-8_12

14. Qian, L., Winfree, E.: Scaling up digital circuit computation with DNA strand displacement cascades. Science **332**(6034), 1196–1201 (2011)

15. Schaeffer, J.M., Thachuk, C., Winfree, E.: Stochastic simulation of the kinetics of multiple interacting nucleic acid strands. In: Phillips, A., Yin, P. (eds.) DNA 2015. LNCS, vol. 9211, pp. 194–211. Springer, Cham (2015). doi:10.1007/978-3-319-21999-8_13

16. Shin, S.W.: Compiling and verifying DNA-based chemical reaction network implementations. Master's thesis, Caltech (2011)

17. Soloveichik, D., Cook, M., Winfree, E., Bruck, J.: Computation with finite stochastic chemical reaction networks. Nat. Comput. **7**(4), 615–633 (2008)

18. Soloveichik, D., Seelig, G., Winfree, E.: DNA as a universal substrate for chemical kinetics. Proc. Natl. Acad. Sci. **107**(12), 5393–5398 (2010)

19. Srinivas, N.: Programming chemical kinetics: engineering dynamic reaction networks with DNA strand displacement. Ph.D. thesis, Caltech (2015)

20. Srinivas, N., Parkin, J., Seelig, G., Winfree, E., Soloveichik, D.: Enzyme-free nucleic acid dynamical systems. bioRxiv (2017). http://biorxiv.org/content/early/2017/05/16/138420

21. Thubagere, A.J., Thachuk, C., Berleant, J., Johnson, R.F., Ardelean, D.A., Cherry, K.M., Qian, L.: Compiler-aided systematic construction of large-scale DNA strand displacement circuits using unpurified components. Nat. Commun. **8**, 14373 (2017)

22. Zhang, D.Y., Seelig, G.: Dynamic DNA nanotechnology using strand-displacement reactions. Nat. Chem. **3**(2), 103–113 (2011)

Thermodynamic Binding Networks

David Doty[1], Trent A. Rogers[2], David Soloveichik[3(✉)], Chris Thachuk[4],
and Damien Woods[5]

[1] University of California, Davis, Davis, USA
doty@ucdavis.edu
[2] University of Arkansas, Fayetteville, USA
tar003@email.uark.edu
[3] University of Texas at Austin, Austin, USA
david.soloveichik@utexas.edu
[4] California Institute of Technology, Pasadena, USA
thachuk@caltech.edu
[5] Inria, Paris, France
damien.woods@inria.fr

Abstract. Strand displacement and tile assembly systems are designed
to follow prescribed kinetic rules (i.e., exhibit a specific time-evolution).
However, the expected behavior in the limit of infinite time—known
as thermodynamic equilibrium—is often incompatible with the desired
computation. Basic physical chemistry implicates this inconsistency as
a source of unavoidable error. Can the thermodynamic equilibrium be
made consistent with the desired computational pathway? In order to
formally study this question, we introduce a new model of molecular
computing in which computation is driven by the thermodynamic driving
forces of enthalpy and entropy. To ensure greatest generality we do not
assume that there are any constraints imposed by geometry and treat
monomers as unstructured collections of binding sites. In this model we
design Boolean AND/OR formulas, as well as a self-assembling binary
counter, where the thermodynamically favored states are exactly the
desired final output configurations. Though inspired by DNA nanotechnology, the model is sufficiently general to apply to a wide variety of
chemical systems.

D. Doty—Supported by NSF grant CCF-1619343.

T.A. Rogers—Supported by the NSF Graduate Research Fellowship Program under
Grant No. DGE-1450079, NSF Grant CAREER-1553166, and NSF Grant CCF-
1422152.

D. Soloveichik—Supported by NSF grants CCF-1618895 and CCF-1652824.

C. Thachuk—Supported by NSF grant CCF-1317694.

D. Woods—Part of this work was carried out at California Institute of Technology.
Supported by Inria (France) as well as National Science Foundation (USA) grants
CCF-1219274, CCF-1162589, CCF-1317694.

R. Brijder and L. Qian (Eds.): DNA 23 2017, LNCS 10467, pp. 249–266, 2017.
DOI: 10.1007/978-3-319-66799-7_16

1   Introduction

Most of the models of computing that have come to prominence in molecular programming are essentially kinetic. For example, models of DNA strand displacement cascades and algorithmic tile assembly formalize desired interaction rules followed by certain chemical systems over time [8,12]. Basing molecular computation on kinetics is not surprising given that computation itself is ordinarily viewed as a *process*. However, unlike electronic computation, where thermodynamics holds little sway, chemical systems operate in a Brownian environment [2]. If the desired output happens to be a meta-stable configuration, then thermodynamic driving forces will inexorably drive the system toward error. For example, *leak* in most strand displacement systems occurs because the thermodynamic equilibrium of a strand displacement cascade favors incorrect over the correct output, or does not discriminate between the two [11]. In DNA tile assembly, we typically must find and exploit kinetic barriers to unseeded growth to enforce that growth happens only from seed assemblies, otherwise thermodynamically favored assemblies will quickly form that are not the intended self-assembly program execution from the seed/input [1,10].

We introduce the Thermodynamic Binding Networks (TBN) model, where information processing is due entirely to the thermodynamic tradeoff between *entropy* and *enthalpy*, and not any particular reaction pathway. In most experimental systems considered in DNA nanotechnology, thermodynamic favorability is determined by a tradeoff between: (1) the number of base pairs formed or broken (all else being equal, a state with more base pairs bound is more favorable); (2) the number of separate complexes (all else being equal, a state with more free complexes is more favorable). We use the terms enthalpy and entropy to describe (1) and (2) respectively (although this use does not perfectly align with their physical definitions, see Sect. 2). Intuitively, the entropic benefit of configurations with more separate complexes is due to additional microstates, each describing the independent three-dimensional positions of each complex. Although the general case of a quantitative trade-off between enthalpy and entropy is complex, we develop an elegant formulation based on the limiting case in which enthalpy is infinitely more favorable than entropy. Intuitively, this limit corresponds to increasing the strength of binding, while diluting (increasing the volume), such that the ratio of binding to unbinding rate goes to infinity. Systems studied in molecular programming can in principle be engineered to arbitrarily approach this limit. Indeed, this is the regime previously studied in the context of leak reduction for strand displacement cascades [11]. Figure 1 shows a simple TBN, which can exist in 9 possible binding configurations. The favored (stable) configuration is the one that, among the maximally bound ones (bottom row), maximizes the number of separate complexes (bottom right).

As a central choice in seeking a general theory, we *dispense with geometry*: formally, we treat monomers simply as multisets of binding sites (domains). Viewed in the context of strand displacement, this abstracts away secondary structure (the order of domains on a strand), allowing us to represent arbitrary molecular arrangements such as pseudoknots [4], and handle non-local error

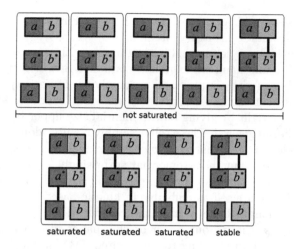

Fig. 1. An example TBN $T = (\mathcal{D}, \mathcal{M})$. $\mathcal{D} = \{a, b\}$ and $\mathcal{M} = \{\mathbf{m}_1, \mathbf{m}_2, \mathbf{m}_3, \mathbf{m}_4\}$, where monomers $\mathbf{m}_1 = \{a, b\}, \mathbf{m}_2 = \{a^*, b^*\}, \mathbf{m}_3 = \{a\}$, and $\mathbf{m}_4 = \{b\}$. Note that the order of domains does not matter (thus, $\{a, b\} = \{b, a\}$). There are nine distinct configurations for the monomer collection $\vec{\mathbf{c}} = \{\mathbf{m}_1, \mathbf{m}_2, \mathbf{m}_3, \mathbf{m}_4\}$ consisting of a single copy of each of these monomers. The five in the top row are *not saturated* meaning that they do not maximize the number of bound domains, whereas the four configurations in the bottom row are all *saturated*. In addition to being saturated, the configuration in the bottom right is *stable* as it maximizes the number of separate complexes (3) among all saturated configurations (the other saturated configurations have 2).

modes such as spurious remote toeholds [5]. In the context of tile self-assembly, we consider configurations in which binding does not follow the typical regular lattice structure. Since the TBN model does not rely on geometric constraints to enforce correct behavior, showing that specific undesired behavior is prevented by enthalpy and entropy alone leads to a stronger guarantee. Thus, for example proving leaklessness in this model would imply that even if pseudoknots, or other typically disallowed structures form, we would still have little leak. Indeed, by casting aside the vagaries of DNA biophysics (e.g., persistence length, number of bases per turn, sequence dependence on binding strength, etc.), our aim is to develop a general theory of programmable systems based on molecular bonds, a theory that will apply to bonds based on other substrates such as proteins, base stacking, or electric charge.

After introducing the TBN model in Sect. 2, we give results on Boolean circuit-based and self-assembly-based computation. In Sect. 3 we show how to construct AND and OR gates where the thermodynamically favored configurations encode the output. We develop provable guarantees on the entropic penalty that must be overcome to produce an incorrect 1 output, showing how the logic gates can be designed to make the penalty arbitrarily large. Although completely modular reasoning seems particularly tough in this model, we develop a proof technique based on logically excising domains to handle the composition of Boolean gates—specifically trees of AND gates. Further work is needed to generalize these results to arbitrary circuits.

In Sect. 4 we look at self-assembly, beginning with questions about large assemblies. On the one hand we exhibit a class of TBNs with thermodynamical stable assemblies (with simple 'tree' connectivity) of size exponential in the number of constituent monomer types. On the other hand, we show that this bound is essentially tight by giving an exponential size upper bound on the size of stable assemblies in general. These self-assembly results, along with the binary counter result below, tell us that *monomer-efficient* self-assembly is indeed possible within this model, but that (somewhat surprisingly for a model that favors enthalpy infinitely over entropy) super-exponential size polymers are necessarily unstable, even if they are self-assemblable in kinetic-based models.

For clarity of thought in separating the computational power of thermodynamics and kinetics, throughout much of this paper we do not identify any particular kinetic pathway leading to the desired TBN stable state. Of course real-world physical systems do not operate at thermodynamic equilibrium, and might take longer than the lifetime of the universe to get there. Thus, for such 'kinetically trapped' systems, encoding desired output in thermodynamic equilibrium is not enough by itself. To address this, in the full version of this paper we give a kinetically and thermodynamically favoured binary counter that assembles in both the abstract Tile Assembly Model and the TBN model. Similarly, the strand displacement AND gate from Ref. [11] can be shown to compute correctly in the TBN model [3]. Nonetheless, more work is needed to come up with TBN schemes that have fast kinetic pathways, in addition to the provable thermodynamic guarantees.

2   Model

Let $\mathbb{N}, \mathbb{Z}, \mathbb{Z}^+$ denote the set of nonnegative integers, integers, and positive integers, respectively. A key type of object in our definitions is a multiset, which we define in a few different ways as convenient. Let $\mathcal{A}$ be a finite set. We can define a multiset over $\mathcal{A}$ using the standard set notion, e.g., $\mathbf{c} = \{a, a, c\}$, where $a, c \in \mathcal{A}$. Formally, we view multiset $\mathbf{c}$ as a vector assigning counts to $\mathcal{A}$. Letting $\mathbb{N}^{\mathcal{A}}$ denote the set of functions $f : \mathcal{A} \to \mathbb{N}$, we have $\mathbf{c} \in \mathbb{N}^{\mathcal{A}}$. We index entries by elements of $a \in \mathcal{A}$, calling $\mathbf{c}(a) \in \mathbb{N}$ the *count of a in $\mathbf{c}$*. Fixing some arbitrary ordering on the elements of $\mathcal{A} = \{a_1, a_2, \dots, a_k\}$, we may equivalently view $\mathbf{c}$ as an element of $\mathbb{N}^k$, where for $i \in \{1, 2, \dots, k\}$, $\mathbf{c}(i)$ denotes $\mathbf{c}(a_i)$. Let $\|\mathbf{c}\| = \|\mathbf{c}\|_1 = \sum_{a \in \mathcal{A}} \mathbf{c}(a)$ denote the *size* of $\mathbf{c}$. For any vector or matrix $\mathbf{c}$, let $\mathrm{amax}(\mathbf{c})$ denote the largest absolute value of any component of $\mathbf{c}$.

We model molecular bonds with precise binding specificity abstractly as binding "domains", designed to bind only to other, specific binding domains. Formally, consider a finite set $\mathcal{D}$ of *primary domain types*. Each primary domain type $a \in \mathcal{D}$ is mapped to a *complementary domain type* (a.k.a., *codomain type*) denoted a^*. Let $\mathcal{D}^* = \{a^* \mid a \in \mathcal{D}\}$ denote the set of codomain types of $\mathcal{D}$. The mapping is assumed 1-1, so $|\mathcal{D}^*| = |\mathcal{D}|$. We assume that domains of type a bind only to those of type a^* and vice versa.[1] The set $\mathcal{D} \cup \mathcal{D}^*$ is the set of *domain types*.

We assume a finite set $\mathcal{M}$ of *monomer types*, where a monomer type $\mathbf{m} \in \mathbb{N}^{\mathcal{D} \cup \mathcal{D}^*}$ is a non-empty multiset of domain types, e.g., $\mathbf{m} = \{a, b, b, c^*, a^*\}$, where primary domain types $a, b, c \in \mathcal{D}$. A *thermodynamic binding network* (TBN) is a pair $\mathcal{T} = (\mathcal{D}, \mathcal{M})$ consisting of a finite set $\mathcal{D}$ of primary domain types and a finite set $\mathcal{M} \subset \mathbb{N}^{\mathcal{D} \cup \mathcal{D}^*}$ of monomer types. A *monomer collection* $\vec{\mathbf{c}} \in \mathbb{N}^{\mathcal{M}}$ of $\mathcal{T}$ is multiset of monomer types; intuitively, $\vec{\mathbf{c}}$ indicates how many of each monomer there are, but not how they are bound.[2]

Since one monomer collection usually contains more than one copy of the same domain type, we use the term *domain* to refer to each copy separately.[3] We similarly reserve the term *monomer* to refer to a particular instance of a monomer type if a monomer collection has multiple copies of the same monomer type.

A single monomer collection $\vec{\mathbf{c}}$ can take on different configurations depending on how domains in monomers are bound to each other. To formally model configurations, we first need the notion of a *bond assignment* M, which is simply a matching[4] on the bipartite graph (U, V, E) describing all possible bonds, where U is the set of all primary domains on all monomers in $\vec{\mathbf{c}}$, V is the set of all codomains on all monomers in $\vec{\mathbf{c}}$, and E is the set of edges between primary domains and their complements $\{\{u, v\} \mid u \in U, v \in V, v = u^*\}$. A *configuration* α of monomer collection $\vec{\mathbf{c}}$ is then the (multi)graph $(U \cup V, E_M)$, where the edges E_M describe both the association of domains within the same monomer, and the bonding due to M. Specifically, for each pair of domains $d_i, d_j \in \mathcal{D} \cup \mathcal{D}^*$ that are part of the same monomer in $\vec{\mathbf{c}}$, let $\{d_i, d_j\} \in E_M$, calling this a *monomer edge*, and for each edge $\{d_i, d_i^*\}$ in the bond assignment M, let $\{d_i, d_i^*\} \in E_M$, calling this a *binding edge*. Let $[\vec{\mathbf{c}}]$ be the set of all configurations of a monomer collection $\vec{\mathbf{c}}$. We say the size of a configuration, written $|\alpha|$, is simply the number of monomers in it.

Another graph that will be useful in describing the connectivity of the monomers, independent of which exact domains are bound, is the *monomer binding graph* $G_\alpha = (V_\alpha, E_\alpha)$, which is obtained by contracting each monomer edge of α. In other words, V_α is the set of monomers in α, with an edge between monomers that share at least one pair of bound domains.

---

[1] That is, we assume *like-unlike* binding such as that found in DNA Watson-Crick base-pairing, as opposed to *like-like* binding such as hydrophobic molecules with an affinity for each other in aqueous solution, or base stacking between the blunt ends of DNA helices [6,13]. It is not clear the extent to which this choice affects the computational power of our model.

[2] Because a monomer collection is a multiset of monomer types, each of which is itself a multiset, we distinguish them typographically with an arrow.

[3] For instance, the monomer collection shown in Fig. 1 has 2 domains of type a, 2 domains of type b, and 1 domain of type a^* and b^* each.

[4] A matching of a graph is a subset of edges that share no vertices in common. In our case this enforces that a domain is bound to at most one other domain.

Which configurations are thermodynamically favored over others depends on two properties of a configuration: its bond count and entropy. The *enthalpy* $H(\alpha)$ of a configuration is the number[5] of binding edges (i.e., the cardinality of the matching M). The *entropy* $S(\alpha)$ of a configuration is the number of connected components of α.[6] Each connected component is called a *polymer*.[7] Note that a polymer is itself a configuration, but of a smaller monomer collection $\vec{\mathbf{c}}' \le \vec{\mathbf{c}}$. As with all configurations, the size of a polymer is the number of monomers in it.

Intuitively, configurations with higher enthalpy $H(\alpha)$ (more bonds formed) and higher entropy $S(\alpha)$ (more separate complexes) are thermodynamically favored. What happens if there is a conflict between the two? One can imagine capturing a tradeoff between enthalpy and entropy by some linear combination of $H(\alpha)$ and $S(\alpha)$. In DNA nanotechnology applications, the tradeoff can be controlled by increasing the number of nucleotides constituting a binding domain (increasing the weight on $H(\alpha)$), or by decreasing concentration (increasing the weight on $S(\alpha)$).[8]

In the rest of this paper, we study the particularly interesting limiting case in which enthalpy is *infinitely* more favorable than entropy.[9] We say a configuration α is *saturated* if it has no pair of domains d and d^* that are both unbound; this

---

[5] We are assuming bonds are of equal strength (although the definition can be naturally generalized to bonds of different strength).

[6] Our use of the terms "enthalpy" and "entropy", and notation H and S is meant to evoke the corresponding physical notions. Note, however, that there are other contributions to physical entropy besides the number of separate complexes. Indeed, the free energy contribution of forming additional bonds typically contains substantial enthalpic and entropic parts.

[7] We are generalizing the convention for the word "polymer" in the chemistry literature. We have no requirement that a polymer be linear, nor that it consist of repeated subunits. We chose "polymer" rather than "complex" to better contrast with "monomer".

[8] In typical DNA nanotechnology applications, the Gibbs free energy $\Delta G(\alpha)$ of a configuration α can be estimated as follows. Bonds correspond to domains of length l bases, and forming each base pair is favorable by ΔG_{bp}°. Thus, the contribution of $H(\alpha)$ to $\Delta G(\alpha)$ is $(\Delta G_{bp}^\circ \cdot l)H(\alpha)$. At 1 M, the free energy penalty due to decreasing the number of separate complexes by 1 is ΔG_{assoc}°. At effective concentration C M, this penalty increases to $\Delta G_{assoc}^\circ + RT \ln(1/C)$. As the point of zero free energy, we take the configuration with no bonds, and all monomers separate. Thus, the contribution of $S(\alpha)$ to $\Delta G(\alpha)$ is $(\Delta G_{assoc}^\circ + RT \ln(1/C))(|\alpha| - S(\alpha))$, where $|\alpha|$ is the total number of monomers. To summarize,

$$\Delta G(\alpha) = (\Delta G_{bp}^\circ \cdot l)H(\alpha) + (\Delta G_{assoc}^\circ + RT \ln(1/C))(|\alpha| - S(\alpha)).$$

Note that, as expected, this is a linear combination of $H(\alpha)$ and $S(\alpha)$, and that increasing the length of domains l weighs $H(\alpha)$ more heavily, while decreasing the concentration C weighs $S(\alpha)$ more heavily. Typically $G_{bp}^\circ \approx -1.5$ kcal/mol, and $G_{assoc}^\circ \approx 1.96$ kcal/mol [9].

[9] Note that the other limiting case, where entropy is infinitely more favorable, is degenerate: the most favorable configuration in that case always has every monomer unconnected to any other.

is equivalent to stating that α has maximal bonding among all configurations in $[\vec{c}]$. We say a configuration $\alpha \in [\vec{c}]$ is *stable* (aka thermodynamically favored) if it is saturated and maximizes the entropy among all saturated configurations, i.e., every saturated configuration $\alpha' \in [\vec{c}]$ obeys $S(\alpha') \leq S(\alpha)$. Let $[\vec{c}]_\square$ denote the set of stable configurations of monomer collection $\vec{c}$. See Fig. 1 for an example thermodynamic binding network that has a single stable configuration. We note that, consistent with our model, in strand displacement cascades "long" domains are assumed to always be paired, and systems can be effectively driven by the formation of more separate complexes [14].

3   Thermodynamic Boolean Formulas

Figure 2 shows an example of a TBN that performs AND computation, based on the CRN strand displacement gate from Ref. [11]. Realized as a strand displacement system, it has a kinetic pathway taking the untriggered (left) to the triggered (right) configuration. The inputs are specified by the presence (logical value 1) or absence (logical value 0) of the input monomers i_1 and i_2. The output convention followed is the following. The output is 1 if and only if *some* stable configuration has the output monomer **o** unbound to any other monomer (free). This can be termed the *weak* output convention. Alternatively, in the *strong* output convention, output 1 implies *every* stable configuration has the output monomer **o** free, and output 0 implies *every* stable configuration has the output

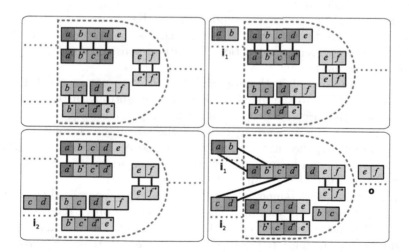

Fig. 2. Basic AND gate: Monomers $i_1 = \{a, b\}$ and $i_2 = \{c, d\}$ represent the input, $o = \{e, f\}$ represents the output, while the remainder are intermediate monomers to implement the logic relating the input to the output. If either or both inputs are missing, then the only stable configuration has the present input monomers free (unbound) and the output monomer **o** not free (bound). If both input monomers are present, then there are two stable configurations: one with inputs free (and **o** not free), or the one depicted with **o** free and both inputs bound.

monomer **o** bound to some other monomer. More complex AND gate designs are compatible with the strong output convention (not shown).

Note that even the weak output convention, coupled with a kinetic pathway releasing the output given the correct inputs, can be used to argue that: (1) if the correct inputs are present the output will be produced (via kinetic argument), (2) if the correct inputs are not present then ultimately little output will be free (thermodynamic argument). In the context of strand displacement cascades, TBNs can explore arbitrary structures (pseudoknots, remote toeholds, etc.) since we do not impose any ordering on domains in a monomer, nor any geometry. This strengthens the conclusion of (2), showing that arbitrary (even unknown) kinetic pathways must lead to a thermodynamic equilibrium with little output.

While individual AND gates can be proven correct with respect to the above output conventions (e.g., through the SAT solver of Ref. [3]), it remains to be shown that these components can be safely composed into arbitrary Boolean circuits. Note that the input and output monomers have orthogonal binding sites. This is important for composing AND gates, where the output of one acts as an input to another. As is typical for strand displacement logic, OR gates can be trivially created when multiple AND gates have the same output. Dual-rail AND/OR circuits are sufficient to compute arbitrary Boolean functions without explicit NOT gates. Nonetheless it is not obvious that the input convention (complete presence or absence of input monomers) matches the output convention (weak or strong). It is also not clear how statements about the stable configurations of the whole circuit can be made based on the stable configurations of the individual modules.

We now show that correct composition can be proven in certain cases. Although we believe that the gate shown in Fig. 2 is composable, the argument below relies on a different construction. We further consider a restricted case of AND gate formulas (trees).

An important concept in the argument below is the notion of "distance to stability". This refers to the difference between the entropy of the stable configurations and the largest entropy of a saturated configuration with incorrect output. The larger the distance to stability, the larger the entropy penalty to incorrectly producing the output. Unlike the simple AND gate from Fig. 2, the constructions below can be instantiated to achieve arbitrary desired distance to stability (by increasing the redundancy parameter n).

Many open questions remain. Can our techniques be generalized to arbitrary circuits, rather than just trees of AND gates? Can we prove these results for logic gates that have a corresponding kinetic pathway (like the AND gates in Fig. 2 which can be instantiated as strand displacement systems)? Finally, in our Boolean gate constructions, we assume that the monomer collection has exactly one copy of certain monomers. It remains open whether these schemes still work if there are many copies of all monomers.

3.1   Translator Cascades

We begin with the simplest of circuits, *translator cascades* $(x_1 \rightarrow x_2 \rightarrow \ldots \rightarrow x_{k+1})$, which simply propagate signal through k layers when the input signal x_1 is present. Logically a translator gate is simply a repeater gate. The input is the presence or absence of the input monomer consisting of n copies of domain x_1. Our analysis below implies that if and only if the input is present, there is a stable configuration with n copies of x_{k+1} domain in the same polymer. The *terminator gadget* converts this output to the weak output convention defined above (whether or not the monomer consisting of n copies of domain x_{k+1} is free). The following Lemma shows that we can exactly compute the distance from stability of a translator cascade shown in Fig. 3. Besides being a "warm-up" for AND gate cascades, the Lemma is used in the proof of Theorem 2.

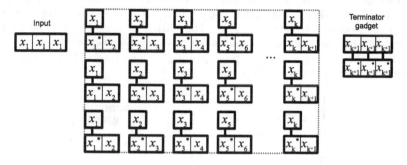

Fig. 3. A cascade of k translator gates discussed in Sect. 3.1, with redundancy parameter $n = 3$. We say that a configuration of a formula has output 1 if the terminator monomer $\{x_{k+1}, \ldots, x_{k+1}\}$ is free, and has output 0 otherwise. Redundancy parameter n specifies the number of copies of monomers and domains as shown.

Observation 1. *The intended configuration α of a monomer collection representing a depth k, redundancy n translator cascade, without input, and with output 0, is saturated and has $S(\alpha) = nk + 1$. (See Fig. 3.)*

Lemma 1. *If γ is a saturated configuration of a monomer collection representing a depth k, redundancy n translator cascade, without input, and with output 1, then $S(\gamma) = n(k - 1) + 2$.*

The proof of Lemma 1 appears in the full version of this paper. Taken together, Observation 1 and Lemma 1 imply that the redundancy parameter (n) guarantees the distance to stability $(n - 1)$ for a translator cascade of any length.

3.2   Trees of AND Gates

In this section we motivate how Boolean logic gates can be composed such that the overall circuit has a guaranteed distance to stability, relative to a redundancy

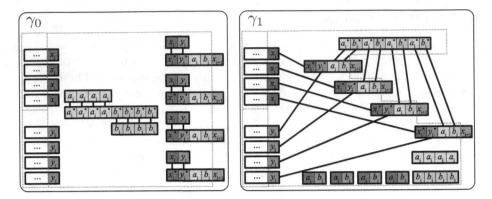

Fig. 4. AND gates used in Sect. 3.2, with redundancy parameter $n = 4$. Two saturated configurations are shown: γ_0 is the intended configuration corresponding to output of 0. γ_1 is the intended configuration corresponding to output of 1. Input domains are x_i y_i, and output domains are x_{i+1}. The output is considered to be 1 in any configuration where all n output domains are in the same polymer, 0 otherwise. Dashed boxes represent that any domain type appearing inside of a box does have have a complement appearing outside of the box.

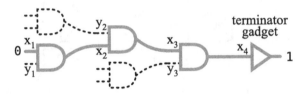

Fig. 5. Shown highlighted is a leak path through a tree of AND gates from a missing input ("0") to erroneous output ("1").

parameter n. Specifically, we start with the AND gate design of Fig. 4, and we give a concrete argument for a tree of these AND gates (e.g., Fig. 5).

Theorem 2. *Consider a TBN for AND gates, with redundancy n, composed into a tree of depth k. If at least one of the inputs is not present, the distance to stability for any saturated configurations with output 1 is at least $n - 2k - 1$.*

Proof. Let γ be any saturated configuration of the TBN with output 1. Consider the missing input and define the *leak path* to be the linear sequence of AND gates from the missing input to and including the terminator gadget. For convenience we imagine relabelling all the domains in the leak path indexed by the position of the AND gate in the leak path. For example, Fig. 5 highlights the leak path through the tree from a missing input ("0") to erroneous output ("1"). Specifically, the domain names as shown in Fig. 4 appear in the ith AND gate (for $1 \leq i \leq k$), where x_{k+1} feeds into the terminator gadget. Domains y_i connect the leak path to the rest of the tree.

Definition 1. *Given a configuration α of a monomer collection $\vec{c}$, we say we excise a domain d if we create a new configuration α' by removing the node corresponding to d and all incident edges. (Note that α' is a configuration of a monomer collection of a different TBN.)*

Manipulation 1. Excise all domains of type y_i and codomains of type y_i^* on monomers of the leak path involved in fan-in, $1 \leq i \leq k$, yielding the new configuration γ'. Note that if domain y_i is on a monomer other than the leak path, then it is not excised.

The leak path in γ' now has no domains in common with the rest of the tree (and thus no bonds). Let γ'_L be the subconfiguration of the leak path, and let γ'_R be the subconfiguration of the rest of the system. (Note $\gamma' = \gamma'_L \cup \gamma'_R$.)

Observation 3. *Given a saturated configuration α, if you excise all domains or codomains of a particular type (or both its domains and codomains) yielding α', then α' is saturated.*

By Observation 3 γ' is saturated since for every domain type y_i and codomain type y_i^*, every instance of y_i^* is excised; $1 \leq i \leq k$. This implies γ'_L and γ'_R are also saturated.

Manipulation 2. Excise all domains of type a_i and b_i and all codomains of type a_i^* and b_i^* in γ'_L, $1 \leq i \leq k$, yielding the new configuration γ''_L. By Observation 3, γ''_L is saturated.

Claim A. $S(\gamma') \geq S(\gamma)$.

Proof of the claim. Entropy can only be decreased via excision if an entire monomer is excised. Since Manipulation 1 only excised domain and codomain types from the set $\mathcal{D}' = \bigcup_{i=1}^{k} \{y_i, y_i^*\}$, and those domain types only appear on monomers which also have domain instances with types not in $\mathcal{D}'$, then no entire monomer was excised. ∎

Claim B. $S(\gamma''_L) \geq S(\gamma'_L) - 3k$.

Proof of the claim. For every layer i, $1 \leq i \leq k$, there are 3 monomers that only contain domain and codomain types in the set $\{a_i, b_i, a_i^*, b_i^*\}$. Therefore, γ''_L contains at most 3 fewer monomers than γ'_L, for each of the k layers. ∎

Claim C. $S(\gamma''_L) = n(k-1) + 2$.

Proof of the claim. Recognize that γ''_L is a saturated configuration of a monomer collection representing a depth k, redundancy n translator cascade, without input, and with output 1. The claim follows by Lemma 1. ∎

Claim D. $S(\gamma) \leq n(k-1) + 2 + S(\gamma'_R) + 3k$.

Proof of the claim.

$$\begin{aligned}
S(\gamma) &\leq S(\gamma') && \text{by Claim A} \\
&= S(\gamma'_L) + S(\gamma'_R) \\
&\leq S(\gamma''_L) + S(\gamma'_R) + 3k && \text{by Claim B} \\
&\leq n(k-1) + 2 + S(\gamma'_R) + 3k && \text{by Claim C}
\end{aligned}$$

∎

Now, take the monomers from the leak path in γ, and configure them into the "untriggered configuration" (see Fig. 4, left), yielding subconfiguration β. Let $\alpha = \beta \cup \gamma'_R$. Note that β is saturated, and therefore α is a saturated configuration of the entire tree (*i.e.*, the same TBN as γ).

Observation 4. $S(\alpha) = S(\gamma'_R) + k(n+1) + 1$.

Finally, consider the entropy gap between α and γ.

$$\begin{aligned}
S(\alpha) - S(\gamma) &\geq S(\gamma'_R) + k(n+1) + 1 - S(\gamma) && \text{by Observation 4} \\
&\geq S(\gamma'_R) + k(n+1) + 1 \\
&\quad - (n(k-1) + 2 + S(\gamma'_R) + 3k) && \text{by Claim D} \\
&= n - 2k - 1
\end{aligned}$$

Therefore, there exists a saturated configuration with output 0 over the same TBN as γ, but with entropy at least $n - 2k - 1$ larger, thus establishing the theorem.   □

Theorem 2 seems to suggest that in order to maintain the bound on distance to stability for incorrect computation, the redundancy parameter n should increase to compensate for an increase in circuit depth k. However, a more sophisticated argument shows that manipulations 1 and 2 can decrease entropy by at most $k + 1$. Following the above argument, the distance to stability is found to be $n - 2$. This is optimal because a single AND gate with redundancy $n = 2$ can be shown to have no entropy gap between output 0 and output 1 configurations.

4   Thermodynamic Self-assembly: Assembling Large Polymers

TBNs can not only exhibit Boolean circuit computation, but they can also be thought of as a model of self-assembly. Here we begin to explore this connection by asking a basic question motivated by the abstract Tile Assembly Model (aTAM) [12]: how many different monomer *types* are required to assemble a large polymer?

Favoring enthalpy infinitely over entropy, on its face, *appears* to encourage large polymers. Perhaps we can imagine designing a single TBN $\mathcal{T}$ that can assemble arbitrarily large polymers where for each $n \in \mathbb{N}$, $\mathcal{T}$ has a stable polymer α composed of at least n monomers. In this section we show that this is impossible: every TBN $\mathcal{T} = (\mathcal{D}, \mathcal{M})$ has stable polymers of size at most exponential in the number of domain types $|\mathcal{D}|$ and monomer types $|\mathcal{M}|$ (Theorem 9). The proof shows that any polymer ρ larger than the bound can be partitioned into at least two saturated (maximally bound) polymers, which implies that ρ is not stable. Figure 6 gives an example. We also show that this upper bound is essentially tight by constructing a family of systems with exponentially large stable polymers (Theorem 5). Taken together, the exponential lower bound of Theorem 5 and upper bound of Theorem 9 give a relatively tight bound on the maximum size achievable for stable TBN polymers.

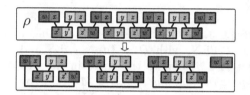

Fig. 6. A polymer ρ composed of several copies of four monomer types, which is not stable since it can be broken into several smaller polymers (bottom panel) such that all domains are bound.

Fig. 7. An example of a TBN from Theorem 5 for $n = 4$ and $k = 2$.

Is it possible to construct algorithmically *interesting* TBN polymers that are stable? In the full version of this paper, we show that a typical binary counter construction from the aTAM model is not stable, but can be modified to become stable in our model. Importantly, this TBN binary counter demonstrates that in principle algorithmically complex assemblies could have effective assembly pathways (aTAM) as well as be thermodynamically stable (TBN).

4.1   Superlative Trees: TBNs with Exponentially Large Stable Polymers

The next theorem shows that there are stable polymers that are exponentially larger than the number of domain types and monomer types required to assemble them.

Theorem 5. *For every $n, k \in \mathbb{Z}^+$, there is a TBN $\mathcal{T} = (\mathcal{D}, \mathcal{M})$ with $|\mathcal{D}| = n-1$ and $|\mathcal{M}| = n$, having a stable polymer of size $\frac{k^n - 1}{k - 1}$.*

Proof. An example of $\mathcal{T}$ for $n = 4$ and $k = 2$ is shown in Fig. 7. Let $\mathcal{D} = \{d_1, \ldots, d_n\}$ and $\mathcal{M} = \{\mathbf{m}_1, \ldots, \mathbf{m}_n\}$, where, for each $j \in \{2, \ldots, n - 1\}$, $\mathbf{m}_j = \{d_{j-1}^*, k \cdot d_j\}$ (i.e., 1 copy of d_{j-1}^* and k copies of d_j), $\mathbf{m}_1 = \{k \cdot d_1\}$, and $\mathbf{m}_n = \{d_{n-1}^*\}$. Define $\vec{\mathbf{c}} \in \mathbb{N}^{\mathcal{M}}$ by $\vec{\mathbf{c}}(\mathbf{m}_j) = k^{j-1}$ for $j \in \{1, \ldots, n\}$. Then $\|\vec{\mathbf{c}}\| = \sum_{j=1}^{n} k^{j-1} = \frac{k^n - 1}{k - 1}$. Observe that $[\vec{\mathbf{c}}]$ has a unique (up to isomorphism) saturated configuration α (which is therefore stable), described by a complete k-ary tree: level $j \in \{1, \ldots, n - 1\}$ of the tree is composed of k^{j-1} copies of $\mathbf{m}_j$, each bound to k children of type $\mathbf{m}_{j+1}$ in level $j + 1$. $\qquad\square$

The remainder of Sect. 4 is devoted to proving that no stable polymer ρ can have size *more* than exponential in $|\mathcal{D}|$ and $|\mathcal{M}|$.

4.2   A Linear Algebra Framework

We prove Theorem 9, the main result of Sect. 4, by viewing TBNs from a linear algebra perspective. Let $\mathcal{T} = (\mathcal{D}, \mathcal{M})$ be a TBN, with $\mathcal{D} = \{d_1, \ldots, d_d\}$ and $\mathcal{M} = \{\mathbf{m}_1, \ldots, \mathbf{m}_m\}$. For a matrix $\mathbf{A}$, let $\mathbf{A}(i, j)$ denote the entry in the i'th row and j'th column. Define the $d \times m$ *positive monomer matrix* $\mathbf{M}_{\mathcal{T}}^+$ of $\mathcal{T}$ by $\mathbf{M}_{\mathcal{T}}^+(i, j) = \mathbf{m}_j(d_i)$. Define the $d \times m$ *negative monomer matrix* $\mathbf{M}_{\mathcal{T}}^-$ of $\mathcal{T}$ by $\mathbf{M}_{\mathcal{T}}^-(i, j) = \mathbf{m}_j(d_i^*)$. Define the $d \times m$ *monomer matrix* $\mathbf{M}_{\mathcal{T}}$ of $\mathcal{T}$ to be $\mathbf{M}_{\mathcal{T}} = \mathbf{M}_{\mathcal{T}}^+ - \mathbf{M}_{\mathcal{T}}^-$. Note that $\mathbf{M}_{\mathcal{T}}^+$ and $\mathbf{M}_{\mathcal{T}}^-$ are matrices over $\mathbb{N}$, but $\mathbf{M}_{\mathcal{T}}$ is over $\mathbb{Z}$.

The rows of the monomer matrix $\mathbf{M}_{\mathcal{T}}$ correspond to domain types and the columns correspond to monomer types. The mapping from a TBN $\mathcal{T}$ to a monomer matrix $\mathbf{M}_{\mathcal{T}}$ is not 1-1: $\mathbf{M}_{\mathcal{T}}(i, j)$ is the number of d_i domains minus the number of d_i^* domains in monomer type $\mathbf{m}_j$, which would be the same, for instance, for monomer types $\mathbf{m}_1 = \{d_1, d_3\}$ and $\mathbf{m}_2 = \{d_1, d_1, d_1^*, d_3\}$. Let $\vec{\mathbf{c}}$ be a monomer collection and let $\mathbf{d} = \mathbf{M}_{\mathcal{T}} \vec{\mathbf{c}} \in \mathbb{N}^d$; for $i \in \{1, \ldots, d\}$, $\mathbf{d}(i)$ is the number of d_i domains minus the number of d_i^* domains in the whole monomer collection $\vec{\mathbf{c}}$.

Let $\alpha \in [\vec{\mathbf{c}}]$ be saturated; α can only have a domain d_i unbound if all copies of its complement d_i^* are bound, and vice versa. If $\mathbf{d}(i) > 0$, in α there is an excess of d_i domains, and all d_i^* domains are bound. If $\mathbf{d}(i) < 0$, in α there is an excess of d_i^* domains, and all d_i domains are bound. This leads to the following observation.

Observation 6. *Let $\mathcal{T} = (\mathcal{D}, \mathcal{M})$ be a TBN and $\vec{\mathbf{c}} \in \mathbb{N}^{\mathcal{M}}$ a monomer collection. Let $\mathbf{d} = \mathbf{M}_{\mathcal{T}} \vec{\mathbf{c}}$. Then for every configuration $\alpha \in [\vec{\mathbf{c}}]$, α is saturated if and only if, for all $i \in \{1, \ldots, d\}$, if $\mathbf{d}(i) \geq 0$ (respectively, if $\mathbf{d}(i) \leq 0$), then $\mathbf{d}(i)$ is the number of unbound d_i (resp., d_i^*) domains in α.*

Let $\mathcal{T} = (\mathcal{D}, \mathcal{M})$ and $\mathcal{T}' = (\mathcal{D}, \mathcal{M}')$ be TBNs with the same set of domain types. Then we call $\mathcal{T}'$ a *relabeling* of $\mathcal{T}$ if there exists a subset $D \subseteq \mathcal{D}$ such that $\mathcal{M}'$ can be obtained from $\mathcal{M}$ by starring any instance of $d_i \in D$ in $\mathcal{M}$ and unstarring any instance of d_i^* in $\mathcal{M}$. Since this corresponds to negating the i'th row of $\mathbf{M}_{\mathcal{T}}$, which negates the i'th entry of the vector $\mathbf{d} = \mathbf{M}_{\mathcal{T}} \vec{\mathbf{c}}$, this gives the following observation.

Observation 7. *Let* $\mathcal{T} = (\mathcal{D}, \mathcal{M})$ *be a TBN and* $\vec{\mathbf{c}} \in \mathbb{N}^{\mathcal{M}}$ *a monomer collection. There exists a relabeling* $\mathcal{T}'$ *of* $\mathcal{T}$ *so that* $\mathbf{M}_{\mathcal{T}'} \vec{\mathbf{c}} \geq 0$.

Combining Observations 6 and 7 results in the following observation, which essentially states that for any given monomer collection $\vec{\mathbf{c}}$, we may assume without loss of generality that domains unbound in saturated configurations $\alpha \in [\vec{\mathbf{c}}]$ are all primary domain types.

Observation 8. *Let* $\mathcal{T} = (\mathcal{D}, \mathcal{M})$ *be a TBN and* $\vec{\mathbf{c}} \in \mathbb{N}^{\mathcal{M}}$ *a monomer collection. There exists a relabeling* $\mathcal{T}'$ *of* $\mathcal{T}$ *so that, letting* $\mathbf{d} = \mathbf{M}_{\mathcal{T}'} \vec{\mathbf{c}}$, *for all configurations* $\alpha \in [\vec{\mathbf{c}}]$, α *is saturated if and only if, for all* $i \in \{1, \ldots, d\}$, $\mathbf{d}(i) \in \mathbb{N}$ *is the number of unbound primary domains of type* $d_i \in \mathcal{D}$ *in* α.

The following lemma is a key technical tool for showing that a polymer is not stable (or equivalently that a stable configuration has entropy greater than 1 and therefore cannot be a single polymer). It generalizes the idea shown in Fig. 6 that if one can find a monomer subcollection $\vec{\mathbf{c}}_1$ in a larger collection $\vec{\mathbf{c}}$, and $\vec{\mathbf{c}}_1$ has a saturated configuration with *no* bonds left unbound, then one can create a saturated configuration $\gamma \in [\vec{\mathbf{c}}]$ with no bonds between $\vec{\mathbf{c}}_1$ and $\vec{\mathbf{c}} - \vec{\mathbf{c}}_1$. (Thus γ has at least two polymers.)

More generally, given a monomer collection $\vec{\mathbf{c}}$ with at least as many d_i as d_i^* domains (under appropriate relabeling this holds for each i by Observation 7), if we can partition $\vec{\mathbf{c}}$ into subcollections $\vec{\mathbf{c}}_1$ and $\vec{\mathbf{c}}_2$, and each of them *also* has at least as many d_i as d_i^* domains for each $i \in \{1, \ldots, d\}$, then every stable configuration $\alpha \in [\vec{\mathbf{c}}]_\square$ has at least two polymers, since there is a saturated configuration of $\vec{\mathbf{c}}$ in which there are no bonds between $\vec{\mathbf{c}}_1$ and $\vec{\mathbf{c}}_2$.[10]

Lemma 2. *Let* $\mathcal{T} = (\mathcal{D}, \mathcal{M})$ *be a TBN, let* $\vec{\mathbf{c}} \in \mathbb{N}^{\mathcal{M}}$ *be a monomer collection of* $\mathcal{T}$ *such that* $\mathbf{M}_{\mathcal{T}} \vec{\mathbf{c}} \geq 0$, *and let* $\alpha \in [\vec{\mathbf{c}}]_\square$ *be a stable configuration. If there exist nonempty subcollections* $\vec{\mathbf{c}}_1, \vec{\mathbf{c}}_2 \in \mathbb{N}^{\mathcal{M}}$ *where 1)* $\vec{\mathbf{c}}_1 + \vec{\mathbf{c}}_2 = \vec{\mathbf{c}}$ *and 2)* $\mathbf{M}_{\mathcal{T}} \vec{\mathbf{c}}_1 \geq 0$ *and* $\mathbf{M}_{\mathcal{T}} \vec{\mathbf{c}}_2 \geq 0$, *then* $S(\alpha) > 1$.

The proof of Lemma 2 appears in the full version of this paper.

4.3 Exponential Upper Bound on Polymer Size

We now show a converse to Theorem 5, namely Theorem 9, showing that stable polymers have size at most exponential in the number of domain and monomer types. The proof of Theorem 9 closely follows Papadimitriou's proof that integer programming is contained in NP [7]. That proof shows, for any linear system

---

[10] Observations 6, 7, and 8 are not really *necessary* for our technique, but simplify the description of the conditions under which $\vec{\mathbf{c}}_1$ and $\vec{\mathbf{c}}_2$ would be saturated: specifically, that if $\mathbf{d} = \mathbf{M}_{\mathcal{T}} \vec{\mathbf{c}}$ is in the nonnegative orthant, then so are $\mathbf{d}_1 = \mathbf{M}_{\mathcal{T}} \vec{\mathbf{c}}_1$ and $\mathbf{d}_2 = \mathbf{M}_{\mathcal{T}} \vec{\mathbf{c}}_2$. If we did not use relabeling (thus could not guarantee that $\mathbf{d}$ is in the nonnegative orthant) then the requisite condition to apply Lemma 2 would be that $\mathbf{d}$, $\mathbf{d}_1$, and $\mathbf{d}_2$ all occupy the same orthant; i.e., for all $i \in \{1, \ldots, d\}$, if any of $\mathbf{d}(i)$, $\mathbf{d}_1(i)$, or $\mathbf{d}_2(i)$ are negative, then the other two are not positive.

$\mathbf{Ax} = \mathbf{b}$, where $\mathbf{A}$ is a given $n \times m$ integer matrix, $\mathbf{b} \in \mathbb{Z}^n$ is a given integer vector, and $\mathbf{x}$ represents the m unknowns, that if the system has a solution $\mathbf{x} \in \mathbb{N}^m$, then it has a "small" solution $\mathbf{x}' \in \mathbb{N}^m$. "Small" means that $\mathrm{amax}(\mathbf{x}')$ is at most exponential in $n + m + \mathrm{amax}(\mathbf{A}) + \mathrm{amax}(\mathbf{b})$. The technique of [7] proceeds by showing that any sufficiently large solution $\mathbf{x} \in \mathbb{N}^m \setminus \{\mathbf{0}\}$ can be split into two vectors $\mathbf{x}_1, \mathbf{x}_2 \in \mathbb{N}^m \setminus \{\mathbf{0}\}$ such that $\mathbf{x}_1 + \mathbf{x}_2 = \mathbf{x}$, where $\mathbf{Ax}_1 = \mathbf{0}$, so $\mathbf{x}_2$ is also a solution: $\mathbf{Ax}_2 = \mathbf{A}(\mathbf{x} - \mathbf{x}_1) = \mathbf{Ax} - \mathbf{Ax}_1 = \mathbf{Ax} = \mathbf{b}$. This is useful because $\mathbf{x}_1$ and $\mathbf{x}_2$ satisfy the hypothesis of Lemma 2, which tells us that all stable configurations $\alpha \in [\mathbf{x}]$ obey $S(\alpha) > 1$, so any single-polymer configuration of $\mathbf{x}$ is not stable.

We include the full proof for three reasons: (1) self-containment, (2) it requires a bit of care to convert our inequality $\mathbf{Ax} \geq \mathbf{0}$ into an equality as needed for the technique,[11] and (3) although the proof of [7] is sufficiently detailed to prove our theorem, the statement of the theorem in [7] hides the details about splitting the vector, which are crucial to obtaining our result.

We require the following discrete variant of Farkas' Lemma, also proven in [7].

Lemma 3 ([7]). *Let $a, d, l \in \mathbb{Z}^+$, $\mathbf{v}_1, \ldots, \mathbf{v}_l \in \{0, \pm 1, \ldots, \pm a\}^d$, and $K = (ad)^{d+1}$. Then exactly one of the following statements holds:*

1. *There exist l integers $n_1, \ldots, n_l \in \{0, 1, \ldots, K\}$, not all 0, such that $\sum_{j=1}^{l} n_j \mathbf{v}_j = \mathbf{0}$.*
2. *There exists a vector $\mathbf{h} \in \{0, \pm 1, \ldots, \pm K\}^d$ such that, for all $j \in \{1, \ldots, l\}$, $\mathbf{h}^\mathsf{T} \cdot \mathbf{v}_j \geq 1$.*

Intuitively, statement (1) of Lemma 3 states that the vectors can be added to get $\mathbf{0}$ (they are "directions of balanced forces" [7]). This is false if and only if statement (1) holds: the vectors all lie on one side of some hyperplane, whose orthogonal vector $\mathbf{h}$ would then have positive dot product with each of the vectors $\mathbf{v}_j$ (thus adding any of them would move positively in the direction $\mathbf{h}$ and could never cancel to get $\mathbf{0}$).

Intuitively, Theorem 9 states that the size of polymers in stable configurations is upper bounded by a function which is exponential in d. We prove this by first defining a constant K which is exponential in d. If each of the m individual monomer counts is less than K, then we are done since no polymer in the configuration can have size bigger than mK. If some of the monomer counts are greater than K (call these *large-count monomers*), we consider two cases.

For the first case, we consider the scenario where the vectors which describe the monomer types with large monomer counts are such that they can "balance" each other out with relatively small linear combination coefficients. If this is

---

[11] In particular, the proof of [7] upper bounds the size of $\mathbf{x}$ in terms of the entries of both $\mathbf{A}$ and $\mathbf{b}$. However, the naïve way to solve a linear inequality $\mathbf{Ax} \geq \mathbf{0}$ using an equality, by introducing slack variables $\mathbf{b}$ and asking for solutions $\mathbf{x} \in \mathbb{N}^m$, $\mathbf{b} \in \mathbb{N}^n$ such that $\mathbf{Ax} = \mathbf{b}$, allows for the possibility that $\|\mathbf{b}\|$ is very large compared to $\|\mathbf{A}\|$, in which case upper bounding $\|\mathbf{x}\|$ in terms of both $\mathbf{A}$ and $\mathbf{b}$ does not help to bound $\|\mathbf{x}\|$ in terms of $\mathbf{A}$ alone.

the case, then we can make a saturated subconfiguration which has at least one polymer using these small linear combination coefficients and large-count monomer types since the domains and codomains completely "balance" each other out. We can then use the rest of the counts of the configuration to make another saturated subconfiguration which has at least one polymer. This is shown mathematically by applying Lemma 3 to show that the monomer counts in the polymer can be split to find a configuration consisting of two separate saturated polymers. This means that there is a saturated configuration that has at least two polymers which contradicts the assumption α is a single stable polymer.

If there exist no such linear combination to "balance out" out the vectors describing the large-count monomers, then Lemma 3 tells us all of these vectors lie on the same side of some hyperplane. In this case, we show that counts of the small-count monomers play a role in bounding the counts of the large-count monomers. Intuitively, if all of the vectors describing the large-count monomers lie on the same side of some hyperplane, they are missing domains and codomains which will allow them to bind together. The domains and codomains they need in order to bind together, then must be found on the small-count monomer. Consequently, this means the size of polymers will be bound by the counts of small-count monomers (which is exponential in K). The proof appears in the full version of this paper.

Theorem 9. *Let* $\mathcal{T} = (\mathcal{D}, \mathcal{M})$ *be a TBN with* $d = |\mathcal{D}|$ *and* $m = |\mathcal{M}|$. *Let* $a = \max\limits_{\mathbf{m} \in \mathcal{M}, d_i \in \mathcal{D} \cup \mathcal{D}^*} \mathbf{m}(d_i)$ *be the maximum count of any domain in any monomer. Then all polymers of every stable configuration* α *of* $\mathcal{T}$ *have size at most* $2(m + d)(ad)^{2d+3}$.

References

1. Barish, R.D., Schulman, R., Rothemund, P.W.K., Winfree, E.: An information-bearing seed for nucleating algorithmic self-assembly. Proc. Natl. Acad. Sci. **106**(15), 6054–6059 (2009)
2. Bennett, C.H.: The thermodynamics of computation—a review. Int. J. Theor. Phys. **21**(12), 905–940 (1982)
3. Breik, K., Prakash, L., Thachuk, C., Heule, M., Soloveichik, D.: Computing properties of stable configurations of thermodynamic binding networks (2017, in preparation)
4. Dirks, R.M., Bois, J.S., Schaeffer, J.M., Winfree, E., Pierce, N.A.: Thermodynamic analysis of interacting nucleic acid strands. SIAM Rev. **49**(1), 65–88 (2007)
5. Genot, A.J., Zhang, D.Y., Bath, J., Turberfield, A.J.: Remote toehold: a mechanism for flexible control of DNA hybridization kinetics. J. Am. Chem. Soc. **133**(7), 2177–2182 (2011)
6. Gerling, T., Wagenbauer, K.F., Neuner, A.M., Dietz, H.: Dynamic DNA devices and assemblies formed by shape-complementary, non-base pairing 3D components. Science **347**(6229), 1446–1452 (2015)
7. Papadimitriou, C.H.: On the complexity of integer programming. J. ACM (JACM) **28**(4), 765–768 (1981)

8. Phillips, A., Cardelli, L.: A programming language for composable DNA circuits. J. R. Soc. Interface **6**(Suppl 4), S419–S436 (2009)
9. SantaLucia Jr., J., Hicks, D.: The thermodynamics of DNA structural motifs. Annu. Rev. Biophys. Biomol. Struct. **33**, 415–440 (2004)
10. Schulman, R., Winfree, E.: Programmable control of nucleation for algorithmic self-assembly. SIAM J. Comput. **39**(4), 1581–1616 (2009)
11. Thachuk, C., Winfree, E., Soloveichik, D.: Leakless DNA strand displacement systems. In: Phillips, A., Yin, P. (eds.) DNA 2015. LNCS, vol. 9211, pp. 133–153. Springer, Cham (2015). doi:10.1007/978-3-319-21999-8_9
12. Winfree, E.: Algorithmic self-assembly of DNA. Ph.D. thesis, California Institute of Technology (1998)
13. Woo, S., Rothemund, P.W.K.: Programmable molecular recognition based on the geometry of DNA nanostructures. Nat. Chem. **3**, 620–627 (2011)
14. Zhang, D.Y., Turberfield, A.J., Yurke, B., Winfree, E.: Engineering entropy-driven reactions and networks catalyzed by DNA. Science **318**(5853), 1121–1125 (2007)

Author Index

Printed in the United States
By Bookmasters